Urban Runoff Control and Sponge City Construction

Urban Runoff Control and Sponge City Construction

Editors

**Haifeng Jia
Jiangyong Hu
Tianyin Huang
Albert S. Chen
Yukun Ma**

MDPI • Basel • Beijing • Wuhan • Barcelona • Belgrade • Manchester • Tokyo • Cluj • Tianjin

Editors

Haifeng Jia
Tsinghua University
China

Jiangyong Hu
National University of
Singapore
Singapore

Tianyin Huang
Suzhou University of Science
and Technology
China

Albert S. Chen
University of Exeter
UK

Yukun Ma
Beijing Normal University
China

Editorial Office
MDPI
St. Alban-Anlage 66
4052 Basel, Switzerland

This is a reprint of articles from the Special Issue published online in the open access journal *Water* (ISSN 2073-4441) (available at: https://www.mdpi.com/journal/water/special_issues/ UrbanRunoff_Control).

For citation purposes, cite each article independently as indicated on the article page online and as indicated below:

LastName, A.A.; LastName, B.B.; LastName, C.C. Article Title. *Journal Name* **Year**, *Volume Number*, Page Range.

ISBN 978-3-0365-4789-3 (Hbk)
ISBN 978-3-0365-4790-9 (PDF)

© 2022 by the authors. Articles in this book are Open Access and distributed under the Creative Commons Attribution (CC BY) license, which allows users to download, copy and build upon published articles, as long as the author and publisher are properly credited, which ensures maximum dissemination and a wider impact of our publications.

The book as a whole is distributed by MDPI under the terms and conditions of the Creative Commons license CC BY-NC-ND.

Contents

About the Editors vii

Haifeng Jia, Jiangyong Hu, Tianyin Huang, Albert S. Chen and Yukun Ma
Urban Runoff Control and Sponge City Construction
Reprinted from: *Water* **2022**, *14*, 1910, doi:10.3390/w14121910 1

Dingkun Yin, Changqing Xu, Haifeng Jia, Ye Yang, Chen Sun, Qi Wang and Sitong Liu
Sponge City Practices in China: From Pilot Exploration to Systemic Demonstration
Reprinted from: *Water* **2022**, *14*, 1531, doi:10.3390/w14101531 7

Zijing Liu, Yuehan Yang, Jingxuan Hou and Haifeng Jia
Decision-Making Framework for GI Layout Considering Site Suitability and Weighted Multi-Function Effectiveness: A Case Study in Beijing Sub-Center
Reprinted from: *Water* **2022**, *14*, 1765, doi:10.3390/w14111765 31

Haozheng Wang, Guanyu Han, Lei Zhang, Yiting Qiu, Juntao Li and Haifeng Jia
Integrated and Control-Oriented Simulation Tool for Optimizing Urban Drainage System Operation
Reprinted from: *Water* **2022**, *14*, 25, doi:10.3390/w14010025 45

Zhouyang Peng, Xi Jin, Wenjiao Sang and Xiangling Zhang
Optimal Design of Combined Sewer Overflows Interception Facilities Based on the NSGA-III Algorithm
Reprinted from: *Water* **2021**, *13*, 3440, doi:10.3390/w13233440 59

Carlos Martínez, Zoran Vojinovic, Roland Price and Arlex Sanchez
Modelling Infiltration Process, Overland Flow and Sewer System Interactions for Urban Flood Mitigation
Reprinted from: *Water* **2021**, *13*, 2028, doi:10.3390/w13152028 75

Luís Mesquita David and Rita Fernandes de Carvalho
Designing for People's Safety on Flooded Streets: Uncertainties and the Influence of the Cross-Section Shape, Roughness and Slopes on Hazard Criteria
Reprinted from: *Water* **2021**, *13*, 2119, doi:10.3390/w13152119 93

Chengyao Wei, Jin Wang, Peirong Li, Bingdang Wu, Hanhan Liu, Yongbo Jiang and Tianyin Huang
A New Strategy for Sponge City Construction of Urban Roads: Combining the Traditional Functions with Landscape and Drainage
Reprinted from: *Water* **2021**, *13*, 3469, doi:10.3390/w13233469 107

Wensheng Tang, Haiyuan Ma, Xinyue Wang, Zhiyu Shao, Qiang He and Hongxiang Chai
Study on the Influence of Sponge Road Bioretention Facility on the Stability of Subgrade Slope
Reprinted from: *Water* **2021**, *13*, 3466, doi:10.3390/w13233466 121

Qian Li, Haifeng Jia, Hongkai Guo, Yunyun Zhao, Guohua Zhou, Fang Yee Lim, Huiling Guo, Teck Heng Neo, Say Leong Ong and Jiangyong Hu
Field Study of the Road Stormwater Runoff Bioretention System with Combined Soil Filter Media and Soil Moisture Conservation Ropes in North China
Reprinted from: *Water* **2022**, *14*, 415, doi:10.3390/w14030415 139

Chia-Chun Ho and Yi-Xuan Lin
Pollutant Removal Efficiency of a Bioretention Cell with Enhanced Dephosphorization
Reprinted from: *Water* **2022**, *14*, 396, doi:10.3390/w14030396 159

Fang Yee Lim, Teck Heng Neo, Huiling Guo, Sin Zhi Goh, Say Leong Ong, Jiangyong Hu, Brandon Chuan Yee Lee, Geok Suat Ong and Cui Xian Liou
Pilot and Field Studies of Modular Bioretention Tree System with *Talipariti tiliaceum* and Engineered Soil Filter Media in the Tropics
Reprinted from: *Water* **2021**, *13*, 1817, doi:10.3390/w13131817 . . . . **179**

Teck Heng Neo, Dong Xu, Harsha Fowdar, David T. McCarthy, Enid Yingru Chen, Theresa Marie Lee, Geok Suat Ong, Fang Yee Lim, Say Leong Ong and Jiangyong Hu
Evaluation of Active, Beautiful, Clean Waters Design Features in Tropical Urban Cities: A Case Study in Singapore
Reprinted from: *Water* **2022**, *14*, 468, doi:10.3390/w14030468 . . . . **209**

Bo Meng, Mingjie Li, Xinqiang Du and Xueyan Ye
Flood Control and Aquifer Recharge Effects of Sponge City: A Case Study in North China
Reprinted from: *Water* **2022**, *14*, 92, doi:10.3390/w14010092 . . . . **227**

About the Editors

Haifeng Jia

Prof. Dr. Haifeng Jia; He conducted his research and teaching in the fields of watershed / urban water environmental planning and management, water quality and hydrologic modelling, urban runoff control and Sponge city, and Environmental Remote Sensing and GIS. He has finished 100 research projects. He has published more 200 peer-reviewed journal papers and conference papers, and 10 books. He has received 39 different level academic and engineering Awards and Honors. He is active in international academic activities and international collaborations. He has organized and attended many international conferences.

Jiangyong Hu

Prof. Dr. Jiangyong Hu; She is specializing in the field of water innovative treatment technology, emerging contaminants detection and removal, water disinfection and biofilm control, stormwater management. She has published more than 170 papers in various international journals and been invited to deliver more than 70 invited speeches. She has been acted as chairs of organizing committee and co-president for a few international conferences and symposium. She currently serves as editors, guest editors and editorial board members for a number of journals and reviewers for funding agencies. She is Fellows of International Water Association and Institute of Engineers, Singapore. She is the President of Environmental Engineering Society of Singapore, Board member of International Ultraviolet Association, Chair of IWA Specialist Group Management Committee on Assessment and Control of Hazardous Substances in Water.

Tianyin Huang

Prof. Dr. Tianyin Huang; His research focused on wastewater control, advanced biological treatment, water environment restoration, municipal engineering planning and design. He has published more 100 peer-reviewed journal papers and 4 books. He established the institute of water environment and safety, worked on development of new technologies for wastewater treatment, ecological restoration, rain and flood management. These developed techniques have been applied in many projects including wastewater treatment and reclamation, sponge city construction and municipal wastewater treatment and reuse.

Albert S. Chen

Prof. Dr. Albert S. Chen; An Associate Professor at the Centre for Water Systems (CWS), University of Exeter with over 20 years of experience in Water and the Human Environment research. his interests include water resources, hydrology and hydraulic modelling, urban drainage, flood forecasting and early warning, innovation technology applications, water-food-energy-ecosystems nexus, climate change impact to critical infrastructure, prediction of water-borne disease, hazard impact assessment, mitigation and resilience strategies. he have participated in more than 40 international projects, and published over 60 peer-reviewed journal articles and 100 peer-reviewed conferences papers.

Yukun Ma

A/Prof. Yukun Ma; Her research interests include urban non-point source pollution, water environment pollution control in river basins, rainwater runoff and pollution control in sponge cities, and relevant results have been published in Journal of Hazardous Materials, Journal of Cleaner Production, Environmental Pollution, Science of the Total Environment, Chemosphere, Ecotoxicology and Environmental Safety and other international high-level journals.

Editorial

Urban Runoff Control and Sponge City Construction

Haifeng Jia [1,*], Jiangyong Hu [2], Tianyin Huang [3], Albert S. Chen [4] and Yukun Ma [5]

1. School of Environment, Tsinghua University, Beijing 100084, China
2. Department of Civil and Environmental Engineering, National University of Singapore, Singapore 117576, Singapore; hujiangyong@nus.edu.sg
3. School of Environment, Suzhou University of Science and Technology, Suzhou 215009, China; huangtianyin111@163.com
4. Centre for Water Systems, University of Exeter, Exeter EX4 4QF, UK; a.s.chen@ex.ac.uk
5. School of Environment, Beijing Normal University, Beijing 100875, China; ykma@bnu.edu.cn
* Correspondence: jhf@tsinghua.edu.cn

Citation: Jia, H.; Hu, J.; Huang, T.; Chen, A.S.; Ma, Y. Urban Runoff Control and Sponge City Construction. *Water* 2022, 14, 1910. https://doi.org/10.3390/w14121910

Received: 9 June 2022
Accepted: 10 June 2022
Published: 14 June 2022

Publisher's Note: MDPI stays neutral with regard to jurisdictional claims in published maps and institutional affiliations.

Copyright: © 2022 by the authors. Licensee MDPI, Basel, Switzerland. This article is an open access article distributed under the terms and conditions of the Creative Commons Attribution (CC BY) license (https://creativecommons.org/licenses/by/4.0/).

1. Introduction

Rapid population growth, urbanization and high-intensity human activities cause a multitude of extremely serious environmental problems all over the world [1]. In the construction of many urban areas, pervious vegetated ground surfaces have been progressively replaced with impervious pavements. Over the years, urbanization has induced floods and, consequently, the deterioration of the urban water environment. To alleviate these problems, the concept of a sponge city was first proposed and constructed in China [2,3].

Along with the onset of sponge city construction, many related studies were conducted, which yielded many positive results, such as the development of the China Sponge City database [4], environmental and economic cost–benefit comparison of sponge city construction [5,6], temporally and spatially adaptive optimal placement of green and grey runoff control infrastructures [7,8], etc. In order to reflect the state-of-the-art advances in urban runoff control and sponge city construction, we organized this Special Issue. We aimed to discuss and present research focused on the theories and technologies of sponge city construction; urban hydrology; methods of quantifying the benefits of a sponge cities; rainwater utilization; practices that mitigate urban flooding and soil erosion; the performance of GI; the impact of media; preferential flow paths; vegetation; climate; the design of hydrological, hydrodynamic and pollutant removal processes; and case studies on sustainable urban design and management using LID-GI principles and practices. We wish to express our gratitude to all the contributors who made this Special Issue so successful.

2. Summary of This Special Issue

In total, 13 papers were published in this Special Issue. The article types, authors, titles, keywords and study areas of these articles are summarized in Table 1. We have categorized these papers by article type in the table below.

Yin et al. presents a review on sponge city practices in China from their inception through to a systematic demonstration [9]. The main contents of the paper include: (1) Source control or a drainage system design for China's sponge city construction. The key element of sponge city construction is to combine various specific technologies to alleviate urban water problems, such as flooding, water environment pollution, shortage of water resources and deterioration of water ecology. (2) The sponge city pilot projects in China are introduced, the achievements obtained and lessons learned are summarized; (3) the objectives, corresponding indicators, key contents and needs of sponge city construction at various scales are identified. Moreover, the paper also describes the obligations of sponge city construction for various stakeholders.

Table 1. Summary of the papers published in the Special Issue "Urban Runoff Control and Sponge City Construction" for the journal *Water* (https://www.mdpi.com/journal/water/special_issues/UrbanRunoff_Control accessed on 13 June 2022).

Article Type	Authors	Title	Keywords	Study Area
General review	Yin, D.; Xu, C.; Jia, H.; et al.	"Sponge City Practices in China: From Pilot Exploration to Systemic Demonstration"	sponge city; low-impact development; pilot exploration; systematic demonstration; construction scale; stakeholders	China
Methods and tools	Liu, Z.; Yang, Y.; Hou, J.; et al.	"Decision-Making Framework for GI Layout Considering Site Suitability and Weighted Multi-Function Effectiveness: A Case Study in Beijing Sub-Center"	multifunctional decision-making framework; cost-effectiveness; site suitability; stakeholders' preference; green infrastructure	Beijing, China
Methods and tools	Wang, H.; Han, G.; Zhang, L.; et al.	"Integrated and Control-Oriented Simulation Tool for Optimizing Urban Drainage System Operation"	control-oriented model; urban drainage system; real-time optimization; Simuwater	/
Methods and tools	Peng, Z.; Jin, X.; Sang, W.; et al.	"Optimal Design of Combined Sewer Overflows Interception Facilities Based on the NSGA-III Algorithm"	combined sewer overflows; optimization; SWMM; NSGA-III	Wuhan, China
Methods and tools	Martínez, C.; Vojinovic, Z.; Price, R.; et al.	"Modelling Infiltration Process, Overland Flow and Sewer System Interactions for Urban Flood Mitigation"	Green-Ampt method; infiltration; overland flow; urban flood modelling; 1D/2D coupled modelling	/
Methods and tools	David, L.; Carvalho, R.	"Designing for People's Safety on Flooded Streets: Uncertainties and the Influence of the Cross-Section Shape, Roughness and Slopes on Hazard Criteria"	dual drainage modelling; extreme rainfall; flooding; safety criteria; urban drainage; uncertainty	/
Methods and tools	Wei, C.; Wang, J.; Li, P.; et al.	"A New Strategy for Sponge City Construction of Urban Roads: Combining the Traditional Functions with Landscape and Drainage"	urban water management; drainage function; permeable pavement; biological retention	Suzhou, China.
Typical source control facility	Tang, W.; Ma, H.; Wang, X.; et al.	"Study on the Influence of Sponge Road Bioretention Facility on the Stability of Subgrade Slope"	sponge city; bioretention facility; rain infiltration; slope stability	Chongqing, China
Typical source control facility	Li, Q.; Jia, H.; Guo, H.; et al.	"Field Study of the Road Stormwater Runoff Bioretention System with Combined Soil Filter Media and Soil Moisture Conservation Ropes in North China"	modified bioretention facility; road stormwater runoff; combined soil filter media; soil moisture conservation rope; field study; microbial diversity	Tianjin, China
Typical source control facility	Ho, C.; Lin, Y.	"Pollutant Removal Efficiency of a Bioretention Cell with Enhanced Dephosphorization"	low impact development; Sustainable Development Goals; non-point source pollution; enhanced dephosphorization bioretention	Hefei, China
Typical source control facility	Lim, F.; Neo, T.; Guo, H.; et al.	"Pilot and Field Studies of Modular Bioretention Tree System with Talipariti tiliaceum and Engineered Soil Filter Media in the Tropics"	urban runoff remediation; Talipariti tiliaceum; modular bioretention tree; field study; tree-pit	Singapore
Typical source control facility	Neo, T.; Xu, D.; Fowdar, H.; et al.	"Evaluation of Active, Beautiful, Clean Waters Design Features in Tropical Urban Cities: A Case Study in Singapore"	urban stormwater runoff management; field monitoring; ABC Waters design features; water quality; bioretention; swales	Singapore
Typical source control facility	Meng, B.; Li, M.; Du, X.; et al.	"Flood Control and Aquifer Recharge Effects of Sponge City: A Case Study in North China"	Sponge City; aquifer recharge; urban stormwater; green infrastructure	Zhengzhou, China

There are six papers that focus on the methods and tools of urban runoff control and sponge city construction. Liu et al. proposed a decision-making framework for GI layout considering site suitability and weighted multi-function effectiveness [10]. A case study in Beijing Sub-Center showed the feasibility of the proposed framework. Wang et al. developed an innovative modeling software that could play a role in the integrated simulation and overflow control application of urban drainage system [11]. The software was utilized in a real-time case–control study in one city of China, and it obtained significant optimized operation results, reducing combined sewer overflow (CSO) by making full use of the storage facilities and actuators. Peng et al. proposed a new simulation optimization method with new features of multithreading individual evaluation and fast data exchange by recoding SWMM with object-oriented programming [12]. These new features can rapidly accelerate optimization processes. The non-dominated sorting genetic algorithm-III (NSGA-III) was selected as the optimization framework for a better performance in dealing with multi-objective optimization. The proposed method was used in the optimal design of a terminal CSO interception facility in Wuhan, China.

Martínez et al. considered the influence of green infrastructure on urban surface runoff generation and proposed a new modelling setup, which includes a rainfall–runoff infiltration process in overland flow and its interaction with a sewer network [13]. The effect of infiltration losses on the overland flow was evaluated through an infiltration algorithm added in a so-called Surf-2D model. Then, the surface flow from a surcharge sewer was also investigated by coupling the Surf-2D model with the SWMM 5.1. An evaluation of two approaches representing urban floods was carried out based on two 1D/2D model interactions. Two test cases were implemented to validate the model. David and Carvalho highlighted how the change in street cross-section profile affected flood characteristics, which pose different levels of risk to pedestrians [14]. They also found that the uncertainty of roughness could be more influential in runoff than in street profiles. This methodology can be applied to improve street and drainage design to better manage urban runoffs. Wei et al. proposed a new strategy to combine roads, green spaces, and drainage systems [15]. The crux of this strategy is to consider the organization of the runoff and the construction of the drainage system (including sponge city source control facilities), so that both the drainage and traffic functions are achieved. This new strategy was implemented in a pilot study of road reconstruction conducted in Zhangjiagang, Suzhou, China.

A further six papers concentrated on research of the structures and performances of typical source control facilities; among these source control facilities, bioretention facilities are the focus. By establishing a three-dimensional finite element model for numerical analysis and combining it with geotechnical tests, Tang et al. studied the effects of bioretention facility on water pressure distribution, seepage path, and slope stability under rainwater seepage conditions [16]. In addition, this study explored the relationship between the parameters of the bioretention facility and the stability of the slope in combination with the effect of runoff pollution control. Li et al. invented a modified bioretention facility that contains soil moisture conservation ropes [17]. Additionally, a modified bioretention facility and a contrasting traditional bioretention facility were constructed in Tianjin Eco-city, China. A redundancy analysis was performed to evaluate the relationships between the variation in media physicochemical properties and microbial communities. It was found that an increase in media moisture could promote an increase in the relative abundance of several dominant microbial communities. Ho and Lin developed a new type of enhanced dephosphorization bioretention cell (EBC), which can not only efficiently remove nitrogen and COD, but also provides excellent phosphorus removal performance [18]. An EBC (length: 45 m; width: 15 m) and a traditional bioretention cell (TBC) of the same size were constructed in Anhui, China to treat rural nonpoint source pollution with high phosphorus concentration levels. After almost 2 years of on-site operation, the results indicate that TBCs and EBCs show similar performances in the removal of ammonium nitrogen and COD, but that the EBC significantly outperforms the TBC in terms of the total phosphorus removed.

In Singapore, Lim et al. developed a modular bioretention tree with a small footprint and a reduced on-site installation time for applications in a tropical environment [19]. The results show that the modular bioretention tree can effectively remove total suspended solids (TSS), total phosphorus (TP), zinc, copper, cadmium, and lead with removal efficiencies of greater than 90%. A field study in Singapore had a very clean effluent quality. Neo et al. characterized the performances of a rain garden and a vegetated swale that were implemented in a 4 ha urban residential precinct and monitored for a period of 15 months [20]. The results show that total suspended solids (TSS), total phosphorus (TP) and total nitrogen (TN) concentrations were low in the new residential precinct runoff. The findings from this study can help us to better understand the performance of source control facilities receiving low influent concentrations and the implications for further investigations that aim to improve stormwater runoff management in the tropics.

In order to evaluate the effect of aquifer recharge on flood control, Meng et al. proposed a sponge city design that highlighted aquifer recharge in a study area in Zhengzhou, China [21]. The stormwater management model of SWMM and the groundwater flow model of MOD-Flow were adopted to evaluate the flood control effect and aquifer-recharge effect, respectively. The results show that the sponge city design has a positive impact on maintaining groundwater level stabilization and even raises the groundwater level in some specific areas where stormwater seepage infrastructure is located.

3. Conclusions

This Special Issue highlights and discusses topics related to urban runoff control and sponge city construction. During the call period, 22 submissions were received. After the peer review process of the journal, a total of 13 papers were published in this Special Issue. Among the published papers, one review paper presents an in-depth review on sponge city practices in China, from its inception to national pilot construction projects and then to systematic demonstration. Six papers focus on the methods and tools of urban runoff control and sponge city construction, including planning strategies, simulation models, and optimization methods. Six papers concentrate on new findings regarding structures and performances of typical source control facilities, especially bioretention facilities. However, sponge city construction is still a new paradigm in city management; therefore, there are many theoretical, technical and practical problems that need to be addressed and solved, and it is expected that groundbreaking and innovative findings.

Funding: This research received no external funding.

Acknowledgments: Thanks to all of the contributions to this Special Issue, the time invested by each author, as well as the anonymous reviewers who contributed to the development of the articles in this Special Issue. All the guest editors are very pleased with the review process and management of the Special Issue and offer their gratitude.

Conflicts of Interest: The authors declare no conflict of interest.

References

1. Jia, H.; Yao, H.; Yu, S. Advances in LID BMPs research and practices for urban runoff control in China. *Front. Environ. Sci. Eng.* **2013**, *7*, 709–720. [CrossRef]
2. Jia, H.; Wang, Z.; Zhen, X.; Clar, M.; Yu, S. China's Sponge City Construction: A Discussion on Technical Approaches. *Front. Environ. Sci. Eng.* **2017**, *11*, 18. [CrossRef]
3. Yin, D.; Chen, Y.; Jia, H.; Wang, Q.; Chen, Z.; Xu, C.; Li, Q.; Wang, W.; Yang, Y.; Fu, G.; et al. Sponge City Practice in China: A Review of Construction, Assessment, Operational and Maintenance. *J. Clean. Prod.* **2021**, *280*, 124963. [CrossRef]
4. Xu, C.; Shi, X.; Jia, M.; Han, Y.; Zhang, R.; Ahmad, S.; Jia, H. China Sponge City database development and urban runoff source control facility configuration comparison between China and the US. *J. Environ. Manag.* **2022**, *304*, 114241. [CrossRef]
5. Zhu, Y.; Xu, C.; Yin, D.; Xu, J.; Wu, Y.; Jia, H. Environmental and economic cost-benefit comparison of Sponge City construction in different urban functional regions. *J. Environ. Manag.* **2022**, *304*, 114230. [CrossRef]
6. Xu, C.; Liu, Z.; Chen, Z.; Zhu, Y.; Yin, D.; Leng, L.; Jia, H.; Zhang, X.; Xia, J.; Fu, G. Environmental and economic benefit comparison between coupled grey-green infrastructure system and traditional grey one through a life cycle perspective. *Resour. Conserv. Recycl.* **2021**, *174*, 105804. [CrossRef]
7. Jia, H.; Liu, Z.; Xu, C.; Chen, Z.; Zhang, X.; Xia, J.; Yu, S. Adaptive pressure-driven multi-criteria spatial decision-making for a targeted placement of green and grey runoff control infrastructures. *Water Res.* **2022**, *212*, 118126. [CrossRef]
8. Xu, T.; Li, K.; Eengel, B.A.; Jia, H.; Leng, L.; Sun, Z.; Yu, S.L. Optimal adaptation pathway for sustainable low impact development planning under deep uncertainty of climate change: A greedy strategy. *J. Environ. Manag.* **2019**, *248*, 109280. [CrossRef]
9. Yin, D.; Xu, C.; Jia, H.; Yang, Y.; Sun, C.; Wang, Q.; Liu, S. Sponge City Practices in China: From Pilot Exploration to Systemic Demonstration. *Water* **2022**, *14*, 1531. [CrossRef]

10. Liu, Z.; Yang, Y.; Hou, J.; Jia, H. Decision-Making Framework for GI Layout Considering Site Suitability and Weighted Multi-Function Effectiveness: A Case Study in Beijing Sub-Center. *Water* **2022**, *14*, 1765. [CrossRef]
11. Wang, H.; Han, G.; Zhang, L.; Qiu, Y.; Li, J.; Jia, H. Integrated and Control-Oriented Simulation Tool for Optimizing Urban Drainage System Operation. *Water* **2022**, *14*, 25. [CrossRef]
12. Peng, Z.; Jin, X.; Sang, W.; Zhang, X. Optimal Design of Combined Sewer Overflows Interception Facilities Based on the NSGA-III Algorithm. *Water* **2021**, *13*, 3440. [CrossRef]
13. Martínez, C.; Vojinovic, Z.; Price, R.; Sanchez, A. Modelling Infiltration Process, Overland Flow and Sewer System Interactions for Urban Flood Mitigation. *Water* **2021**, *13*, 2028. [CrossRef]
14. David, L.; Carvalho, R. Designing for People's Safety on Flooded Streets: Uncertainties and the Influence of the Cross-Section Shape, Roughness and Slopes on Hazard Criteria. *Water* **2021**, *13*, 2119. [CrossRef]
15. Wei, C.; Wang, J.; Li, P.; Wu, B.; Liu, H.; Jiang, Y.; Huang, T. A New Strategy for Sponge City Construction of Urban Roads: Combining the Traditional Functions with Landscape and Drainage. *Water* **2021**, *13*, 3469. [CrossRef]
16. Tang, W.; Ma, H.; Wang, X.; Shao, Z.; He, Q.; Chai, H. Study on the Influence of Sponge Road Bioretention Facility on the Stability of Subgrade Slope. *Water* **2021**, *13*, 3466. [CrossRef]
17. Li, Q.; Jia, H.; Guo, H.; Zhao, Y.; Zhou, G.; Lim, F.; Guo, H.; Neo, T.; Ong, S.; Hu, J. Field Study of the Road Stormwater Runoff Bioretention System with Combined Soil Filter Media and Soil Moisture Conservation Ropes in North China. *Water* **2022**, *14*, 415. [CrossRef]
18. Ho, C.; Lin, Y. Pollutant Removal Efficiency of a Bioretention Cell with Enhanced Dephosphorization. *Water* **2022**, *14*, 396. [CrossRef]
19. Lim, F.; Neo, T.; Guo, H.; Goh, S.; Ong, S.; Hu, J.; Lee, B.; Ong, G.; Liou, C. Pilot and Field Studies of Modular Bioretention Tree System with Talipariti tiliaceum and Engineered Soil Filter Media in the Tropics. *Water* **2021**, *13*, 1817. [CrossRef]
20. Neo, T.; Xu, D.; Fowdar, H.; McCarthy, D.; Chen, E.; Lee, T.; Ong, G.; Lim, F.; Ong, S.; Hu, J. Evaluation of Active, Beautiful, Clean Waters Design Features in Tropical Urban Cities: A Case Study in Singapore. *Water* **2022**, *14*, 468. [CrossRef]
21. Meng, B.; Li, M.; Du, X.; Ye, X. Flood Control and Aquifer Recharge Effects of Sponge City: A Case Study in North China. *Water* **2022**, *14*, 92. [CrossRef]

Review

Sponge City Practices in China: From Pilot Exploration to Systemic Demonstration

Dingkun Yin [1], Changqing Xu [1,*], Haifeng Jia [1,*], Ye Yang [2], Chen Sun [2], Qi Wang [2] and Sitong Liu [3]

1. School of Environment, Tsinghua University, Beijing 100084, China; ydk19@mails.tsinghua.edu.cn
2. China Urban Construction Design Research Institute Co., Ltd., Beijing 100120, China; yangye@cucd.cn (Y.Y.); sunchen@cucd.cn (C.S.); wangqiyuanlin@cucd.cn (Q.W.)
3. Capital Urban Planning & Design Consulting Development Co., Ltd., Beijing 100031, China; liuchao@bmicpd.com.cn
* Correspondence: xuchangqing@tsinghua.edu.cn (C.X.); jhf@tsinghua.edu.cn (H.J.)

Abstract: In recent years, China has been committed to strengthening environmental governance and trying to build a sustainable society in which humans and nature develop in harmony. As a new urban construction concept, sponge city uses natural and ecological methods to retain rainwater, alleviate flooding problems, reduce the damage to the water environment, and gradually restore the hydrological balance of the construction area. The paper presents a review of sponge city construction from its inception to systematic demonstration. In this paper, research gaps are discussed and future efforts are proposed. The main contents include: (1) China's sponge city construction includes but is not limited to source control or a drainage system design. Sponge city embodies foreign experience and the wisdom of ancient Chinese philosophy. The core of sponge city construction is to combine various specific technologies to alleviate urban water problems such as flooding, water environment pollution, shortage of water resources and deterioration of water ecology; (2) this paper also introduces the sponge city pilot projects in China, and summarizes the achievements obtained and lessons learned, which are valuable for future sponge city implementation; (3) the objectives, corresponding indicators, key contents and needs of sponge city construction at various scales are different. The work at the facility level is dedicated to alleviating urban water problems through reasonable facility scale and layout, while the work at the plot level is mainly to improve the living environment through sponge city construction. The construction of urban and watershed scales is more inclined to ecological restoration and blue-green storage spaces construction. Besides, the paper also describes the due obligations in sponge city construction of various stakeholders.

Keywords: sponge city; low-impact development; pilot exploration; systematic demonstration; construction scale; stakeholders

1. Introduction

The rapid urbanization process in China has effectively driven the development of the national economy [1]. However, it has also exposed the risk of urban water issues due to the increase in impervious underlying surface and a decrease in green space and water areas [2–5]. This led to a significant reduction in the amount of rainfall runoff absorbed in the processes of plant interception, infiltration, depression detention, and evapotranspiration [6–8], and then increased the flooding risk [9–11]. When the rainfall runoff is large enough and exceeds the capacity of the drainage networks, it will bring more serious urban water safety and water environment issues [12,13]. Many cities in China have suffered a variety of water related problems such as frequent flooding, water environment pollution, water resources shortage and water ecology deterioration, which have seriously impacted the quality of people's life [14–16].

In order to improve the status quo, on the basis of learning from the stormwater management experiences in developed countries, China initiated the development of sponge city in 2013. This became a new solution for urban stormwater management. Whereafter, the government issued a series of related policies and guidelines for the sponge city development in an effort to improve the sponge city construction [17,18]. In addition, China central government selected 30 pilot cities considering their different natural and social conditions (with the average construction area of 31.3 km^2 for each city) for the sponge city construction exploration in 2015 and 2016, and all of them have completed performance assessment in the end of 2019 [19]. Furthermore, in 2021, based on the experiences of pilot cities, China began to systematically promote the sponge city demonstration on a national scale.

The construction of sponge city emphasis the full utilization of the natural absorption and infiltration capacity of pervious areas to effectively control stormwater runoff, thereby minimizing water system problems caused by the damage of urbanization-induced hydrological effects. The philosophy of sponge city is to transform the traditional "fast drainage" principle to a systematic implementation of "infiltration, detention, retention, purification, utilization and discharge" [13]. It aims to achieve stormwater runoff control from source reduction and process control to systematic remediation through planning, design, construction, operation, and management, which would lead to a sustainable approach for urban development.

At present, various sponge city related studies have become more and more extensive, including the analysis of various green infrastructures' performance, the interpretation of policies, and the optimization of green-gray infrastructure layout at the planning level. Based on the analysis of relevant references, government reports and actual sponge city projects, we present a comprehensive review of the sponge city from the inception to the development. In particular, construction modes, achievements, lessons learned during the pilot exploratory phase are all identified and analyzed. Then, we try to provide an outlook of the next stage work, which can be a guide for the promotion of sponge city systematic demonstration, especially the working contents at different scales and obligations of various stakeholders.

2. Inception of Sponge City

2.1. Foreign Advanced Experiences

The urban drainage concept can be traced back to the 3000 BC, with the most important goal being to quickly discharge the rainfall runoff from the urban area to the downstream channel or other receiving water bodies [20]. However, with the complexity of urban development and the frequent occurrence of extreme stormwater events, a series of relatively 'novel' concepts have appeared in the different developed countries. These concepts mainly include the Low Impact Development (LID), Green Infrastructure (GI) and Best Management Practices (BMPs) [21] in US, the Sustainable Urban Drainage System (SUDS) in UK [22], the Water Sensitive Urban Design (WSUD) in Australia [23] and Nature-Based Solution (NBS) [24]. These concepts attempted to combine urban drainage with natural processes to reduce rainfall runoff through various nature-based solutions and therefore, achieved a benign urban water cycle. All of these concepts are closely related, but different terms represent different technical systems, which also have certain differences in the field of application scale, technical measures and control objectives [25]. This section describes the concepts related to drainage proposed by the US, the UK and Australia, and explores the relationship and connection between these concepts and sponge city construction.

2.1.1. LID-BMP and GI for Source Control in US

BMP first appeared in the "Clean Water Act", enacted by the US Congress in 1972 and was first applied mainly in the field of sewage discharges or point sources [26]. After 15 years, the BMP for stormwater runoff or nonpoint pollution control was implemented. The main technical measures included different low-cost engineering measures. Besides, it also emphasized the non-engineering measures, such as facility maintenance rules. Since then, the concept of LID has been used in the reports related to urban stormwater management issued by the US EPA and in the related design manuals of various states.

The application of source runoff control facilities has been promoted by taking a "nature design approach", such as green roofs and rain gardens. Moreover, the US also promoted urban drainage design by using LID-BMPs, which represented all of the BMPs for urban stormwater runoff control using the LID strategy, and the frequency of this concept rapidly increased in international literature in recent years [27]. Subsequently, green infrastructure or GI, which covers traditional BMP and typical LID measures, was evolved as the term represented source control infrastructure for urban runoff. GI can bring multiple ecological benefits, such as alleviating the urban heat island (UHI) effect, increasing biological habitat, and improving biodiversity [20]. In the sponge city construction, source runoff control is also a top priority since it can effectively reduce the total amount of runoff and absorb part of the runoff on-site. However, the source control of sponge city includes not only small, decentralized infiltration and retention facilities (green roofs, grass swales and bioretention), but also large-scale storage facilities, such as stormwater ponds and wetlands. It is important to select appropriate facilities according to the scale and characteristics of runoff quantity and quality of the specific region.

2.1.2. SUDS for Multifunctional Drainage System Design in UK

In UK, a sustainable drainage concept was proposed in 2007, which not only includes the concepts of LID-BMPs and GI in the US, it also diversifies the design of the drainage system to avoid the traditional sewer network being the only drainage outlet [28,29]. Meanwhile, the filtering effect of drainage facilities was taken into account to reduce the discharge of pollutants into the receiving water body. In addition, rainfall collection and utilization were also emphasized [30–32]. Thus, strategies for the urban stormwater management became more functional rather than focusing solely on rapid runoff discharge. Besides, various corresponding environment, social and economic benefits were also obtained [33,34]. These benefits were reflected not only in the overall reduction of urban runoff, the improvement of air quality, and the CO_2 storage, but also in the burden reduction of the stormwater fees and the energy consumption [33,35–37]. It is not difficult to see that SUDS rose from a traditional "rapid drainage" system to a more sustainable and multifunctional drainage system that maintains a high level of benign water circulation. Meanwhile, it began to optimize the entire water system including urban drainage, sewage, and reclaimed water system rather than that of only urban drainage facilities. This also coincided with the concept of sponge city construction. In the sponge city design, the water quantity and quality, potential landscape and ecological value of runoff are all comprehensively considered.

2.1.3. WSUD for Urban Water System Optimization in Australia

Around the technical core of urban stormwater management, Australia put forward the concept of WSUD in 1994 through a whole understanding of the water cycle in the local physical and environmental context [38,39]. It was also the first time that stormwater, groundwater, drinking water, sewage and reclaimed water system were comprehensively considered together. WSUD was described as "a philosophical approach to urban planning and design aimed at reducing the hydrological impact of urban development on the surrounding environment" [23]. It emphasized the consideration of stormwater management issues within an integrated framework of the entire urban water cycle [40]. Different with LID-BMPs, WSUD used integrated method to achieve stormwater management rather than only micro-scale landscape stormwater control [41]. All of these had a higher overlap with sponge city construction [42]. For example, the fragmentation of management and the discretization of related departments might hinder the evolution of WSUD. Thus, WSUD promoted urban water management through institutional construction and administrative measures to build a long-term mechanism for sustainable urban design [41]. As far as the sponge city construction is concerned, it is still necessary to learn from WSUD and conduct various studies to provide scientific construction guidance, including the runoff regulation capacity of different GIs, long-term tracking monitoring, and comprehensive performance evaluation [31,43–45]. All of these actions play a vital role in the development of sponge city construction and its promotion in the future.

2.2. Chinese Historical Inheritance

The idea of "natural storage, natural infiltration, and natural purification" in sponge city is derived from the wisdom of the ancient Chinese people using nature approaches to discharge and collect stormwater. As early as the Qin and Han dynasties (221 BC–220 CE), China began to build strip-shaped or wave-shaped terraced fields along the contour lines on the hills for farming [46]. This was also an effective measure for controlling soil erosion on sloping farmland. The terraced fields has been listed as a United Nations Educational, Scientific and Cultural Organization (UNSCEO) heritage since 2013 [47]. We can see that people in ancient times have been able to combine the living environment with the natural environment to realize the recycling of stormwater. In the settlement development, Chinese ancients also took advantage of natural power to harvest stormwater for utilization, to drainage stormwater for safety.

2.2.1. Ancient Stormwater System of Courtyards and Villages

The domestic studies on ancient Chinese drainage system are mostly including structure, composition, and operation mode. In China, the quadrangle is one of the most common buildings in ancient times, but the styles are slightly different to adapt local climate. The ancient wisdom and concepts contained in the ancient courtyard drainage system also can be a valuable reference for the current sponge city construction. In northern China, the rainfall depth is much less than that in the southern region. However, the rainfall is more turbulent, and the instantaneous rainfall intensity is stronger, which requires the drainage system to have a good drainage capacity. The quadrangles are all built with walls. Along these walls, flowers and trees are planted (Figure 1a), so that people can enjoy the natural scenery while utilizing rainwater. Usually, as the rain falling down the eaves, part of the runoff is absorbed on the permeable surface in the courtyard, and the rest is discharged out of the courtyard along the ditches into the drainage system.

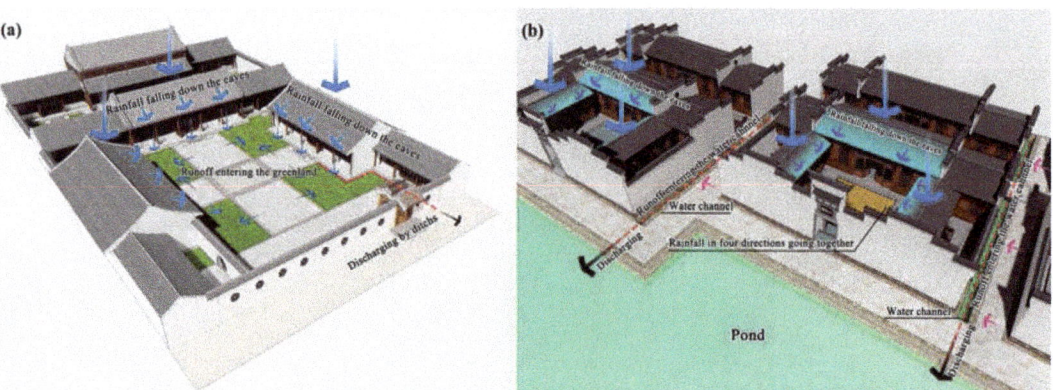

Figure 1. Drainage systems in ancient southern and northern China: (**a**) Typical courtyard buildings in northern China; (**b**) Ancient villages in southern China.

In the southern China, courtyard is more restrained and smaller compared with northern region due to the scarce land resources. Therefore, patios are often used as a substitute for courtyards. The average annual rainfall in the southern region exceeds 800 mm, which is generally higher than that in northern area.

Taking Hongcun which located at the south of Huangshan City, Anhui Province, as an example, it was built in the Southern Song Dynasty (1127–1279 CE) and was also selected as an UNESCO heritage. Based on historical records, Hongcun was unscathed and unaffected by heavy rainfall events in history [48,49]. The reason is that Hongcun includes a smart drainage system that combines storage and drainage facilities together, so that it can retain stormwater runoff on-site first and drain the extra stormwater to downstream water bodies safely. As can be seen in Figure 1b, when rainfall events occur, the stormwater is flowing down the eaves, entering the courtyard, and then drains into the river from the trenches around the patio. Thus, stormwater runoff in Hongcun can be merged into the channel and spread throughout the village. The runoff from the channel then flows into the pond for midway regulation and storage along the terrain, and finally flows into the receiving water body. Hongcun solves the problem of water shortage in the dry season and can reduce flood peaks and runoff flow volume through the rational use of channel, pond and receiving water body at the raining time. The villagers use part of the collected stormwater for production and living. In addition, the management system in Hongcun clearly states that domestic sewage water cannot be directly discharged into the channel and needs to be infiltrated through the soil. This measure ensures the water environment of the stormwater system is not affected by domestic sewage water.

There are also examples for larger areas to deal stemwater rationally, such as Ganzhou in Jiangxi province, China [48]. The urban drainage system of Ganzhou also make reasonable use of ditches, ponds and city walls to achieve source reduction and resource utilization of a large amount of runoff generated by rainfall events, and quickly discharge excess runoff into downstream receiving water bodies [50]. All of the above are good references for sponge city construction and modern stormwater management.

2.2.2. Ancient Drainage System of Architectural Complex

In addition to the smart stormwater management in courtyard and village, the stormwater system in architectural complex was also well designed in ancient China. Taking Tuancheng (Figure 2), Beijing, which was constructed during the Ming Dynasty (1368–1644 CE), as an example, its area is about 0.5 hectares with an average annual rainfall depth of 560 mm [48]. On 21 July 2012, Beijing suffered the extraordinary stormwater event in the past 60 years, with an average rainfall depth of 210.7 mm. According to data, approximately 1.602 million people were affected, and the economic loss was about 11.64 billion Yuan. However, the drainage system in Tuancheng was still in service, and there was no report of flooding there. In Tuancheng, there are no open ditches. The ground of Tuancheng is paved by bricks with good water permeability, and the shape of these bricks is an inverted trapezoid. When rainfall occurs, the stormwater runoff infiltrates into the ground through the gaps between adjacent bricks (Figure 2). When the runoff quantity is large enough that cannot be absorbed locally, it will flow into the surrounding water holes from north to south according to the terrain. The vertical shafts are directly below the water holes, and connected by culverts with a height of 80–150 cm. Therefore, the runoff flows into the water holes can be stored among the culverts, this design cleverly solves the local drainage problem, which can be used for reference when dealing with urban flooding issues.

There are countless historical sites similar to Tuancheng scattered in China, which contain extremely rich scientific and technological value. They are the concrete reflection of ancient scientific thinking, water culture, and technological progress. Besides, these architectures also embody the ingenuity of ancient people and demonstrate historical process of social science and technology development. Traditional culture is the source of modern culture. In order to better understand the modern urban stormwater management system, traditional culture must be learned. The historical site faithfully records the traditional way of stormwater utilization with high cohesion of traditional culture and is also a window for scholars to explore the symbiosis of human and water. At the same time, it provides the possibility for people to experience the broad, profound, and splendid traditional culture.

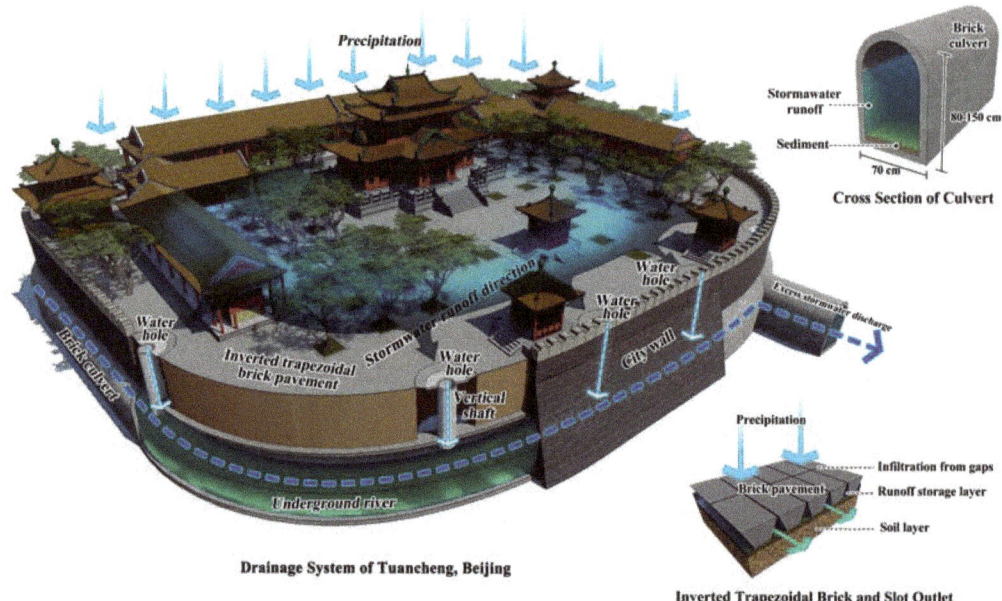

Figure 2. Typical ancient drainage system of Tuancheng, Beijing in northern region.

However, with the onset of rapid urbanization, some ancient drainage systems have been gradually replaced by a large number of engineering pipe systems in urban area which led to urban water quantity and quality issues. In order to avoid the aggravation of such problems, it is urgent to carry out sponge city constructions in urban area to reverse these kinds of situations.

3. Sponge City Pilot Exploration

3.1. Sponge City Construction Implementation Mode in China

Based on the several years' exploration of sponge city construction, the Chinese implementation mode was formed (Figure 3). In the implantation mode, the government carries the main responsibility. Central government acts as the promoter and local municipal governments are organizers of sponge city construction and management. Usually, a sponge city construction office or committee would be setup, which includes the officers from the related municipal bureau or agencies of urban planning, construction, landscaping, transportation, environment protection, water resources, and so on. The sponge city construction office or committee organizes all of the related issues on sponge city construction and management.

During the planning stage, all of the different levels of planning are oriented by technology. For overall city planning level, the concept and the target of sponge city construction should be included, and the implementation strategy should be proposed. For special planning level, the main principles focus on "flooding control, water environment improvement, water resources conservation and water ecology rehabilitation". The objectives are "no ponding in light rain, no flooding in heavy rain, no black and odorous urban water, and mitigating the heat island effect". The schemes which can meet the above objectives should be proposed related with infrastructure space layout, water system, green space and road system.

As of the design, construction and operation stage, the role of different stakeholders in sponge city construction are emphasized. The Public-Private Partnership (PPP) mode is encouraged. Typically, it involves private capital financing government projects and services up-front, and then drawing profits from taxpayers and/or users over the course of the PPP contract. In sponge city construction, usually the related projects are bundled up by municipal government and sign PPP contract with the qualified enterprises. The enterprise will have responsibility for design, construction, and operations.

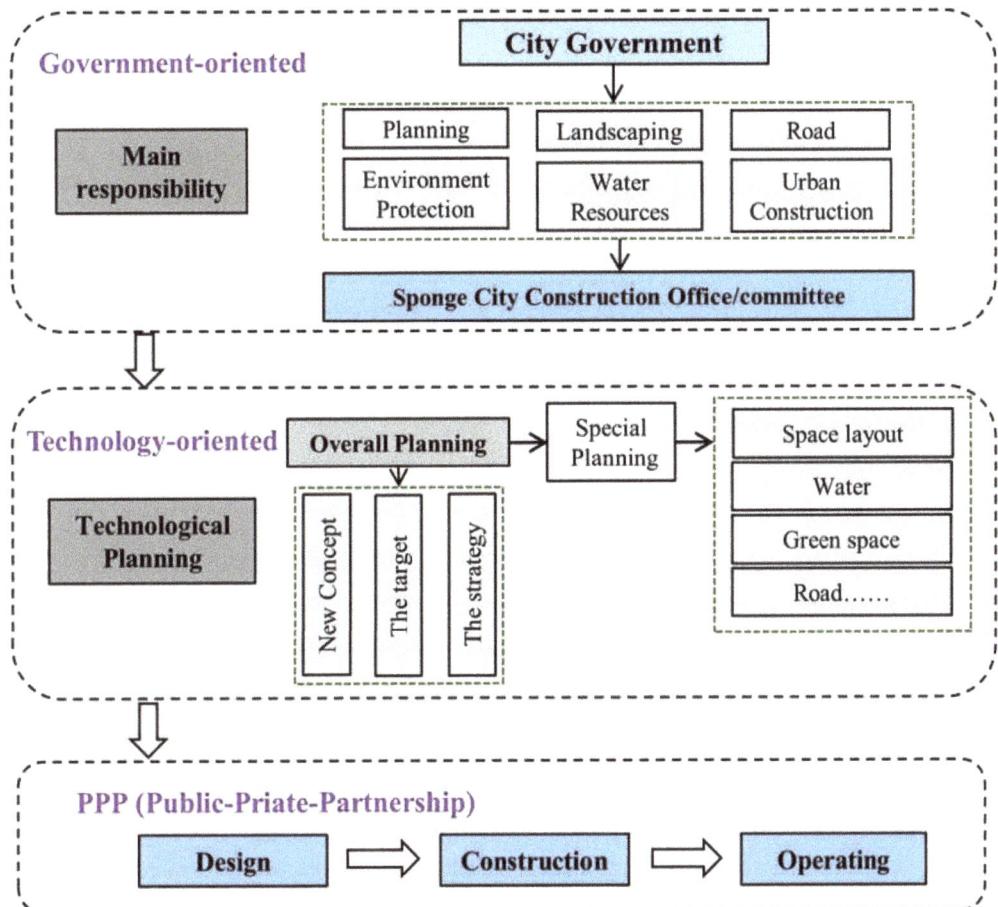

Figure 3. Chinese implementation mode of sponge city pilot construction.

3.2. National and Local Pilot Sponge Cities

To explore the specific development model of sponge cities construction in different regions of China, the Ministry of Finance (MF), the Ministry of housing and urban-rural development (MHURD) and the Ministry of water resources (MW) of China coordinately promoted the national pilot sponge cities construction. According to the characteristics of China's geographical climate, average annual rainfall, and urban development intensity, 30 cities were selected as national pilot sponge cities with different annual rainfall volume capture ratio targets in 2015 and 2016 (Figure 4). Each pilot city constructed a pilot region with no less than 15 km^2 in 3 years. The main task of pilot cities was to explore a development model which is suitable for the construction of sponge cities in the specific region, and to form a set of practices, experiences, policies, and systems which can be promoted in similar cities. In addition to the national pilot construction, provinces and cities have also carried out their own sponge city pilot construction. According to statistics, 13 provinces have carried out local pilot programs in 90 cities, 28 provinces have issued requirements of sponge city construction, and two-thirds of the cities in China have formulated special plans for sponge city construction [51].

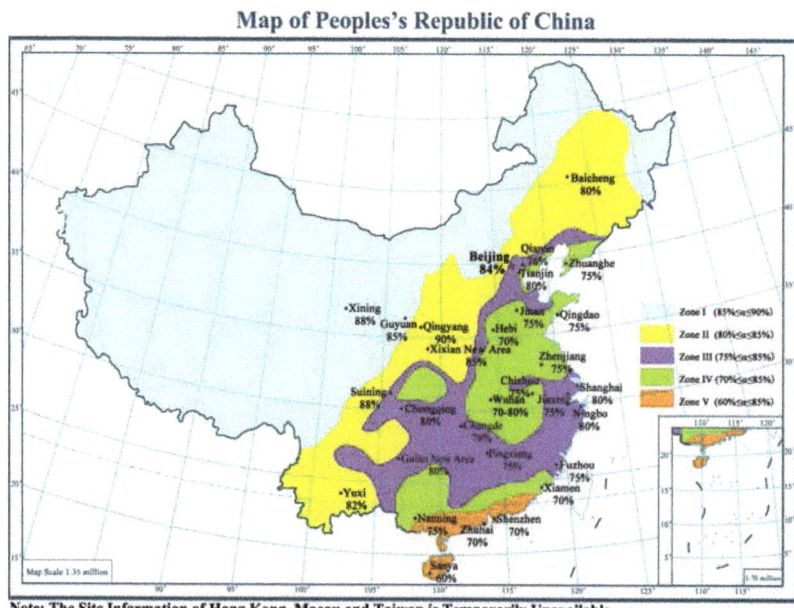

Figure 4. Location of 30 national pilot sponge cities in China (reprinted with permission from Ref. [13]. 2020, Haifeng Jia).

3.3. Achievements

The 30 national pilot sponge cities have passed the joint acceptance check of MHURD, MF, and MW in 2019. After summarizing the national pilot experiences and practices, it was found that many impressive achievements have been obtained.

3.3.1. Worldwide Influence

Sponge city has gained worldwide attention due to its innovative concepts, huge implementation plan and strong performance in improving water quality and controlling flooding situation. To strengthen the academic communication, there are many books, journal papers, international conferences, workshops, and other academic activities on sponge city. The most attractive international conferences are the 2016 International LID Conference held in Beijing and the 2018 International sponge city conference held in Xi'an. Statistical results showed that more than 1200 attendants from more than 20 countries and regions attended the 2016 International LID Conference, and more than 2000 attendants (more than 800 thousand persons online) attended the 2018 International sponge city conference. The internationally renowned journal, *Nature* and *Science*, also reported that sponge city is crucial for many cities which suffer severe flooding and water shortage [52,53]. These studies pointed out that the green sponge infrastructure should be combined with conventional drainage systems, particularly in areas of medium- and high intensity urbanization.

In recent years, related literatures have been rapidly growing. The literature of China National Knowledge Network (CNKI) from 1995 to 2021 showed that sponge city related (for example, sponge city, permeable pavement, low impact development, rain garden, LID, sunken green space, green roof) has reached more than 2000 papers. These papers cover engineering cases, reviews, experimental studies, planning schemes, efficiency evaluation, and so on. Moreover, many books, such as typical cases of sponge city construction, sponge city construction and operation technology system, theory and practice of urban sponge green space planning and design, etc. have been published to help people further understand sponge city [54–56].

The International Water Association (IWA) published many articles (for example, getting to Climate Resilient and Low Carbon Urban Water) about the performance of sponge city in controlling urban runoff and avoiding water scarcity. Major international journals, such as Water Research, Resources, Conservation & Recycling, Journal of Cleaner Production, Science of the Total Environment, Journal of Hydrology, Journal of Environmental Management, and etc. published a large number of papers which are related to sponge city [13,57–60].

3.3.2. Sponge City Performance

In recent years, a large amount of data on sponge city performance has been accumulated. However, these data have various sources, diverse formats, and uneven quality, making it difficult to fulfill relevant research and design needs directly. Thus, Xu et al. establishes a China sponge city database with a clear structure and convenient management schemes [61]. It includes facility size and cost information under various environmental conditions. At present, 1066 urban runoff source control facilities parameters from 30 pilot sponge cities are included in the database. The database can provide useful guidance information to other countries with similar environmental and fiscal conditions for the construction of urban runoff source control facilities.

Sponge city urban runoff source control facilities can achieve good water quality control performance. The average pollutant removal rates of urban runoff source control facilities are presented in Table 1. Results showed that these facilities have good removal performance in COD, SS, NH_3-N, TP, TN, Pb, and Zn.

Table 1. Pollutants average removal rate of urban runoff source control facilities (reprinted with permission from Ref. [51]. 2020, Changqing Xu).

Urban Runoff Source Control Facilities	Pollutants Average Removal Rate (%)						
	COD	SS	NH_3-N	TP	TN	Pb	Zn
Concaved green space	51.65	-	60.39	54.88	33	-	-
Constructed wetland	86.23	71.18	67.07	70.56	85.33	62.71	-
Bioretention	59.10	79.15	65.45	72	73.90	-	-
Permeable pavement	62	34.93	39	57	53	60	60
Detention pond	41.88	59.32	21.62	20.05	15	-	-
Buffer strip	77.97	90	-	85.11	69.93	-	-
Grassed swale	26.70	46.25	44.70	51.40	-	98	97

As mentioned before, the 30 national pilot sponge cities have passed the joint acceptance check by MHURD, MF, and MW. The 30 cities all achieved the pilot objectives of water environment, water ecology, water resources, and water security. For example, performance evaluation of sponge city construction in Qian'an city is good [62]. The scores of water resources, water security, water ecology and economic benefits are high, indicating Qian'an has made positive progress in rational utilization of water resources, water security, and economic benefits after sponge city implementation. Similar to Qian'an, Jiaxing's sponge city construction performance is at a "relatively high" level [63]. From the performance evaluation results, the sponge city construction in Jiaxing has achieved remarkable results, indicating the rationality and feasibility of its overall implementation plan and related policies. Specifically, the environmental performance, which includes water ecology, water environment and water resources indicators (for example, volume capture ratio of annual rainfall, groundwater level, water resource utilization rate, water environmental quality), has higher index weight coefficient than social performance and management performance. Therefore, Jiaxing should focus on the control of the above factors in the future construction of sponge city. The Science and Education Channel of China Central Television (CCTV-10) reported that Nanning City implemented permeable pavement, rain garden, green roof and other practices in sponge city pilot areas, these made Nakao River change from a gutter to an ecological wetland park. The citizens living in Nanning have experienced the significant positive changes in the urban water environment around them.

3.3.3. Education and Talent Training

To further promote the sponge city construction sustainably, strengthening professional expertise and social publicity are necessary. Apart from building information promotion platforms and implementing community demonstration projects, letting sponge city enter the campus and training sponge city talented people are key strategies. At present, there are some teaching materials (for example, Designing Our Sponge Community, Sponge Castle Adventure (primary school version), Sponge city Exploration (middle school version) and activities have been promoted in schools [64–66].

The comprehensive practice series instruction book "Designing Our Sponge Community" was based on the Primary Science Curriculum Standards for Compulsory Education issued by the Ministry of Education. This textbook becomes the "China's first STEAM (Science, Technology, Engineering, Arts, and Mathematics) project teaching guide for primary and secondary schools". Besides the textbooks, many schools launched activities to help students have a comprehensive understanding of sponge city and feel the great changes sponge city has brought to the city. In Suzhou, sponge city course taught on campus has aroused the students' interest in knowing the theory of sponge city. Field trips were organized to guide students to understand the design of bioretention, a permeable pavement, and a green roof etc. In Jinan, the students carry out some experiments in campus by themselves to help them better understand the principle and function of SPONGE CITY. In fact, the active promotion of sponge city construction allows everyone to develop an environmental protection and sustainable thinking mode.

Sponge city construction is a multi- and cross-disciplinary project that involve the majors of water supply and drainage engineering, environmental engineering, water conservancy engineering, urban planning, land use, landscape, transportation and ecology [67]. In the process of sponge city construction, close cooperation across all these disciplines is very important. Wide academic discussion and collaboration are needed. The summary of achievements, such as books, handouts, papers, atlas, manuals, and etc. are significant in education and talent training.

3.3.4. Public Awareness

As a national scale public project to address environmental issues, sponge city project is subject to public financial support and perception. Willingness to pay (WTP) is an effective tool to explore public behavioral intention and evaluate integrated benefits of a project. Wang et al. examined public perceptions of sponge city construction, as well as the public's willingness to support sponge cities in Zibo and Dongying City, Shandong Province [68]. A total of 1800 questionnaires were distributed with 900 each in Zibo and Dongying City, finally 1443 were valid returns. Results indicated that most respondents knew about sponge city projects and supported sponge city construction in residential areas. Respondents also accepted 17% of the domestic water price as a surcharge to be used for sponge city construction. Results also showed that educational level, income, and occupation were main factors affecting respondents' WTP to support sponge city initiatives. Wang et al. made a questionnaire survey in the flood affected communities with 656 respondents in Guyuan City [69]. Survey results showed that most respondents accepted an 8.3% surcharge of domestic water tariff for sponge city development. The results provide practical implications for government and developers to optimize financing and operation of sponge city developments and thus can improve the sustainable performance of sponge city.

During the process of sponge city construction, public dissent has arisen over the effectiveness of sponge cities, the most common dissent is "Omnipotence" and "Uselessness" of sponge city [70]. This is mainly because the public holds unrealistic expectations to the effect of sponge city construction, hoping to solve all water problems in one way, once and for all. In addition, some sponge city construction projects exaggerate the implementation effect in the early publicity, undoubtedly contributing to the public misjudgment. Besides, the public has a partial understanding of the nature of sponge city construction, thinking only source reduction LID facilities represent sponge city construction. Another reason is that the public only cares about sponge city construction projects when flooding occurs, and they are more concerned about the quality of outdoor built environment. Whether sponge facilities are effective depends largely on construction quality, which is consistent with the public's perception of built environment quality. For some sponge facilities, the poor engineering quality has been criticized for a long time. sponge city construction should focus on improving residents' well-being. Only when residents feel the urban water problem is alleviated and observe the improvement of living environment quality, can they support the continuation of more such work, and sponge city construction can be promoted systematically.

3.4. Some Lessons Learned

Although pilot sponge city construction presented much experience for future implementation, there are still some problems that need to be addressed. To further promote sponge city construction, the following lessons learned need to be emphasized.

3.4.1. Lacking Local Sponge City Technical Parameter for Planning & Design

The theory research and practice of urban runoff control in China started late. In the Technical Guide for Sponge city Construction issued in 2014, the definition, construction requirements, typical structures, scale calculation methods, usability, advantages, and disadvantages of individual facilities are given, but the descriptions are relatively simple and general. Urban runoff characteristics are very site-specific. In practice, the technical guide of sponge city should be formulated according to local conditions.

However, since sponge city practice has only begun in China in recent years, there are few literature reports on monitoring data of sponge city in the actual operation process, and most of them are for individual facilities. Domestic studies on optimization of design parameters of runoff control facilities are also in the initial stage, some of which are based on model simulation. For example, Xu et al. simulated optimization of design parameters of low impact development facilities on urban roads based on SWMM model [71]. Meng et al. used SWMM model to simulate the performance of grassed swale and found that the rainfall return period and slope ratio were negatively correlated with the hysteresis capacity of grassed swale [72].

Studies that analyzed the influence of design parameters on the effectiveness of runoff control facilities through experiments are also quite few. Only a few researchers made some attempts, such as Sang et al. used a green roof test device composed of six different substrates to conduct a simulated rainfall test, and found that the green roof composed of light materials had a better control effect on TN, and different substrates had obvious leaching loss of TP and COD [73]. According to the special environment of the red soil area in Southern China, Li optimized the design of the rain garden from the aspects of filler ratio, plant selection, internal water storage area setting, and obtained the appropriate filler ratio and plant type [74]. Zhang determined artificial rainfall parameters according to the special hydrological characteristics and water quality of Xuzhou, selected experimental groups with different ratio of cushion layer, base layer and permeable material, and compared the permeability of permeable pavements in different experimental groups [75].

Although many pilot cities had issued some so-called local guidelines, most lacked solid research but a copy of national technical guideline or foreign experience. Lacking in-depth analysis of design, monitoring, and performance evaluation system is a main issue in sponge city construction of China.

3.4.2. Lacking Professional Managers and Technicians

From the previous sponge city pilot construction experiences, many mistakes were caused by misunderstanding of sponge city. Many misconceptions still exist, for example, LID can solve all urban flooding problems, or LID is useless for flooding control. The main reason is that current education and training do not provide the necessary skills for designing and implementing sponge city. To successfully implement an sponge city project, knowledge from various disciplines is required. For example, planning/design of LID facilities would need skills in stormwater management, urban hydrology and hydraulics (scales from site to region and to watershed), water quality modeling, optimization techniques, landscaping, etc. However, the specific system education and outreach programs are still lacking in universities and colleges. Professional training is also missing.

To implement the sponge city construction nationwide smoothly, a large number of qualified professionals in all stages of sponge city construction are essential. Enough managers, planners, designers and construction workers are required to support this colossal initiative [52]. Firstly, there is a shortage of "specialized technicians" who can provide depth into a certain field. At present, sponge city construction is short of qualified professionals in every post, whether it is design, construction or operation and maintenance. Secondly, there is shortage of general managers who can carry out in-depth professional communication and realize the knowledge and technology links of various professional processes and fields. In future, as to various practices related to sponge city construction, the problems to be solved will become more complex. Cross-field and cross-professional team collaboration will become much more common. Therefore, it is necessary and urgent to strengthen interdisciplinary exchange and cooperation and cultivate comprehensive talents in professional education.

3.4.3. Lacking Policy Coordination

Sponge city is a new paradigm of urban construction and governance; it has multi-benefits in many aspects. Besides runoff control and water environment improvement, sponge city construction can alleviate the urban heat island effect and promote building energy consumption reduction. It also contributes to carbon dioxide reduction and is consistent with the China's "dual carbon" target. In context, nearly all of the policies are related and can impact each other. For example, design of the bioretention needs input from both stormwater and landscape architect professionals. Such a coordinated effort has not been the norm because the responsibilities for stormwater management and roadside vegetation management are belonging to different agencies. Currently, there are close coordination attempts among the key ministries (MHURC, MF and MWR) for implementing the sponge city plan at the national level. However, conflicts still exist among many current policies issued by difference ministries.

At the city level, many agencies are involved in sponge city construction, such as the urban planning, construction, water conservancy, and environment protection bureaus, etc. A smooth and efficient sponge city implementation requires a great effort for inter-agency coordination. To facilitate such efforts, some sponge city pilot cities have created the "Sponge city Offices," which include representatives from all bureaus related. However, during the real operation of the sponge city Offices, there are still many inconsistencies exist because of a difference in interests and responsibility.

The normalization of sponge city construction needs to seek institutional breakthrough under the support of national and local policies. The kernel is the harmony among all of the related national and local technical standards and management regulations issued by different agencies. Therefore, it is important to review all of these current technical standards and management regulations. Then all the inconsistencies and conflicts among them should be identified. After that, a cross-field and cross-professional team is needed to improve these standards and regulations.

4. Sponge City Systematic Demonstration

During the 14th Five-year Plan period (from 2021 to 2025) in China, the MF, MHURD, and MWR planned to facilitate systematic demonstrations at various "representative" cities. In 2021, 20 sponge city demonstration cities had been selected according to their basic conditions. For these selected sponge city demonstration cities, a fixed subsidy range from 700 million to 1.1 billion Yuan were provided by the central government [76]. Another 25 sponge city demonstration cities will be selected in 2022.

4.1. Multi Sponge City Scale Implementation

For the systematic sponge city demonstration, the focus of implementation strategies varies at different scales. The objectives, corresponding indicators, key contents and needs under difference construction scale are illustrated in Figure 5. The construction of sponge city is committed to comprehensively strengthen the flooding mitigation, the water environment improvement, the water resources recycling and the water ecology restoration. These four objectives can be quantified by several indicators, such as non-point source pollution control ratio and volume capture ratio. In order to achieve these goals, different engineering contents and requirements need to be implemented at various scales. Besides, in order to systematically establish and promote a long-term operation and maintenance mechanism for the overall construction period, demonstration cities should also make full use of the previous working experiences and achievements of sponge city pilot construction [76]. In this way, the sponge city strategy can be effectively implemented, thereby promoting the construction of national sponge cities to a new level. This section gives an overview of the engineering content and requirements that need to be performed at each scale.

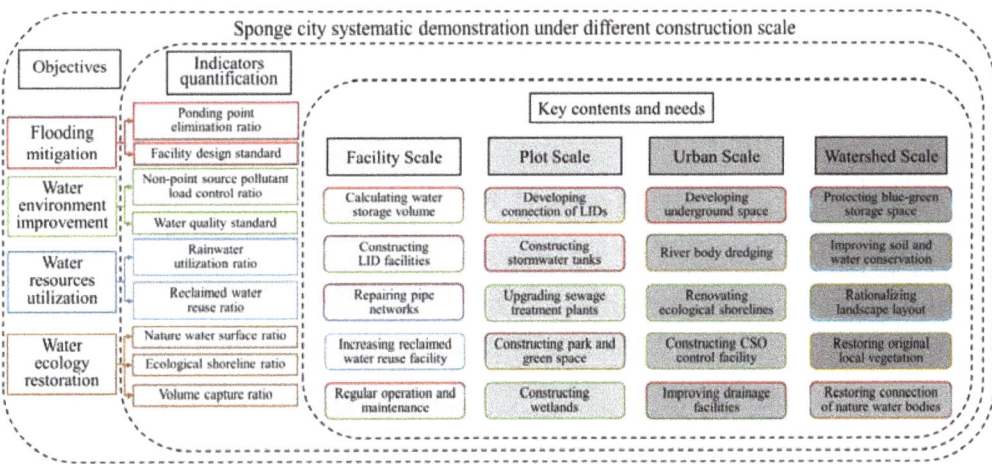

Figure 5. Objectives, indicators, key contents and needs for systematic demonstration of sponge city. Note: The colors of the boxes with different content and needs indicate that they can be used to improve the objectives or indicators under the same corresponding colors.

4.1.1. Facility Scale

The facilities mainly include green infrastructure (LID facilities at the source) such as rain gardens and green roofs, gray infrastructure (transit facilities at mid-way) such as drainage pipe networks and pumps, and blue infrastructure such as receiving water bodies (terminal storage facilities) [44]. The planning layout of green, gray and blue facilities should fully consider the needs of the storage volumes and how to deal with urban water issues [57]. Besides, the renovation and construction of drainage pipe networks should be coordinated with the construction of LID facilities to achieve better performance [77–79]. A full use of green and blue infrastructures is more sustainable and environmental-friendly, it also enhance the resilience of the drainage system [80]. Therefore, how to reconcile the proportion and scale of green-gray-blue infrastructures requires further analysis based on local conditions. Compared with flooding and water quality control, it is also important to increase unconventional water resources (rainwater and reclaimed water) utilization [81]. Currently at facility scale, the most common unconventional water reuse facilities in cities are rain barrels. Li showed that a 25.74 m^2 roof can bring potential benefits of about 1.0903 to 1.2474 million Yuan per year in an area with an annual rainfall depth of about 520 mm–600 mm [82]. Therefore, a lot of economic and environmental benefits can be brought about by rationally arranging rain barrels.

After the facility construction, it is necessary to conduct long-term real-time monitoring, operations, and maintenance of typical facilities for better understanding the relationship between facility capacity and multiple influencing factors [13,45,83,84]. The operation and maintenance of infrastructures is indispensable in sponge city construction. Better operation and maintenance methods can greatly improve the construction effect of sponge city [13]. The operation and maintenance management system of sponge city needs to clarify the operation and maintenance content, risk management, funding source and supervision method. To understand the operation status of a facility, monitoring data is essential. Therefore, the monitoring of typical infrastructures is very important for sponge city effect evaluation since it can provide real-time feedback on facilities' runoff control capacities [85]. the monitoring content of these facilities should include total runoff volume, peak flow, runoff pollution, soil medium infiltration rate, moisture content, and emptying time [86]. Besides, the monitoring data also can be used to calibrate and validate certain urban hydrologic and hydraulic models [87]. Only after the model has been calibrated and validated through monitoring data can it be used to evaluate the effectiveness of sponge city construction [45].

4.1.2. Plot Scale

At plot scale, how to improve the living environment through sponge city construction is one of the most important problems that need to be considered. Measures such as "infiltration, retention, storage, purification, utilization and discharge" should be fully implemented as a priority to solve the problem related to the stormwater and sewage networks [88]. Meanwhile, the vacant land in the plot also needs to be utilized to increase the public activity space, such as parks and green space. The land use types of the plots are mainly divided into community, commercial land, industrial land, green space, and square land.

Communities can be further divided into old and new communities [89,90]. The old community renewal is an important issue in the process of sponge city construction [88]. At present, the common renewal arrangement of old community is problem-oriented, putting residents' demands first to solve local specific problems. Taking an old community in Beijing, China as an example, the overall guiding ideology is to repair the old infrastructures and make full control of the runoff (Figure 6). In accordance with the concept of sponge city construction, the existing stormwater infrastructure were fully used to avoid excessive modification. According to the vertical terrain in the plot, the connection of different types of LID facilities, such as bioretention or sunken green spaces were designed, so that the runoff can be discharged in an organized manner. Meanwhile, the damaged infrastructure was renovated, and the road was reorganized. Some damaged roads or parking spaces were converted into a permeable pavement. Then, several small stormwater tanks were built to store excess stormwater runoff and reuse. It is easier to construct a new sponge community than an old one which is usually goal oriented. New communities often construct under strict sponge city target and combine with the district or city planning. For example, Beijing has incorporated the permeable area and storage volume requirements into local standards of newly built communities [91].

Industrial land is one of the main types of urban construction land and an important area for sponge city construction [92]. Distinct from residential areas, it typically has lower greening rates and a higher risk of runoff pollution. Therefore, the sponge city construction method should be different from that in communities. Firstly, green roofs can be used when the load requirements of building roofs are met. The green roof can not only absorb and store roof runoff and use the soil infiltration process to purify some pollutants, but also reduce the overall stormwater runoff coefficient of the underlying surface of the industrial site. This can reduce the intensity of stormwater discharge and slow down the speed of stormwater concentration. It should be noted that green roofs can only be built on flat roofs since sloping roof buildings with greater inclination may cause the structural layer of green roofs to slip [93]. As for the more polluted industrial land, sewage treatment plants or wetlands can be considered to use for regulation, storage and collection of initial stormwater.

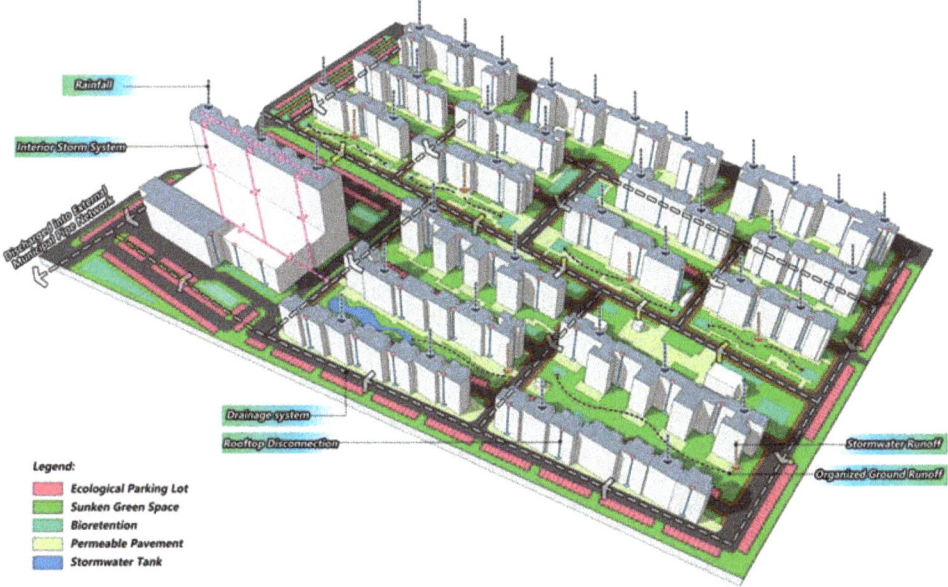

Figure 6. Community renovation under sponge city concept in Beijing, China.

With the continuous acceleration of economic development and urban expansion, commercial complexes have become an important part of urban construction. At present, commercial lands not only require to fully meet the needs of aesthetics and functions, but also needs to consider ecological functions, such as the effective use of natural resources and the promotion of the overall urban environment [94]. During sponge city construction, permeable pavements cannot be used in areas with high traffic loads, such as highways. However, they can be widely used in commercial pedestrian streets. It can not only increase the penetration rate of stormwater runoff, but also reduce the pollutants. In addition, it is also a good runoff control measure to build some high-level flower beds and ecological tree boxes in the commercial square. Although such a flower bed or tree box occupies a small area, its capacity to purify runoff should not be underestimated. The combined use of ecological tree boxes, permeable pavement and the municipal pipe network in the commercial land partly solves the contradiction between rapid discharge and sewage interception.

As for the land type of green and square space, their own conditions enable them to achieve better infiltration, retention, or storage of rainfall runoff than other types of plots. Therefore, for this type of plots, the requirements for runoff quantity and quality control capacity are higher than other types of plots in sponge city construction [18]. The main method is to combine green space, surrounding buildings and roads through vertical terrain adjustment, so that stormwater runoff can be absorbed locally. For the stormwater runoff that exceeds the storage capacity of the facility, it should be merged into the municipal stormwater pipe network through the transfer facilities.

To sum up, in the construction of sponge city at the plot scale, the joint utilizations of different types of green, gray and blue infrastructures can greatly improve the runoff quantity and quality control capacity and enhance people's satisfaction degree. Different types of plots need to be systematically arranged according to specific problems. With the construction of sponge cities in multiple plots, the sponge effect of the whole urban can be improved.

4.1.3. Urban Scale

At the urban scale, it is necessary to build an ecological, safe, and sustainable urban water circulation system to improve the overall level of water resources security, disaster prevention and mitigation capabilities. In the new and renewal projects of urban green space, buildings, roads, squares, etc., not only should various stormwater infiltration and storage facilities be constructed according to local conditions, but urban permeable pavement also need to be promoted to expand the urban permeable area.

The focus of flooding mitigation at the urban scale mainly focuses on the drainage system and the layout of LIDs [18]. Urban drainage system construction needs to be systematically managed on the premise of reducing external influences for the existing storage and drainage channels, such as the main rivers and lakes. Besides, such rivers and lakes are also the important link between the drainage facilities and the plot-scale sponge project [95]. The drainage capacity can be improved by means of diversion, interception, regulation, storage, etc., [96]. A proper GI layout needs to refine the sub-catchment of the region and achieve a certain degree of runoff reduction and drainage efficiency improvement through sponge projects at facilities and plot scale. It is also important to develop underground space in addition to the layout of LID facilities. Reasonable underground space construction can be used to store excess stormwater when a storm event occurs, and then slowly discharge or reuse it after the event.

The improvement of the urban water environment needs to be managed through a variety of engineering or non-engineering measures from two aspects: reduction of pollutants and sediment and increasing water environment capacity. Pollutants may come from urban non-point source pollution or combined sewer overflow (CSO) pollution, which can be alleviated by improving urban drainage facilities and constructing CSO control facilities [97,98]. The urgent need of sponge city construction at urban scale is to alleviate the urban non-point source pollution and CSO problems [18].

Moreover, regular dredging and ecological shoreline rehabilitation of urban rivers and lakes can be implemented for better water ecological restoration. Sediment dredging, however, aims to solve the problem of internal source pollution [99]. Through ecological shoreline rehabilitation, a diversified living environment and natural habitats for living things can be created. Many cities have carried out the transformation and restoration of ecological shorelines in the process of water ecological restoration. For example, during the pilot sponge city construction period, Wuhan, China has restored more than 50% of ecological shoreline. Besides, it is also essential to implement comprehensive management, monitoring, surveillance, and early warning of the ecological environment of the entire ecosystem to achieve better performance.

In addition, along with the global climate change, the frequent extreme storm and drought events have threatened human survival [13]. The consideration of climate change in sponge city construction is first reflected in the resilience of the city to control stormwater. Extreme rainfall events increase the drainage pressure of the city. The sponge city construction at the urban scale can reduce the threat of storm events to the city through the linkage of various green, gray, and blue infrastructures, so as to increase the resilience of the city. Another impact caused by climate change is UHI effect. Studies have shown that the larger the green area, the higher the temperature reduced by transpiration [100]. So, green infrastructures used in sponge city construction also can effectively alleviate the UHI effects [35]. Studies also found that green infrastructures such as green roofs can not only increase urban greening rates, but also reduce greenhouse gases [101]. The use of permeable pavements instead of impervious ground covers can also greatly reduce the UHI effect [102]. In the evaluation system of sponge city construction, the mitigation of UHI effect is also one of the important evaluation indicators.

4.1.4. Watershed Scale

At a watershed scale, it is necessary to aim at protecting nature blue-green storage space to build an ideal spatial pattern of landscape layout and improving connection of natural water bodies to increase the nature storage and drainage capacity. The spatial distribution of mountain, forest, field, river, lake, and grass needs to be accurately identified to protect the natural features. Besides, it is also important to improve soil and water conservation to protect the existing stormwater storage space and expand the natural storage space outside the urban built-up area.

Taking the capital city of Inner Mongolia Autonomous Region, Hohhot, as an example, an urban water system with healthy circulation has been constructed. First, the important ecological elements, such as surface water system and local vegetation in the watershed, are identified and protected. It is important to strengthen the connection of the natural water bodies and protect the blue-green storage space in the watershed to maximum the water storage and drainage capacity and, therefore, prevent flooding or non-point sources pollution. Secondly, considering the actual water shortage problems, the existing green space resources in Hohhot are fully utilized to build an LID system so that they maximize the utilization of stormwater resources. Then, by restoring the drainage pipe and pump networks, the current drainage system can be revitalized. Through the synergy of the above green, gray, and blue systems, a healthy and sustainable water resource utilization system can be created. Although the sponge city is promoted as a whole in the watershed, the natural and historical conditions of each area are different. It is very important to drive the overall construction of sponge city in the watershed through small typical demonstration areas. There are a total of 9 demonstration areas in Hohhot with an area of about 81.21 km^2.

On the whole, the sponge city construction at the watershed scale focuses more on the coordination of macroscopic blue-green storage spaces to improve water and soil resources conservation and enhance purification capabilities. The entire sponge city planning can provide mutual guidance and feedback to jointly ensure the healthy operation of the water system and, thus, maximize the overall benefits of the whole watershed.

4.2. The Roles of Different Stakeholders

In order to promote the sponge city construction in an orderly manner, the main responsibility of different stakeholders must be brought into full play (Figure 7). Stakeholders can be divided into promoters, implementers, and protectors. The central government act as the promotor. Currently, the related ministries of central government are changing their thinking and fully realizing that the sponge city is a new way of urban development. In the implementation stage, local governments, planners, designers, and constructors assume corresponding responsibilities, and no gaps should appear in each section. After the construction is completed, maintenance and the protection of the facilities by the public have also become the necessary conditions for the continuous and normal operation of various facilities in sponge city. This section discusses the obligations of each stakeholder.

Firstly, the central government has a guiding role in the formulation of planning ideas and the selection of directions. The central government has integrated the concept of sponge city into the process of urban planning and development to play an effective role in the work of systematically promoting sponge city construction since 2013. Besides, relevant policies were also issued since the central government can promote the construction and transformation of sponge city through policies and corresponding laws and regulations. Urban infrastructure is usually an important part of financial investment. Therefore, doing a good job in fundraising and investment control is of great significance for the sponge city construction. On the one hand, the central government absorbed social capital to participate in the sponge city construction, and on the other hand, it increased financial investment to ensure the smooth implementation of the project during sponge city construction. In terms of investment control, the central government strengthened not only the process control to ensure the normal payment of funds, but also the role of audit supervision to ensure the rational and efficient use of funds. National technical standards were also issued so that the concept and top-level design of sponge cities can be deeply rooted in the hearts of implementers and protectors. Finally, it is necessary to establish a sound supervision system for the whole process of sponge city construction to ensure that various problems can be discovered and solved in time.

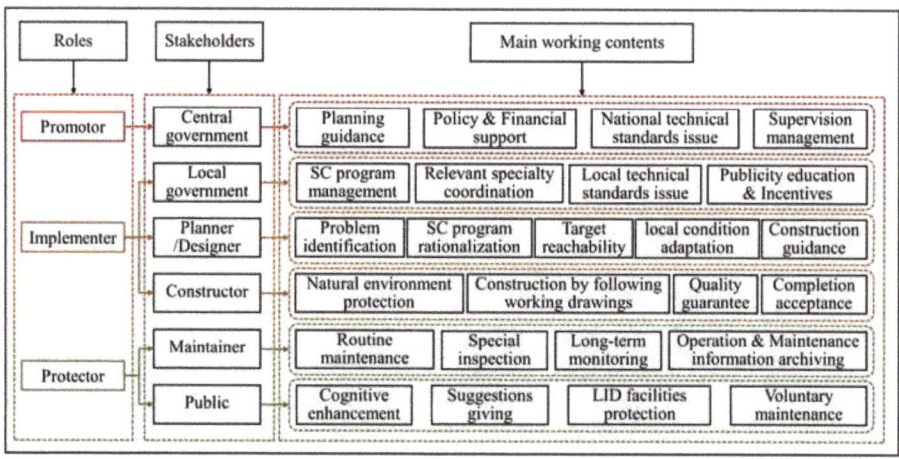

Figure 7. Main working contents for different stakeholders in sponge city construction.

Secondly, local governments, planners, designers, and constructors act as the implementers to construct sponge cities. Local governments formed a normalized sponge city program management and control workflow during the whole process including project establishment, planning, design, construction, acceptance and assessment, etc., so that relevant specialties can be better coordinated. Further, local governments also incorporated the compliance of sponge city construction into the local urban development performance assessment system and issued corresponding technical, management, and operation and maintenance standards adapted to their own regions on the premise of fully understanding the national technical standards. Carrying out publicity education activities and issuing corresponding incentives also played a positive role in the smooth development of sponge city construction. At the same time, local government explored diversified and multi-channel investment and financing modes, attracted social capital input, and encouraged relevant financial institutions to increase credit support for sponge city construction. It is also necessary to use demonstrative projects as a carrier for experience promotion for other regions and local citizens.

Planners and designers then fully identified the local problems before sponge city related planning and designing to ensure rationalization and adaptation of all sponge city programs. After completing the design scheme, it is also necessary to carefully analyze the feasibility. More importantly, planners and designs were supposed to guide construction scheme. Therefore, designers and constructors can coordinate and cooperate with each other to reduce conflicts. This requires designers to understand the actual situation of construction, listen to relevant opinions and suggestions humbly, and figure out if the design need improve, in order to make design economical and practical. Construction following design drawing was also one of the most important jobs because it was the embodiment of the designer's intention. After construction is completed, a systematic acceptance check was implemented.

Finally, maintainers carry out routine maintenance of the sponge city facilities, and special inspections were required for the key parts before and after the occurrence of storm. The operational effect of the facilities can be judged by long-term online or manual monitoring. According to long-term monitoring, facilities should be checked whether it meets the design requirements. If met, the next routine maintenance would continue as usual; if not, special maintenance would be required to make sure the design requirements are met [13]. In addition to relevant maintainers, public also needs to be brought into play to raise the awareness of sponge city construction, and then, guiding them to develop environmental-friendly living habits, and mobilizing the enthusiasm of all social parties to participate in sponge city construction. In daily life, the public strived to improve their own cognition, strengthened their awareness of environmental protection, and actively participated in the construction of sponge cities. At present, many people live in the sponge city renovated communities have already benefited from sponge city construction. Besides, the public can also protect the facilities in the community, for example, no littering, trampling on the lawn, and not destroying the sponge city facilities, etc. At the same time, public can also carry out voluntary activities spontaneously or in an organized manner to help maintaining the facilities in the community.

In summary, the sponge city construction is a large program, which also involving many fields, and it needs the joint efforts of many parties to achieve better results. Only on the premise that all stakeholders enhance their sense of responsibility and strengthen organizational management, it is possible to promote systematic demonstration of sponge city.

5. Conclusions

With the national scale exploration of sponge city pilot projects, we have summarized experiences and lessons learned to help improve future sponge city implementation. The sponge city initiative has made some achievements so far, in the long-term, it still needs more efforts in the urban water system management. Future directions can be summarized as follow:

Increasingly serious water security, water ecology and water environment problems add much more pressure to the traditional gray engineering infrastructure. Sponge city construction (specifically refers to green infrastructures), in some extent, can alleviate part of pressure on traditional gray infrastructure. However, when considering safety during extreme storm events, traditional gray infrastructures is still an essential part in sponge city construction. Appropriately controlling construction scale of gray infrastructure is necessary to avoid excessive artificial interference to the ecological environment. Accordingly, the green and gray infrastructures should be coupled to play an effective role in urban water system management.

Many studies reported that sponge city performed well in flood control, stormwater harvesting, and water quality purification. However, sponge city implementation includes many stages (for example, production and transportation of raw materials, operation, maintenance, labor, and decommissioning). Each stage can have different environmental and economic impacts or benefits. To provide useful and reliable information for policy- and decision-makers with regard to sponge city construction, a quantitative evaluation for both environmental and economic burdens through life cycle perspective is highly needed.

A large amount of data has been accumulated during the implementation phase of sponge city in recent years. To effectively organize, store, and apply these datasets in the assessment of urban runoff control is a great challenge.

Author Contributions: Writing—Original Draft Preparation, D.Y., C.X. and H.J. Writing—Review & Editing, D.Y., C.S., Y.Y., Q.W. and S.L.; Supervision, H.J.; Funding Acquisition, C.X. and H.J. All authors have read and agreed to the published version of the manuscript.

Funding: This work was funded by China Postdoctoral Science Foundation (Grant No. 2021M691769), Shuimu Scholar Programme of Tsinghua University (No. 2019SM125), and the National Nature Science Foundation of China (Grant No. 52070112 and 41890823). The research is also supported by Ningxia Hui Autonomous Region Key R&D Project "Research and Demonstration of Key Technologies of Full Digital Water Control Based on Water Network" (2020BCF01002) and the GEF Hai Basin Integrated Water and Environment Management Project (GEF ID 5561/WB ID P145897).

Institutional Review Board Statement: Not applicable.

Informed Consent Statement: Not applicable.

Conflicts of Interest: The authors declare no conflict of interest.

References

1. Jiang, C.; Li, J.; Liu, J. Does urbanization affect the gap between urban and rural areas? Evidence from China. *Socio-Economic Plan. Sci.* 2022, *101271*. in process. [CrossRef]
2. Yin, D.; Evans, B.; Wang, Q.; Chen, Z.; Jia, H.; Chen, A.S.; Fu, G.; Ahmad, S.; Leng, L. Integrated 1D and 2D model for better assessing runoff quantity control of low impact development facilities on community scale. *Sci. Total Environ.* 2020, *720*, 137630. [CrossRef] [PubMed]
3. Wu, J.; Cheng, D.; Xu, Y.; Huang, Q.; Feng, Z. Spatial-temporal change of ecosystem health across China: Urbanization impact perspective. *J. Clean. Prod.* 2021, *326*, 129393. [CrossRef]
4. Xu, C.; Jia, M.; Xu, M.; Long, Y.; Jia, H. Progress on environmental and economic evaluation of low-impact development type of best management practices through a life cycle perspective. *J. Clean. Prod.* 2018, *213*, 1103–1114. [CrossRef]
5. Cheng, M.; Qin, H.; Fu, G.; He, K. Performance evaluation of time-sharing utilization of multi-function sponge space to reduce waterlogging in a highly urbanizing area. *J. Environ. Manag.* 2020, *269*, 110760. [CrossRef]
6. Pumo, D.; Arnone, E.; Francipane, A.; Caracciolo, D.; Noto, L.V. Potential implications of climate change and urbanization on watershed hydrology. *J. Hydrol.* 2017, *554*, 80–99. [CrossRef]
7. Mei, C.; Liu, J.; Wang, H.; Yang, Z.; Ding, X.; Shao, W. Integrated assessments of green infrastructure for flood mitigation to support robust decision-making for sponge city construction in an urbanized watershed. *Sci. Total Environ.* 2018, *639*, 1394–1407. [CrossRef]
8. Guan, X.; Wang, J.; Xiao, F. Sponge city strategy and application of pavement materials in sponge city. *J. Clean. Prod.* 2021, *303*, 127022. [CrossRef]
9. Liu, K.; Li, X.; Wang, S. Characterizing the spatiotemporal response of runoff to impervious surface dynamics across three highly urbanized cities in southern China from 2000 to 2017. *Int. J. Appl. Earth Obs. Geoinf. ITC J.* 2021, *100*, 102331. [CrossRef]
10. Mahmoud, S.; Gan, T.Y. Urbanization and climate change implications in flood risk management: Developing an efficient decision support system for flood susceptibility mapping. *Sci. Total Environ.* 2018, *636*, 152–167. [CrossRef]
11. Chan, F.K.S.; Griffiths, J.A.; Higgitt, D.; Xu, S.; Zhu, F.; Tang, Y.-T.; Xu, Y.; Thorne, C.R. "Sponge city" in China—A breakthrough of planning and flood risk management in the urban context. *Land Use Policy* 2018, *76*, 772–778. [CrossRef]
12. Qi, Y.; Chan, F.K.S.; Griffiths, J.; Feng, M.; Sang, Y.; O'Donnell, E.; Hutchins, M.; Thadani, D.R.; Li, G.; Shao, M.; et al. Sponge city Program (SCP) and Urban Flood Management (UFM)—The Case of Guiyang, SW China. *Water* 2021, *13*, 2784. [CrossRef]
13. Yin, D.; Chen, Y.; Jia, H.; Wang, Q.; Chen, Z.; Xu, C.; Li, Q.; Wang, W.; Yang, Y.; Fu, G.; et al. Sponge city practice in China: A review of construction, assessment, operational and maintenance. *J. Clean. Prod.* 2020, *280*, 124963. [CrossRef]
14. Wang, C.; Hou, J.; Miller, D.; Brown, I.; Jiang, Y. Flood risk management in sponge cities: The role of integrated simulation and 3D visualization. *Int. J. Disaster Risk Reduct.* 2019, *39*, 101139. [CrossRef]
15. Hasan, H.H.; Razali, S.F.M.; Zaki, A.Z.I.A.; Hamzah, F.M. Integrated Hydrological-Hydraulic Model for Flood Simulation in Tropical Urban Catchment. *Sustainability* 2019, *11*, 6700. [CrossRef]
16. Li, J.; Zhang, B.; Mu, C.; Chen, L. Simulation of the hydrological and environmental effects of a sponge city based on MIKE FLOOD. *Environ. Earth Sci.* 2018, *77*, 32. [CrossRef]
17. MHURD. *Technical Guide for Sponge Cities-Construction of Low Impact Development*; China Architecture Publishing & Media Co., Ltd.: Beijing, China, 2014.
18. MHURD. *Assessment Standard for Sponge city Effect*; China Architecture Publishing & Media Co., Ltd.: Beijing, China, 2019.
19. Jia, H.; Yin, D. Green Infrastructure for Stormwater Runoff Control in China. In *Oxford Research Encyclopedia of Environmental Science*; Oxford University Press: Oxford, UK, 2021. [CrossRef]
20. Fletcher, T.D.; Shuster, W.; Hunt, W.F.; Ashley, R.; Butler, D.; Arthur, S.; Trowsdale, S.; Barraud, S.; Semadeni-Davies, A.; Bertrand-Krajewski, J.-L.; et al. SUDS, LID, BMPs, WSUD and more—The evolution and application of terminology surrounding urban drainage. *Urban Water J.* 2015, *12*, 525–542. [CrossRef]
21. DoER. *Low Impact Development Design Strategies: An Integrated Design Approach*; U.S. Environmental Protection Agency: Washington, DC, USA, 1999.
22. Woods-Ballard, B.; Kellagher, K.; Martin, P.; Jefferies, C.; Bray, R.; Shaffer, P. *The SUDS Manual*; CIRIA: London, UK, 2007.

23. Lloyd, S.D.; Tony, H.F.W.; Chesterfield, C.J. *Water Sensitive Urban—A Stormwater Management Perspective*; Cooperative Research Centre for Catchment Hydrology: Melbourne, Australia, 2002.
24. European Commission. Policy Topics: Nature-Based Solutions. Available online: https://ec.europa.eu/research/environment/index.cfm?Pg=nbs (accessed on 21 April 2017).
25. Yang, Z.; Li, J.; Wang, W.; Che, W.; Ju, C.; Zhao, Y. The advanced recognition of low impact development and sponge city construction. *Environ. Eng.* **2020**, *38*, 10–15. (In Chinese) [CrossRef]
26. Keller, B.D. Griffin, Georgia's Stormwater Utility "A Non Structural Best Management Practice (BMP)". In Proceedings of the 1999 Georgia Water Resources Conference, the University of Georgia, Athens, Georgia, 30–31 March 1999.
27. Jia, H.; Yao, H.; Yu, S.L. Advances in LID BMPs research and practice for urban runoff control in China. *Front. Environ. Sci. Eng.* **2013**, *7*, 709–720. [CrossRef]
28. Riechel, M.; Matzinger, A.; Pallasch, M.; Joswig, K.; Pawlowsky-Reusing, E.; Hinkelmann, R.; Rouault, P. Sustainable urban drainage systems in established city developments: Modelling the potential for CSO reduction and river impact mitigation. *J. Environ. Manag.* **2020**, *274*, 111207. [CrossRef]
29. Eckart, K.; McPhee, Z.; Bolisetti, T. Performance and implementation of low impact development—A review. *Sci. Total Environ.* **2017**, *607–608*, 413–432. [CrossRef] [PubMed]
30. Lim, H.S.; Lim, W.; Hu, J.Y.; Ziegler, A.; Ong, S.L. Comparison of filter media materials for heavy metal removal from urban stormwater runoff using biofiltration systems. *J. Environ. Manag.* **2015**, *147*, 24–33. [CrossRef] [PubMed]
31. Gong, Y.; Yin, D.; Li, J.; Zhang, X.; Wang, W.; Fang, X.; Shi, H.; Wang, Q. Performance assessment of extensive green roof runoff flow and quality control capacity based on pilot experiments. *Sci. Total Environ.* **2019**, *687*, 505–515. [CrossRef]
32. Zubelzu, S.; Rodríguez-Sinobas, L.; Andrés-Domenech, I.; Castillo-Rodríguez, J.; Perales-Momparler, S. Design of water reuse storage facilities in Sustainable Urban Drainage Systems from a volumetric water balance perspective. *Sci. Total Environ.* **2019**, *663*, 133–143. [CrossRef] [PubMed]
33. Johnson, D.; Geisendorf, S. Are Neighborhood-level SUDS Worth it? An Assessment of the Economic Value of Sustainable Urban Drainage System Scenarios Using Cost-Benefit Analyses. *Ecol. Econ.* **2019**, *158*, 194–205. [CrossRef]
34. Joshi, P.; Leitão, J.P.; Maurer, M.; Bach, P.M. Not all SuDS are created equal: Impact of different approaches on combined sewer overflows. *Water Res.* **2020**, *191*, 116780. [CrossRef] [PubMed]
35. Li, J.; Gong, Y.; Li, X.; Yin, D.; Shi, H. Urban stormwater runoff thermal characteristics and mitigation effect of low impact development measures. *J. Water Clim. Chang.* **2018**, *10*, 53–62. [CrossRef]
36. Guptha, G.C.; Swain, S.; Al-Ansari, N.; Taloor, A.K.; Dayal, D. Assessing the role of SuDS in resilience enhancement of urban drainage system: A case study of Gurugram City, India. *Urban Clim.* **2022**, *41*, 101075. [CrossRef]
37. Maqbool, R.; Wood, H. Containing a sustainable urbanized environment through SuDS devices in management trains. *Sci. Total Environ.* **2021**, *807*, 150812. [CrossRef]
38. Sharma, A.K.; Cook, S.; Tjandraatmadja, G.; Gregory, A. Impediments and constraints in the uptake of water sensitive urban design measures in greenfield and infill developments. *Water Sci. Technol.* **2012**, *65*, 340–352. [CrossRef]
39. Tjandraatmadja, G. Chapter 5—The Role of Policy and Regulation in WSUD Implementation. In *Approaches to Water Sensitive Urban Design*; Sharma, A.K., Gardner, T., Begbie, D., Eds.; Woodhead Publishing: Cambridge, UK, 2019; pp. 87–117.
40. Wong, T.H.F. Water sensitive urban design—The journey thus far. *Australas. J. Water Resour.* **2006**, *10*, 213–222. [CrossRef]
41. Liu, D. A Comparative Research on LID and WSUD from the Perspective of Rainwater Management. *Archit. Cult.* **2021**, 93–96. (In Chinese) [CrossRef]
42. Sun, X.; Qin, H.; Lu, W. Evolution of Water Sensitive Urban Design in Australia and Its Enlightenment to Sponge city. *Chin. Landsc. Archit.* **2019**, *35*, 67–71. (In Chinese) [CrossRef]
43. Mitchell, V.; Deletic, A.; Fletcher, T.; Hatt, B.; McCarthy, D. Achieving multiple benefits from stormwater harvesting. *Water Sci. Technol.* **2007**, *55*, 135–144. [CrossRef] [PubMed]
44. Leng, L.; Mao, X.; Jia, H.; Xu, T.; Chen, A.S.; Yin, D.; Fu, G. Performance assessment of coupled green-grey-blue systems for Sponge city construction. *Sci. Total Environ.* **2020**, *728*, 138608. [CrossRef] [PubMed]
45. Yin, D.; Chen, Z.; Yang, M.; Jia, H.; Xu, K.; Wang, T. Evaluation of runoff control effect in sponge city construction based on online monitoring + simulation modeling. *Environ. Eng.* **2020**, *38*, 151–157. (In Chinese) [CrossRef]
46. Zhang, Y. *Inspiration from Traditional Wisdom of Rainwater Utilization to Contemporary Urban Landscape Design*; Nanchang University: Nanchang, China, 2018. (In Chinese)
47. UNSCEO. Honghe Hani Rice Terraces Inscribed on UNESCO's World Heritage alongside an Extension to the uKhahlamba Drakensberg Park. Available online: https://www.unesco.org/en/articles/honghe-hani-rice-terraces-inscribed-unescos-world-heritage-alongside-extension-ukhahlamba (accessed on 22 June 2013).
48. Cun, C.; Zhang, W.; Che, W.; Sun, H. Review of urban drainage and stormwater management in ancient China. *Landsc. Urban Plan.* **2019**, *190*, 103600. [CrossRef]
49. UNSCEO. Ancient Villages in Southern Anhui-Xidi and Hongcun. Available online: https://whc.unesco.org/en/list/1002 (accessed on 30 November 2020).
50. Xu, Y.-S.; Shen, S.-L.; Lai, Y.; Zhou, A.-N. Design of sponge city: Lessons learnt from an ancient drainage system in Ganzhou, China. *J. Hydrol.* **2018**, *563*, 900–908. [CrossRef]

51. Xu, C. *Research on Life-Cycle Environmental and Economic Benefits Assessment of Sponge city Source Control Facilities*; Tsinghua University: Beijing, China, 2020.
52. Liu, D. China's sponge cities to soak up rainwater. *Nature* **2016**, *537*, 307. [CrossRef]
53. Larsen, T.A.; Hoffmann, S.; Lüthi, C.; Truffer, B.; Maurer, M. Emerging solutions to the water challenges of an urbanizing world. *Science* **2016**, *352*, 928–933. [CrossRef]
54. Zhang, L. *Typical Cases of Sponge city Construction*; China Architecture Publishing: Beijing, China, 2017.
55. Huang, M.; Shen, R.; Yan, D.; Huo, D. *Sponge City Construction and Operation Technology System*; China Architecture Publishing: Beijing, China, 2019.
56. Xu, H. *Theory and Practice of Urban Sponge Green Space Planning and Design*; Southeast University Press: Nanjing, China, 2021.
57. Jia, H.; Liu, Z.; Xu, C.; Chen, Z.; Zhang, X.; Xia, J.; Yu, S.L. Adaptive pressure-driven multi-criteria spatial decision-making for a targeted placement of green and grey runoff control infrastructures. *Water Res.* **2022**, *212*, 118126. [CrossRef] [PubMed]
58. Liu, Z.; Xu, C.; Xu, T.; Jia, H.; Zhang, X.; Chen, Z.; Yin, D. Integrating socioecological indexes in multiobjective intelligent optimization of green-grey coupled infrastructures. *Resour. Conserv. Recycl.* **2021**, *174*, 105801. [CrossRef]
59. Xu, C.; Tang, T.; Jia, H.; Xu, M.; Xu, T.; Liu, Z.; Long, Y.; Zhang, R. Benefits of coupled green and grey infrastructure systems: Evidence based on analytic hierarchy process and life cycle costing. *Resour. Conserv. Recycl.* **2019**, *151*, 104478. [CrossRef]
60. Zhu, Y.; Xu, C.; Yin, D.; Xu, J.; Wu, Y.; Jia, H. Environmental and economic cost-benefit comparison of sponge city construction in different urban functional regions. *J. Environ. Manag.* **2021**, *304*, 114230. [CrossRef]
61. Xu, C.; Shi, X.; Jia, M.; Han, Y.; Zhang, R.; Ahmad, S.; Jia, H. China Sponge city database development and urban runoff source control facility configuration comparison between China and the US. *J. Environ. Manag.* **2021**, *304*, 114241. [CrossRef]
62. Chen, Z.; Tang, Q.; Liang, C. Performance Evaluation of Sponge city Construction in Qian'an City. *Hebei Province Water Conservancy* **2021**, 33–37. (In Chinese)
63. Cao, D.; Wei, Y.; Xu, R.; Chen, Q. Performance Evaluation of sponge city Construction—A case study of Jiaxing City. Spatial Governance for High-quality Development. In Proceedings of the 2021 China Urban Planning Annual Conference (08 Urban Ecological Planning), Chengdu, China, 25–30 September 2021. [CrossRef]
64. Cheng, J.; Wang, Y.; Guo, X. *Sponge Castle Adventure*; Jinan Publishing Press: Jinan, China, 2017.
65. Cheng, J. *Sponge city Exploration*; Jinan Publishing Press: Jinan, China, 2017.
66. Lu, H.; Chen, Y.W. *Design Our Sponge Community*; Shanghai Education Press: Jinan, China, 2018.
67. Zhang, W.; Che, W. Connotation and multi-angle analysis of sponge city construction. *Water Resour. Prot.* **2016**, *32*, 19–26. (In Chinese)
68. Wang, Y.; Sun, M.; Song, B. Public perceptions of and willingness to pay for sponge city initiatives in China. *Resour. Conserv. Recycl.* **2017**, *122*, 11–20. [CrossRef]
69. Wang, Y.; Liu, X.; Huang, M.; Zuo, J.; Rameezdeen, R. Received vs. given: Willingness to pay for sponge city program from a perceived value perspective. *J. Clean. Prod.* **2020**, *256*, 120479. [CrossRef]
70. She, N.; Xie, Y.; Li, D. Reflections and Suggestions on China's Sponge city Construction. *Landsc. Archit. Front.* **2021**, *9*, 82–91. (In Chinese)
71. Xu, P.; He, J.; Ren, X.; Tang, Z.; Zhang, Y.; Huang, J. Optimization of LID facility design parameters of urban roads based on SWMM. *Water Resour. Power* **2016**, *34*, 21–25. (In Chinese)
72. Meng, Y.; Chen, M.; Zhang, S. Experiment and simulation of stagnant Stormwater storage in urban road by planting grass ditch. *Adv. Water Sci.* **2018**, *29*, 32–40. (In Chinese)
73. Sang, M.; Zhang, W.; Zhong, X. Effects of common substrates on water quality characteristics of rainwater retention and outflow of extensive green roofs. *Water Conserv. Hydropower Technol.* **2018**, *49*, 166–173. (In Chinese)
74. Li, Y. *Research on Design OPTIMIZATION and Application of Rain Garden in Red Soil Region of South China*; Nanchang University: Nanchang, China, 2017.
75. Zhang, W. Test on permeable pavement technology of sponge green space. *Hortic. Abstr. China* **2018**, *34*, 37–39. (In Chinese)
76. MF. *Notice on Carrying Out Demonstration Work of Systematic Global Promotion of Sponge City Construction*; Ministry of Finance: Beijing, China, 2021.
77. Leng, L.; Jia, H.; Chen, A.S.; Zhu, D.Z.; Xu, T.; Yu, S. Multi-objective optimization for green-grey infrastructures in response to external uncertainties. *Sci. Total Environ.* **2021**, *775*, 145831. [CrossRef]
78. Chen, W.; Wang, W.; Huang, G.; Wang, Z.; Lai, C.; Yang, Z. The Capacity of Grey Infrastructure in Urban Flood Management: A Comprehensive Analysis of Grey Infrastructure and the Green-Grey Approach. *Int. J. Disaster Risk Reduct.* **2021**, *54*, 102045. [CrossRef]
79. Bakhshipour, A.E.; Dittmer, U.; Haghighi, A.; Nowak, W. Hybrid green-blue-gray decentralized urban drainage systems design, a simulation-optimization framework. *J. Environ. Manag.* **2019**, *249*, 109364. [CrossRef]
80. Fu, X.; Hopton, M.E.; Wang, X. Assessment of green infrastructure performance through an urban resilience lens. *J. Clean. Prod.* **2020**, *289*, 125146. [CrossRef]
81. Karimidastenaei, Z.; Avellán, T.; Sadegh, M.; Kløve, B.; Haghighi, A.T. Unconventional water resources: Global opportunities and challenges. *Sci. Total Environ.* **2022**, *827*, 154429. [CrossRef]
82. Li, F. *Comprehensive Benefit Evaluation of Roof Rainwater Reuse System Based on the Perspective of Sponge city—A Case Analysis in Fengxixincheng District*; Xi'an University of Architecture and Technology: Xi'an, China, 2021.

83. Razzaghmanesh, M.; Borst, M. Monitoring the performance of urban green infrastructure using a tensiometer approach. *Sci. Total Environ.* **2018**, *651*, 2535–2545. [CrossRef]
84. Cheng, Y.-Y.; Lo, S.-L.; Ho, C.-C.; Lin, J.-Y.; Yu, S.L. Field Testing of Porous Pavement Performance on Runoff and Temperature Control in Taipei City. *Water* **2019**, *11*, 2635. [CrossRef]
85. Geberemariam, T.K. Post Construction Green Infrastructure Performance Monitoring Parameters and Their Functional Components. *Environments* **2016**, *4*, 2. [CrossRef]
86. MHURD. *Standard for Sponge city Effect Monitoring*; China Architecture Publishing & Media Co., Ltd.: Beijing, China, 2020; *in preparation and to be submitted*.
87. Alizadehtazi, B.; Montalto, F.A. Precipitation and soil moisture data in two engineered urban green infrastructure facilities in New York City. *Data Brief* **2020**, *32*, 106225. [CrossRef] [PubMed]
88. Liu, W.; Zhen, K.; Cai, D.; Yuan, H.; Li, X. Reconstruction Method of Old Residential Quarters in North China under the Concept of Sponge city—Taking Tongzhou District of Beijing as an Example. *Constr. Sci. Technol.* **2019**, 54–58. (In Chinese) [CrossRef]
89. Li, D.; Du, B.; Zhu, J. Evaluating old community renewal based on emergy analysis: A case study of Nanjing. *Ecol. Model.* **2021**, *449*, 109550. [CrossRef]
90. Chang, Y. Analysis on the Scheme Design of the Sponge city in the Newly-built Community of Xixian. *Sci. Technol. Innov.* **2021**, 152–153. (In Chinese)
91. BMCoPNR. *Code for Design of Stormwater Management and Harvest Engineering*; Beijing Municipal Bureau of Quality and Technical Supervision: Beijing, China, 2013.
92. Xie, S.; Wu, C.; Lv, Y. Standards and technical points of sponge city construction in industrial project. *Water Purif. Technol.* **2021**, *40*, 118–121. (In Chinese) [CrossRef]
93. Jin, Y.; Wang, H.; Shi, C. Application of Sponge city Design in Different Underlying Surfaces of Industrial Factories. *Constr. Des. Proj.* **2021**, 62–65. (In Chinese) [CrossRef]
94. Xia, Q. The "Sponge city" Concept in the Landscape Design Practice of Commercial Office Areas-Taking the Community Service Center Project of Airport 3rd Road, Nanfang Xincheng, Nanjing as an example. *Creat. Living* **2021**, 18–19. (In Chinese)
95. Li, X. *Research on Urban Vertical Planning Based on Major Drainage System*; Beijing University of Civil Engineering and Architecture: Beijing, China, 2020.
96. Zhang, Z.; Miao, Y.; Li, J.; Liu, D.; Fang, X. Improvement effect of rainfall source control facilities on urban drainage capacity in different regions of China. *J. Hydrol.* **2019**, *579*, 124127. [CrossRef]
97. Yang, W.; Zhang, J.; Mei, S.; Krebs, P. Impact of antecedent dry-weather period and rainfall magnitude on the performance of low impact development practices in urban flooding and non-point pollution mitigation. *J. Clean. Prod.* **2021**, *320*, 128946. [CrossRef]
98. Rizzo, A.; Tondera, K.; Pálfy, T.; Dittmer, U.; Meyer, D.; Schreiber, C.; Zacharias, N.; Ruppelt, J.; Esser, D.; Molle, P.; et al. Constructed wetlands for combined sewer overflow treatment: A state-of-the-art review. *Sci. Total Environ.* **2020**, *727*, 138618. [CrossRef] [PubMed]
99. Yin, D.; Xu, T.; Li, K.; Leng, L.; Jia, H.; Sun, Z. Comprehensive modelling and cost-benefit optimization for joint regulation of algae in urban water system. *Environ. Pollut.* **2021**, *296*, 118743. [CrossRef] [PubMed]
100. Sanchez, L.; Reames, T.G. Cooling Detroit: A socio-spatial analysis of equity in green roofs as an urban heat island mitigation strategy. *Urban For. Urban Green.* **2019**, *44*, 126331. [CrossRef]
101. Moghbel, M.; Salim, R.E. Environmental benefits of green roofs on microclimate of Tehran with specific focus on air temperature, humidity and CO_2 content. *Urban Clim.* **2017**, *20*, 46–58. [CrossRef]
102. Ferrari, A.; Kubilay, A.; Derome, D.; Carmeliet, J. The use of permeable and reflective pavements as a potential strategy for urban heat island mitigation. *Urban Clim.* **2019**, *31*, 100534. [CrossRef]

Article

Decision-Making Framework for GI Layout Considering Site Suitability and Weighted Multi-Function Effectiveness: A Case Study in Beijing Sub-Center

Zijing Liu [1], Yuehan Yang [2], Jingxuan Hou [3] and Haifeng Jia [1,*]

[1] School of Environment, Tsinghua University, Beijing 100084, China; liuzj17@mails.tsinghua.edu.cn
[2] Department of Environmental Systems Science, ETH Zurich, 8092 Zurich, Switzerland; y.yuehan16@gmail.com
[3] School of Architecture, Tsinghua University, Beijing 100084, China; houjx19@mails.tsinghua.edu.cn
* Correspondence: jhf@tsinghua.edu.cn

Abstract: The effectiveness of runoff control infrastructure depends on infrastructure arrangement and the severity of the problem in the study area. Green infrastructure (GI) has been widely demonstrated as a practical approach to runoff reduction and ecological improvement. However, decision-makers usually consider the cost-efficacy of the GI layout scheme as a primary factor, leading to less consideration of GI's environmental and ecological functions. Thus, a multifunctional decision-making framework for evaluating the suitability of GI infrastructure was established. First, the study area was described by regional pollution load intensity, slope, available space, and constructible area. Then, to assess the multifunctional performance of GI, a hierarchical evaluation framework comprising three objectives, seven indices, and sixteen sub-indices was established. Weights were assigned to different indices according to stakeholders' preferences, including government managers, researchers, and residents. The proposed framework can be extended to other cities to detect GI preference.

Keywords: multifunctional decision-making framework; cost-effectiveness; site suitability; stakeholders' preference; green infrastructure

1. Introduction

Flooding, water pollution, urban heat island effects, and ecological degradation have necessitated the development of multifunctional infrastructures for adjusting the urban layout. Green infrastructure (GI) effectively boosts cities' sustainability and resilience as it expands a nature-based solution [1,2]. GI is frequently used to enhance the water retention and infiltration capability of urban underlying and can hence regulate urban runoff [3–5]. Additionally, GI can provide ecological functions such as habitat improvement, biodiversity compensation [6], and energy conservation [7]. The effectiveness of GI is highly dependent on the application site and the urgency of runoff-related problems [8,9]. In this context, a hierarchical and multifunctional evaluation of GI is critical for ensuring runoff control efficiency [10–12]. Past GI practice shows that GI is a site-specific runoff management strategy [13]. For example, the cost-effectiveness of GI is impacted by pollution severity, and site conditions constrain the GI construction scale. As a result of urban growth and ecological endowments [14,15], spatial heterogeneity affects the quantitative identification of regional characteristics and the suitable site for GI [10,16]. In detail, the intensity of the pollutant load, the catchment slope, and the constructible area are all important factors for quantifying site limits on runoff control infrastructure [17,18]. Balancing the restrictions of natural endowment and the inherent benefits of GI can facilitate evaluating the viability of runoff management techniques in specific sites. In addition, the preferences of different stakeholders are important for GI arrangements. For example, local managers take the responsible role in regional development, scholars are well-versed in the mechanics underlying runoff control infrastructure, and local citizens benefit directly from GI's multiple functions.

Recently, GIs have been given more weight to urban development because of their multiple benefits. Along with controlling urban floods, GIs can help mitigate non-point source pollution and improve the quality of the aquatic environment [3,18,19]. Additionally, GIs offer significant ecological and aesthetic benefits [10], which improve residents' well-being. Although multi-functionality is commonly assumed, only stormwater runoff management or aquatic environment improvement are considered benefits when implementing GIs [20]. Multiple functions of GI in runoff control, economy, and ecology urgently require joint assessment within a unified evaluation system.

By innovatively incorporating ecological benefits into the unified evaluation system, this study overcomes the limitation that traditional GI effectiveness evaluations focus exclusively on runoff control function and economic cost. The feasibility of the site for GI layout was thoroughly assessed in terms of pollutant load intensity, slope, available floor space, and GI constructible areas. Local stakeholders, such as environmental experts, architect experts, managers, and residents, were consulted regarding their desire for the multi-function of GI. Thus, a multi-objective decision-making framework for GI was developed that takes runoff control function, economic, and ecological considerations into account to balance the region's natural endowments and stakeholder interests.

2. Method and Data

2.1. Study Area

The research area is in Beijing's sub-center (Figure 1), Figure 1a illustrates the location of Beijing in China, and Figure 1b depicts the case area of Beijing sub-center. Its elevations range between 9.5 and 26.9 meters. The mild slope allows for adequate retention time for GI, which makes the study area suitable for GI deployment. The primary local soil type is chalky soil, and the groundwater depth is between 5 and 10 m. Beijing's sub-center is in the warm-temperate monsoon climate zone of the continental monsoon. The average temperature is 11.65 °C, and the relative mean humidity is 56.8%. The annual rainfall is 535.88 mm on average. The flood season lasts from June to September and accounts for approximately 80% of annual precipitation. The research area covers around 155 km², with 27% covered by permeable land.

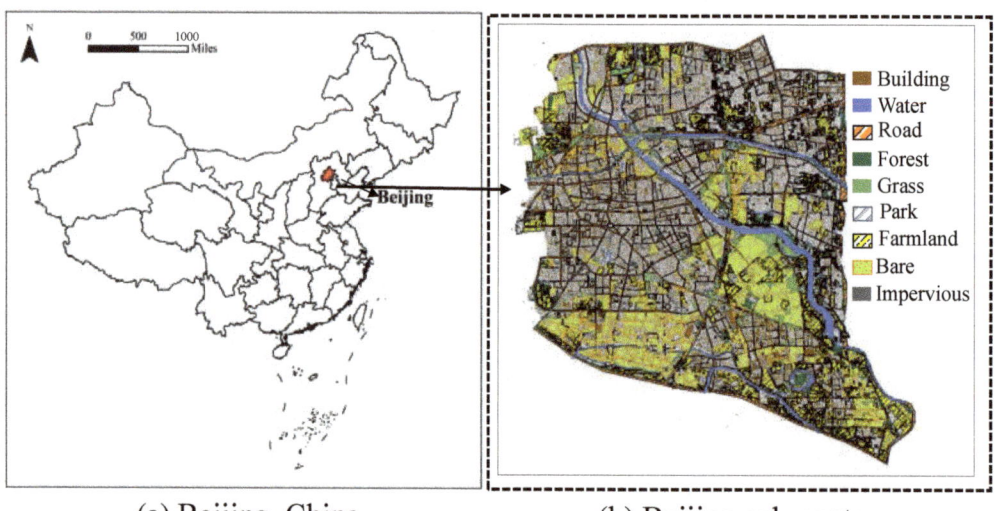

(a) Beijing, China   (b) Beijing sub-center

Figure 1. Study area.

2.2. Methodology

Methodological steps were taken as follows to establish an adaptive GI layout decision-making strategy, as illustrated in Figure 2: (I) The examination of site suitability is the first obstacle in the selection of GI. The intensity of the pollutant load, the catchment slope, the accessible space, and the GI constructible area are essential factors for evaluating the viability of GI site locations. (II) A brief explanation of a typical GI, including its functioning mechanisms, facility characteristics, operational and maintenance requirements, and how residents interact with it, is provided. Three-dimensional evaluation is proposed. This technique considers GI efficacy, cost, and social benefits. The approach for determining GI's effectiveness uses three primary indicators, eight subsidiary indicators, and sixteen tertiary indicators. (III) Local urban managers, relevant professionals, and residents are the critical GI decision-makers and experiencers. This study analyzes stakeholder interests and collects construction intentions from city administrators, architects, environmentalists, and residents. Their preferences are represented hierarchically as weights. The weights correspond to the infrastructure effectiveness indicators. (IV) The selection of GI subjects depends on the suitability of the site. Then, a hierarchical evaluation of the GI's intrinsic efficacy in its various domains and a weighing of the indicators according to local stakeholders are conducted. The decision-making framework for GI layout considered site suitability and weighted malfunction effectiveness are established.

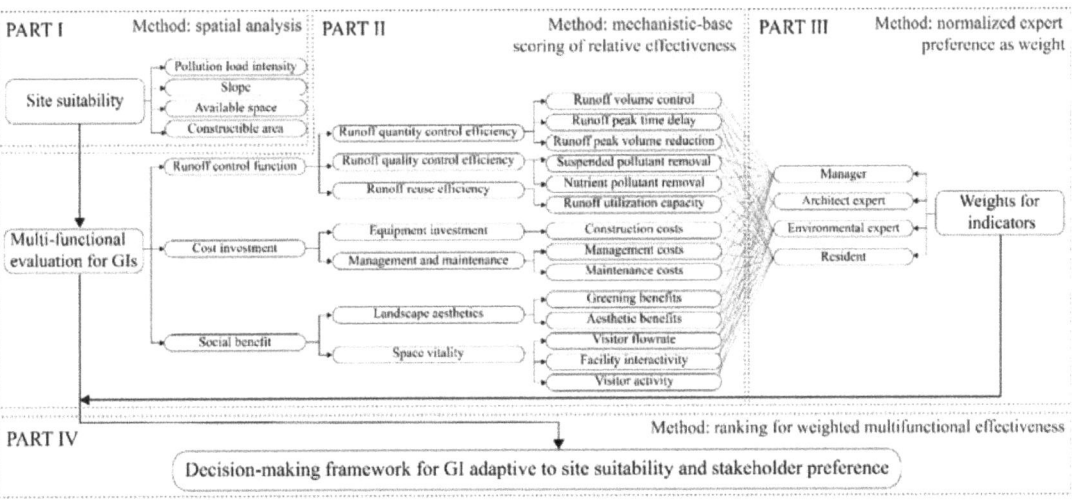

Figure 2. Methodological framework.

2.3. Typical GIs for Evaluation

In the multifunctional decision-making framework, eight commonly used GIs were analyzed. To quantitatively assess the runoff control capacity and the effect of GI, we categorized typical GIs into three types below, based on previous practice and runoff control mechanisms.

(1) Source-oriented runoff control GIs: primarily focused on in-situ runoff dissipation and control. The runoff quantity and quality are regulated by modifying runoff infiltration, retention, and in-situ storage processes. Typical source-oriented runoff reduction measures include bioretention facilities, permeable pavement, green roofs, and sunken green spaces.

(2) Transmission process control GIs: these facilities change runoff flows from sources to sinks, reducing runoff control pressure for the sources. Vegetation swales and infiltration trenches are two common facilities of transmission process control.

(3) Terminal GIs: these are centralized runoff control facilities that are focused on comprehensive management. They are space constrained. Dry ponds and wet ponds are the two most common terminal facilities.

2.4. Site Suitability Evaluation System of GIs

The effectiveness of GI depends on its fitness for the site's features. By incorporating site factors into the GI multifunctional decision-making system, the benefits of GI deployment can possibly be optimized. Table 1 summarizes the various site suitability indices [21]. Four indices serve as decision-making factors for site suitability: pollutant load intensity, catchment slope, available space, and GI constructable areas. The GI site suitability parameters (listed in Table 1) are based on in part on the authors' team's previous research foundation [21,22] and in part on recent documents and literature [20,23–26]. The pollution load intensity was calculated by multiplying the runoff–wash-off pollution concentration in different land use (as shown in Table 2) and the corresponding land use area [27,28]. The pollutant load was normalized for comparison and was described as high, medium, and low-level pollution intensity. Owing to differences in structure and function, the GI's ability to cope with pollutants varies, and tailored installation can improve the efficiency of system runoff control. Catchment slope indices influence the duration and rate of runoff overflow in GI-related pollution capture and removal efficiency. The limited available area constrains the size of GIs, and GIs with low space utilization are not desirable in land-shorted regions. The layout of the GI is constrained by indices such as ecological and aquatic reserve zones.

Table 1. Site suitability indices for GI.

Site Characteristics	Pollution Load Intensity	Slope (%)	Available Space	Constructible Area (Buffer Distance)
Infiltration trench (IT)	Medium	<15	Medium	building > 3 m
Dry pond (DP)	Medium	<17	Large	River > 30 m
Wet pond (WP)	Medium	<10	Large	River > 30 m
Sunken green spaces (SGS)	High	<5	Medium	Road < 30 m
Vegetation swales (VS)	Medium	0.5–5	Medium	Road < 30 m
Green roof (GR)	Low	<4	Medium	Flat roof slope
Permeable pavement (PP.)	Low	<1	-	Road < 30 m
Bioretention facilities (BF)	Low	<15	Small	Road < 30 m, River > 30 m, Building > 3 m

Table 2. Site suitability indices for GI.

	Building	Road	Forest	Grass	Park	Farmland	Bare	Impervious
EMC (g/L)	0.35	1.25	0.03	0.02	0.15	0.07	0.05	0.55

Note: EMC denotes the median event mean concentrations.

2.5. Establishment of a Multifunctional Evaluation System for GI

The multifunctional benefits of GI were assessed. GI aims to manage the quantity and quality of runoff. In terms of runoff control, GI enables the restoration of the source's natural underlying, promotes infiltration and rapid discharge of runoff throughout the transfer process, and enables efficient centralized regulation of runoff quality. The indicators were created to examine the alleviation of strain on urban drainage networks, reduce the pollution of receiving water bodies, and limit peak flooding and pollutant impact on water bodies. The economic costs are associated with the necessity for financial assistance to create and maintain the efficacy of the GI. Efficient investment allocation is possible based on the close correlation between GI and site suitability. In addition, GIs provide various ecological benefits, including improving landscape aesthetics and resident well-being. As a result, a synergistic evaluation system was built for GI regarding functions including runoff control, investment, and ecological benefits. The descriptions of indices are displayed in Table 3.

Table 3. Multi-function evaluation system for GI.

Function	Indicators	Sub-Indicators	Indicator Implication
Runoff control function	Runoff quantity control efficiency	Runoff volume control	Rainfall volume capture rate
		Runoff peak time delay	Delay in the occurrence of flood peaks
		Runoff peak volume reduction	Runoff peak volume control rate
	Runoff quality control efficiency	Suspended pollutant removal	Effectiveness of suspended pollutant removal by GI, counted by suspended solid matter
		Oxygen-consuming pollutant removal	Effectiveness of COD, BOD_5 pollutant removal by GI.
		Nutrient pollutant removal	Effectiveness of nitrogen and phosphorus pollutant removal by GI.
		Toxic pollutant removal	Effectiveness of toxic pollutant removal by GI.
	Runoff reuse efficiency	Runoff utilization capacity	The capacity of runoff harvesting and reuse through GI, including centralized collection, in-situ reuse, and groundwater recharge
Costs investment	Equipment investment	Construction costs	Initial equipment asset investment for the construction of GI.
	Management and maintenance	Management costs	Consider the investment of depreciation and replacement over the life span of GI.
		Maintenance costs	Maintenance costs to ensure the proper functioning of GI such as dredging, renovation, etc.
Social benefit	Landscape aesthetics	Greening benefits	Calculated by greenery and vegetation stereo
		Aesthetic benefits	The landscape effect of the pebbles and paving colors, along with the facilities
	Space vitality	Visitor flowrate	The total number of passengers through the GI is divided by the space.
		Facility Interactivity	The extent to which the facility interacts with the surrounding visitor flow
		Visitor activity	The level of activity is characterized by the frequency of people entering and leaving the GI and its surrounding space

Three indexes are included in the runoff control function of the GI evaluation system. They describe separately the release of runoff volume control pressure in the urban drainage system, the effectiveness of runoff pollutant reduction, and the capacity to increase rainwater collection and reuse via GI. The cost investment in GI refers to the structural costs associated with the construction process and the maintenance costs associated with keeping normal regular operation. The social benefit metrics for GI quantify the extent to which the facilities improve the comfort and liveliness of residents. Urban inhabitants are the primary GI users and quantitative assessment of their perceptions serves as the foundation for assessing the social advantages of GI.

2.6. Quantification of the Multifunctional Effectiveness of GI

The GI functions are based on practical examples, mechanistic studies (Ying, 2010), and expert opinions regarding runoff control, economic costs, and ecological advantages. The values are derived based on each GI's structure and technical parameters, and primarily reflect its intrinsic properties. Each GI was assigned a comparison score, indicating its relative performance to the corresponding index. Each GI indicator's performance was graded as inappropriate, low, low-moderate, moderate, moderate-high, and high. The performance was quantified as 0, 1, 2, 3, 4, and 5, allowing for a mechanism-based assessment of the effects of GIs.

The GI runoff control function considers various structural characteristics and indexes that are influenced by the corresponding mechanism. Source-oriented GIs are based on an in-situ infiltration, detention, and storage mechanism with a hydraulic retention time of several hours. Transmission process control GIs rapidly convey runoff from the source to centralized facilities, alleviating pressure on drainage networks; nevertheless, their storage capacity, hydraulic residence time, and storage volume are limited. Systematically managed facilities focus on centralized runoff control. It is the primary mechanism for achieving quantitative and qualitative runoff control, with hydraulic retention times often lasting several days. The retention volume of GI affects the volume and quality of runoff. Their hydraulic retention techniques allow time for runoff quality enhancement mechanisms such as adsorption and degradation. The typical GI cost was calculated based on data from current research conducted both nationally and globally [29,30]. A higher score for a cost investment index corresponds to lower investment requirements, fewer management efforts, and more excellent operational stability in the evaluation system. Field monitoring was used to calculate the social benefit indices. Greening advantages were evaluated by calculating the green view rate [31].

The aesthetic benefits indexes quantify the GI's capacity to attract occupants. Total visitor flow and the frequency of resident-facility contact were used to quantify spatial vitality. Wi-Fi probes, GoPro photography, and artificial observation were used for the indexes. Wi-Fi monitoring equipment was set up to scan the profusion of Wi-Fi signals emanating from mobile phones within a 30-m radius to measure the interaction between the GI and visitors. Table 4 summarizes the multifunctional evaluation scores for GI. The functions and costs of GI runoff control are based on the process shown in Supplementary Materials Tables S1–S3.

Table 4. The multifunctional evaluation scores for GI.

GI	Runoff Control Function							
	Runoff Quantity Control Efficiency			Runoff Quality Control Efficiency				Runoff Reuse Efficiency
	Runoff Volume Control	Runoff Peak Time Delay	Runoff Peak Volume Reduction	Suspended Pollutant Removal	Oxygen-Consuming Pollutant Removal	Toxic Pollutant Removal	Nutrient Pollutant Removal	Runoff Utilization Capacity
IC	3	5	2	5	4	4	4	2
DP	2	1	3	2	1	1	1	4
WP	5	5	5	4	4	5	4	5
SGS	1	1	1	1	2	1	2	1
VS	3	3	3	2	3	3	3	1
GR	2	2	3	2	3	2	2	1
PP	3	5	2	5	3	4	3	2
BF	3	4	4	4	5	5	5	3

GI	Capital Investment			Social Habitat Benefits				
	Equipment Investment	Maintenance		Landscape Aesthetics		Space Vitality		
	Construction Costs	Management Costs	Maintenance Costs	Greening Benefits	Aesthetic Benefits	Visitor Flowrate	Facility Interactivity	Visitor Activity
IC	4	2	1	1	1	1	0	0
DP	4	5	5	3	2	2	1	2
WP	2	1	1	4	4	3	2	2
SGS	5	5	5	4	3	3	2	3
VS	5	1	2	4	3	3	2	4
GR	2	5	4	4	3	1	0	0
PP	1	4	2	2	2	5	5	5
BF	1	1	1	5	5	3	4	3

2.7. Weight of Multifunctional Indexes for GI Decision-Making

Weights for the GI multifunctional indexes were quantified based on stakeholders' preferences with different occupations. The opinions of experts and stakeholders were tallied and summarized to determine the weights for indicators. The process for acquiring and quantifying ideas was as follows: (1) Select typical stakeholders, including officers responsible for constructing GI projects, scholars of environment, scholars of architecture, and residents. (2) Explain to stakeholders the GI decision-making framework and the meaning of the indexes, and elicit their preferences for the indexes. (3) A comparative scoring system is applied, in which stakeholders assign relative importance to several indexes within the same category. After normalization, the weights for each indication were determined. The weights of indexes at each level are added together to a final performance score.

The GI multifunctional combined score is calculated by multiplying the weights by the index function values and then summing them. The total score was utilized to determine the most appropriate GI at different sites. The GIs with the highest total scores are listed first, with the highest total scores indicating the most recommended GI for the local conditions.

$$I_i = \sum_{j=1}^{16} w_{ij} \times r_{ij}, i = 1, 2, ...8$$

where I_i denotes the weighted total score of GI multifunctional performance, w_{ij} denotes the weight of a specific index, r_{ij} denotes the score of a function for GI, i denotes eight types of GI that are considered in this study, j denotes an evaluation index.

3. Application of GI Decision-Making System in Beijing's Urban Sub-Center

3.1. Site Suitability Indexes of Beijing's Urban Sub-Center for GI

The concentration of suspended solid pollutants in runoff was used as a proxy for the level of site contamination. The land use distribution and runoff coefficients were considered by calculating the intensity of the pollutant load. Pollution load intensity in different blocks was compared. As seen in Figure 3a, the cumulative runoff pollution of each block in Beijing's urban sub-center was statistically represented as low-medium-high. The study area is relatively flat, with an average slope of less than 10%. Blocks were categorized, as illustrated in Figure 3b, according to the slope indexes. The GI scale is constrained by available floor area, and the space use efficiency of GI facilities is an essential factor for heavily impervious underlying terrains. Owing to the structural and functional variances, it is vital to assure GI performance within the space of local sites. As seen in Figure 3c, blocks are categorized into three categories based on the area for GI construction and the site appropriateness evaluation criteria. Source-oriented GI is well-suited to small spaces. Transportation process regulation facilities provide mesoscale runoff control and are well suited for locations with medium available space. Systematic detention and regulation facilities provide centralized runoff regulation and are well suited to sites with large available space. The constructible area for GI development must include buffers from buildings, roads, and waters. As seen in Figure 3d, identifying suitable places for GI construction considers both site appropriateness criteria and the buffer distribution of the underlying surface.

3.2. Weights for the GI Multifunctional Indexes of Beijing's Urban Sub-Center

The weights reflect the decision-making preferences of the construction managers, technical experts, and residents for the GI multifunctional indexes. The examination was conducted by stakeholders from four fields, including architect experts, environmental and ecological experts, local government administrators, and residents of Beijing's sub-center. Table 5 summarizes the weights derived from the opinions of the four stakeholder groups. Stakeholders typically regarded the relevance of GI in the following order: runoff control function > cost input > societal benefit. According to all four stakeholder groups, runoff control is a dominant function for GIs. Environmental experts and urban planners make similar judgments on the critical nature of GI's numerous functions. Among the four expert groups, architects are the only group that believes the social benefits of GI outweigh the expense. Residents choose GI for its social benefits. Because stakeholders assessed the indexes and sub-indexes in the evaluation method equally, only the average aggregate weighted results are provided in Table 5.

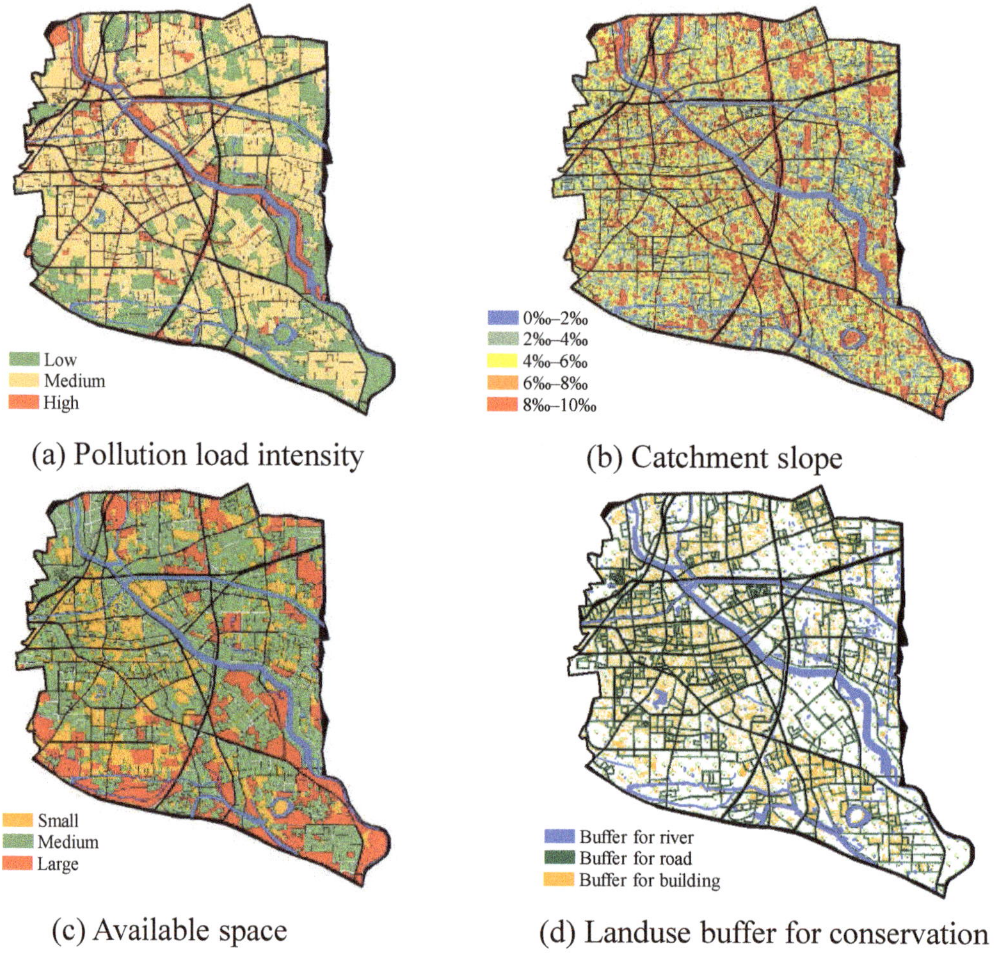

(a) Pollution load intensity
- Low
- Medium
- High

(b) Catchment slope
- 0‰–2‰
- 2‰–4‰
- 4‰–6‰
- 6‰–8‰
- 8‰–10‰

(c) Available space
- Small
- Medium
- Large

(d) Landuse buffer for conservation
- Buffer for river
- Buffer for road
- Buffer for building

Figure 3. The distribution of site suitability indexes for GI.

GI's runoff control effects are weighted similarly and are highly recognized by stakeholders. But the weight of runoff reuse is not as high as runoff quality and quantity control efficiency. The equipment investment and maintenance costs have relative weights in terms of GI cost. Landscape and space vitality plays a similar role in the social advantages of GI. GIs are given equal importance in landscape aesthetics and spatial vitality indicators. The weight values in Table 5 demonstrate that trade-offs between runoff control, cost input, and social benefit are required for GI layout.

Table 5. Weights for the GI multifunctional indexes.

Multi-function	Environmental Expert	Architect Expert	Manager	Resident	Average	Index	Average Weight	Sub-Index	Average Weight
Runoff control function	0.45	0.47	0.42	0.38	0.43	Runoff quantity control efficiency	0.18	Runoff volume control	0.07
								Runoff peak time delay	0.06
								Runoff peak volume reduce	0.04
						Runoff quality control efficiency	0.15	Suspended pollutant removal	0.05
								Oxygen-consuming pollutant removal	0.04
								Toxic pollutant removal	0.04
								Nutrient pollutant removal	0.03
						Runoff reuse efficiency	0.10	Runoff utilization capacity	0.10
Cost investment	0.34	0.20	0.35	0.34	0.31	Equipment investment	0.15	Construction costs	0.15
						Maintenance	0.16	Management costs	0.08
								Maintenance costs	0.08
Social benefit	0.21	0.33	0.23	0.28	0.26	Landscape aesthetics	0.14	Greening benefits	0.08
								Aesthetic benefits	0.06
						Space vitality	0.12	Visitor flowrate	0.04
								Facility interactivity	0.04
								Visitor activity	0.04

3.3. Comprehensive Effectiveness Score Ranking for GI Decision-Making

Figure 4 illustrates the combined effectiveness score ranking for GI, which considers diverse stakeholder perspectives and the inherent multifunctional benefits of GI. The study's findings indicated that WP was the primary GI facility in the study area, followed by BF and VS. The highest-scoring GI is a systemic detention and regulation facility constrained by site space. The necessary hydraulic retention period can ensure a high runoff quantity and quality control and a considerable rainwater resource utilization capacity. Since WPs are highly self-healing during regular operation, management and maintenance need can be moderately eased, improving cost-effectiveness. Because WPs are primarily located in suburban regions with minimal population activity, they perform poorly in visitor flow and engagement with residents, resulting in a low social benefit score. The second-ranked BF is a source-oriented facility that is highly successful in regulating the quantity and quality of runoff. Owing to the expensive initial investment in equipment and ongoing management costs, its cost-effectiveness is compromised. In the core urban area, BF is chosen due to the high volume of visitors and the consequent opportunity to interact effectively with neighboring residents, resulting in more excellent social benefits. At a transport process control facility VS is the third-rated GI. VS requires less initial capital expenditure and minor maintenance and performs well in cost-effectiveness.

The region's suitable GI facilities were selected based on the pollution load intensity, slope, available area, and reserved area. GIs with the highest scores in the multifunctional evaluation system are regarded as the most suited GI facilities in the study area, as shown in Figure 5. WP is preferable in places with a large area in the suburbs, where runoff control pressure arises from centered upstream. The most significant hurdle for WP is the space. However, affordable land property in the suburbs provides chances for WP and ecosystems. End-of-system wetland regulation of runoff quantity and quality has also been extensively shown in previous research [32,33]. According to the site appropriateness evaluation matrix, BF is the most recommended in blocks with little available space and significant pollutant loads. BF provides exceptional runoff quality control but is costly [7]. Despite eliminating budgetary constraints, as illustrated in Figure 5, some areas strongly need runoff quality reduction. In densely built-up places, VS is favored. Length impacts the performance of VS. Owing to its form and purpose, VS is utilized widely [4,34,35], as indicated in this study and previous research, along pathways, riverfronts, and major roads. Many GI systems in urban areas use SGS [36,37] because they store runoff economically and work with the landscape. SGS can control runoff, save money, and boost social benefits. SGS became the widest preferred mode of runoff control, as proved in this study.

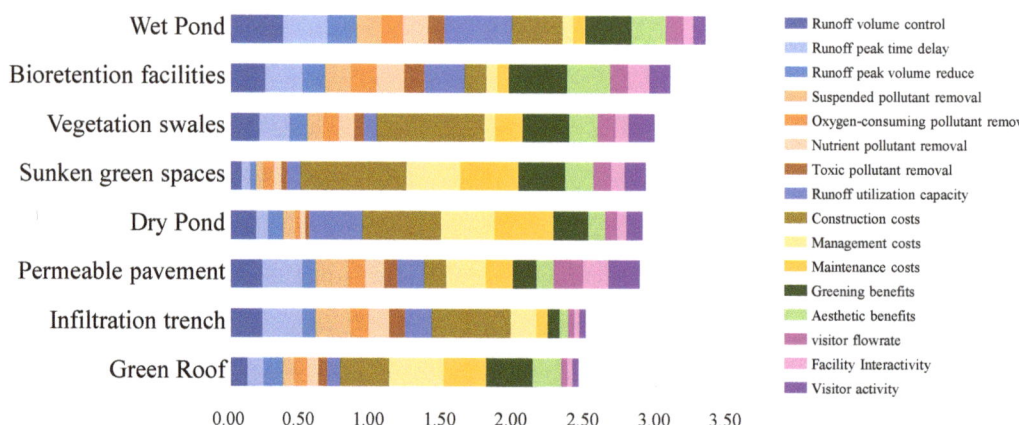

Figure 4. Ranking for weighted multi-functional effectiveness combined score.

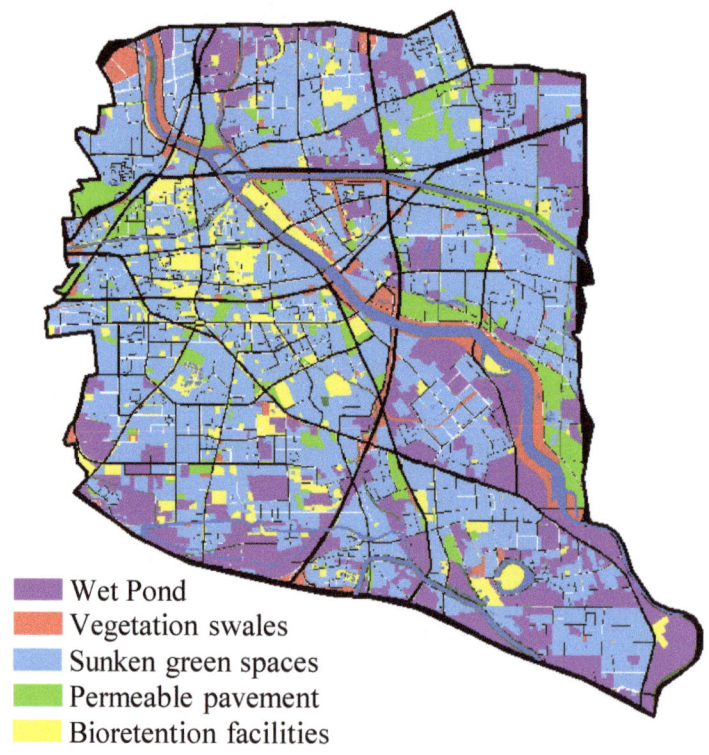

Figure 5. Regional most preferred GI distribution.

3.4. Discussion

This research presents a decision-making framework for the spatial layout of GI that considers aspects such as site suitability, multifunctional effectiveness, and weight assignment based on the desires of stakeholders. Promoting the technique requires highlighting three aspects. First, select and evaluate GI sites. GIs were traditionally allocated based only on managers' opinions [2,21], ignoring the diverse stakeholders who directly profit from them. In GI layout planning, opportunistic site selection is insufficient without systematic analysis, robust data, and in-depth investigation [38]. This study considers the spatial heterogeneity offered by natural conditions, pollutant concentration characteristics, and constructible area of the region. These features are analyzed as prerequisites for an appropriate GI layout, ensuring that GI effectiveness can be maximized. Second, local monitoring data and parameters are integrated. This study's GI layout decision-making frameworks could be replicated in other cities. The sophisticated site localization and GI efficacy evaluation system involves various indicators. Generalizing optimization insights depends mainly on geography, requiring localized experiments and data processing. It is also possible to streamline the evaluation system's indicators by retaining only those essential indicators. During the systematic examination, consideration must be given to the synergistic application of data from numerous sources. Third, stakeholders are involved in decision-making. Residents, architects, and the government are GI stakeholders. Architects and residents are subject to government-imposed constraints [39]. Urban amenity as a resident's objective has been disregarded. This study demonstrates in a novel manner the preferences of residents and industry academics for GI multifunctional effectiveness, ensuring that implementation benefits satisfy essential stakeholders.

4. Conclusions

An assessment index approach was developed to make it easier to identify a GI layout plan that meets the site's characteristics. First, the usefulness of GI for a particular site was determined by the pollution load intensity, slope, available area, and constructible area. Then, the multifunctional benefits of a typical GI were quantified in terms of runoff control function, cost investment, and social benefits. The case study was conducted in the sub-center of Beijing. To determine the index weights for decision-making, we examined the GI multifunctional preferences of local stakeholders, including administrators, experts, and residents. BF is the most preferred in densely built-up areas with limited available space and significant pollutant loads. In the most urbanized region, VS is favored. WP is preferable in places with a large area in the suburbs, where runoff control pressure arises upstream. The optimal layout outcomes are consistent with the region's natural resources and stakeholder interests. The GI is designed with a specific layout to maximize multifunctional benefits.

Supplementary Materials: The following supporting information can be downloaded at: https://www.mdpi.com/article/10.3390/w14111765/s1, Table S1: Control effectiveness of GI; Table S2: Runoff control mechanisms and effectiveness for common structural GIs; Table S3: Capital, operational, and maintenance cost of structural GIs.

Author Contributions: Z.L. Conceptualization, methodology, software, writing—original draft preparation; Y.Y. validation, writing—review and editing; J.H. writing—review and editing; H.J. supervision, project administration, funding acquisition. All authors have read and agreed to the published version of the manuscript.

Funding: This work was supported by the National Nature Science Foundation of China (Grant No. 41890823, 52070112, 7181101209).

Conflicts of Interest: The authors declare no conflict of interest.

References

1. Choi, C.; Berry, P.; Smith, A. The climate benefits, co-benefits, and trade-offs of green infrastructure: A systematic literature review. *J. Environ. Manag.* **2021**, *291*, 112583. [CrossRef]
2. Yin, D.; Chen, Y.; Jia, H.; Wang, Q.; Chen, Z.; Xu, C.; Li, Q.; Wang, W.; Yang, Y.; Fu, G.; et al. Sponge city practice in China: A review of construction, assessment, operational and maintenance. *J. Clean. Prod.* **2021**, *280*, 124963. [CrossRef]
3. Gong, Y.; Zhang, X.; Li, J.; Fang, X.; Xie, P.; Nie, L. Factors affecting the ability of extensive green roofs to reduce nutrient pollutants in rainfall runoff. *Sci. Total Environ.* **2020**, *732*, 139248. [CrossRef]
4. Li, H.; Li, K.; Zhang, X. Performance Evaluation of Grassed Swales for Stormwater Pollution Control. *Procedia Eng.* **2016**, *154*, 898–910. [CrossRef]
5. Liu, J.; Yan, H.; Liao, Z.; Zhang, K.; Schmidt, A.R.; Tao, T. Laboratory analysis on the surface runoff pollution reduction performance of permeable pavements. *Sci. Total Environ.* **2019**, *691*, 1–8. [CrossRef]
6. Martin, D.M.; Piscopo, A.N.; Chintala, M.M.; Gleason, T.R.; Berry, W. Developing qualitative ecosystem service relationships with the Driver-Pressure-State-Impact-Response framework: A case study on Cape Cod, Massachusetts. *Ecol. Indic.* **2018**, *84*, 404–415. [CrossRef]
7. Wang, M.; Zhang, D.; Adhityan, A.; Ng, W.J.; Dong, J.; Tan, S.K. Assessing cost-effectiveness of bioretention on stormwater in response to climate change and urbanization for future scenarios. *J. Hydrol.* **2016**, *543*, 423–432. [CrossRef]
8. Zhang, X.; Chen, L.; Zhang, M.; Shen, Z. Prioritizing Sponge City Sites in Rapidly Urbanizing Watersheds Using Multi-Criteria Decision Model. *Environ. Sci. Pollut. Res.* **2021**, *28*, 63377–63390. [CrossRef]
9. Zischg, J.; Zeisl, P.; Winkler, D.; Rauch, W.; Sitzenfrei, R. On the sensitivity of geospatial low impact development locations to the centralized sewer network. *Water Sci. Technol.* **2018**, *77*, 1851–1860. [CrossRef]
10. Liu, Z.; Xu, C.; Xu, T.; Jia, H.; Zhang, X.; Chen, Z.; Yin, D. Integrating socioecological indexes in multiobjective intelligent optimization of green-grey coupled infrastructures. *Resour. Conserv. Recycl.* **2021**, *174*, 105801. [CrossRef]
11. Raei, E.; Reza Alizadeh, M.; Reza Nikoo, M.; Adamowski, J. Multi-objective decision-making for green infrastructure planning (LID-BMPs) in urban storm water management under uncertainty. *J. Hydrol.* **2019**, *579*, 124091. [CrossRef]
12. Wang, J.; Liu, J.; Mei, C.; Wang, H.; Lu, J. A multi-objective optimization model for synergistic effect analysis of integrated green-gray-blue drainage system in urban inundation control. *J. Hydrol.* **2022**, *609*, 127725. [CrossRef]
13. Martin-Mikle, C.J.; de Beurs, K.M.; Julian, J.P.; Mayer, P.M. Identifying priority sites for low impact development (LID) in a mixed-use watershed. *Landsc. Urban Plan.* **2015**, *140*, 29–41. [CrossRef]
14. Patault, E.; Ledun, J.; Landemaine, V.; Soulignac, A.; Richet, J.-B.; Fournier, M.; Ouvry, J.-F.; Cerdan, O.; Laignel, B. Analysis of off-site economic costs induced by runoff and soil erosion: Example of two areas in the northwestern European loess belt for the last two decades (Normandy, France). *Land Use Policy* **2021**, *108*, 105541. [CrossRef]
15. Zhang, N.; Luo, Y.-J.; Chen, X.-Y.; Li, Q.; Jing, Y.-C.; Wang, X.; Feng, C.-H. Understanding the effects of composition and configuration of land covers on surface runoff in a highly urbanized area. *Ecol. Eng.* **2018**, *125*, 11–25. [CrossRef]
16. Yao, L.; Wu, Z.; Wang, Y.; Sun, S.; Wei, W.; Xu, Y. Does the spatial location of green roofs affects runoff mitigation in small urbanized catchments? *J. Environ. Manag.* **2020**, *268*, 110707. [CrossRef]
17. Eckart, K.; McPhee, Z.; Bolisetti, T. Performance and implementation of low impact development—A review. *Sci. Total Environ.* **2017**, *607–608*, 413–432. [CrossRef]
18. Wang, X.; Tian, Y.; Zhao, X. The influence of dual-substrate-layer extensive green roofs on rainwater runoff quantity and quality. *Sci. Total Environ.* **2017**, *592*, 465–476. [CrossRef]
19. Yang, W.; Wang, Z.; Hua, P.; Zhang, J.; Krebs, P. Impact of green infrastructure on the mitigation of road-deposited sediment induced stormwater pollution. *Sci. Total Environ.* **2021**, *770*, 145294. [CrossRef]
20. Xu, C.; Jia, M.; Xu, M.; Long, Y.; Jia, H. Progress on environmental and economic evaluation of low-impact development type of best management practices through a life cycle perspective. *J. Clean. Prod.* **2019**, *213*, 1103–1114. [CrossRef]
21. Jia, H.; Yao, H.; Tang, Y.; Yu, S.L.; Zhen, J.X.; Lu, Y. Development of a multi-criteria index ranking system for urban runoff best management practices (BMPs) selection. *Environ. Monit. Assess.* **2013**, *185*, 7915–7933. [CrossRef] [PubMed]
22. Tang, Y. SUSTAIN-Supported BMP Planning Study for Optimal Management of Urban Rainfall Runoff. Master's Thesis, Tsinghua University, Beijing, China, 2010.
23. Gwak, J.H.; Lee, B.K.; Lee, W.K.; Sohn, S.Y. Optimal location selection for the installation of urban green roofs considering honeybee habitats along with socio-economic and environmental effects. *J. Environ. Manag.* **2017**, *189*, 125–133. [CrossRef] [PubMed]
24. Jia, H.; Wang, Z.; Zhen, X.; Clar, M.; Yu, S.L. China's sponge city construction: A discussion on technical approaches. *Front. Environ. Sci. Eng.* **2017**, *11*, 18. [CrossRef]
25. Xu, C.; Hong, J.; Jia, H.; Liang, S.; Xu, T. Life cycle environmental and economic assessment of a LID-BMP treatment train system: A case study in China. *J. Clean. Prod.* **2017**, *149*, 227–237. [CrossRef]
26. Xu, C.; Tang, T.; Jia, H.; Xu, M.; Xu, T.; Liu, Z.; Long, Y.; Zhang, R. Benefits of coupled green and grey infrastructure systems: Evidence based on analytic hierarchy process and life cycle costing. *Resour. Conserv. Recycl.* **2019**, *151*, 104478. [CrossRef]
27. Ji, H.; Peng, D.; Fan, C.; Zhao, K.; Gu, Y.; Liang, Y. Assessing effects of non-point source pollution emission control schemes on Beijing's sub-center with a water environment model. *Urban Clim.* **2022**, *43*, 101148. [CrossRef]

28. Shajib, M.T.I.; Hansen, H.C.B.; Liang, T.; Holm, P.E. Rare earth elements in surface specific urban runoff in Northern Beijing. *Sci. Total Environ.* **2020**, *717*, 136969. [CrossRef]
29. Muthukrishnan, S.; Field, R. *The Use of Best Management Practices (BMPs) in Urban Watersheds*; EPA/600/R-04; DEStech Publications, Inc.: Lancaster, PA, USA, 2004.
30. Pradhan, S.; Al-Ghamdi, S.G.; Mackey, H.R. Greywater recycling in buildings using living walls and green roofs: A review of the applicability and challenges. *Sci. Total Environ.* **2019**, *652*, 330–344. [CrossRef]
31. Hou, J.; Chen, L.; Zhang, E.; Jia, H.; Long, Y. Quantifying the usage of small public spaces using deep convolutional neural network. *PLoS ONE* **2020**, *15*, e0239390. [CrossRef]
32. Boucher-Carrier, O.; Brisson, J.; Abas, K.; Duy, S.V.; Sauvé, S.; Kõiv-Vainik, M. Effects of macrophyte species and biochar on the performance of treatment wetlands for the removal of glyphosate from agricultural runoff. *Sci. Total Environ.* **2022**, *838*, 156061. [CrossRef]
33. Kill, K.; Grinberga, L.; Koskiaho, J.; Mander, Ü.; Wahlroos, O.; Lauva, D.; Pärn, J.; Kasak, K. Phosphorus removal efficiency by in-stream constructed wetlands treating agricultural runoff: Influence of vegetation and design. *Ecol. Eng.* **2022**, *180*, 106664. [CrossRef]
34. Huang, C.-L.; Hsu, N.-S.; Liu, H.-J.; Huang, Y.-H. Optimization of low impact development layout designs for megacity flood mitigation. *J. Hydrol.* **2018**, *564*, 542–558. [CrossRef]
35. Tang, J.; Wang, W.; Feng, J.; Yang, L.; Ruan, T.; Xu, Y. Urban green infrastructure features influence the type and chemical composition of soil dissolved organic matter. *Sci. Total Environ.* **2021**, *764*, 144240. [CrossRef] [PubMed]
36. Du, S.; Wang, C.; Shen, J.; Wen, J.; Gao, J.; Wu, J.; Lin, W.; Xu, H. Mapping the capacity of concave green land in mitigating urban pluvial floods and its beneficiaries. *Sustain. Cities Soc.* **2019**, *44*, 774–782. [CrossRef]
37. Liu, W.; Chen, W.; Peng, C. Influences of setting sizes and combination of green infrastructures on community's stormwater runoff reduction. *Ecol. Model.* **2015**, *318*, 236–244. [CrossRef]
38. Starkl, M.; Brunner, N.; López, E.; Martínez-Ruiz, J.L. A planning-oriented sustainability assessment framework for peri-urban water management in developing countries. *Water Res.* **2013**, *47*, 7175–7183. [CrossRef]
39. Chen, Y.; Chen, H. The Collective Strategies of Key Stakeholders in Sponge City Construction: A Tripartite Game Analysis of Governments, Developers, and Consumers. *Water* **2020**, *12*, 1087. [CrossRef]

Article

Integrated and Control-Oriented Simulation Tool for Optimizing Urban Drainage System Operation

Haozheng Wang [1,*], Guanyu Han [1], Lei Zhang [1], Yiting Qiu [1], Juntao Li [1] and Haifeng Jia [2]

1. North China Municipal Engineering Design & Research Institute Co., Ltd., Tianjin 300070, China; hanguanyu9301@126.com (G.H.); mumu.zhang@outlook.com (L.Z.); qytt0309@163.com (Y.Q.); wyljtmail@163.com (J.L.)
2. School of Environment, Tsinghua University, Beijing 100084, China; jhf@tsinghua.edu.cn
* Correspondence: haozheng_bnu@hotmail.com

Abstract: With the management and operation of urban drainage systems (UDS) becoming more complicated and difficult, integrated models aiming to control and manage the entire drainage system are under enormous demand. Ideally, integrated models, as a potential tool for meeting the increasing demands, should combine both conceptual and mechanistic models that merge all UDS components and balance simulation accuracy with time constraints. Within this context, our study introduces an innovative modeling software, Simuwater, which couples multiple principles, simulates multiple components, and combines optimized control functions, playing a role in the integrated simulation and overflow control application of UDS. The software has been utilized in a real-time case-control study in one city of China, and it obtained significant optimized operation results to reduce combined sewer overflow (CSO) by making full use of the storage facilities and actuators. As the Simuwater model continues to improve in depth and breadth, it will play an increasingly important role in more application scenarios of UDS.

Keywords: control-oriented model; urban drainage system; real-time optimization; Simuwater

Citation: Wang, H.; Han, G.; Zhang, L.; Qiu, Y.; Li, J.; Jia, H. Integrated and Control-Oriented Simulation Tool for Optimizing Urban Drainage System Operation. *Water* 2022, 14, 25. https://doi.org/10.3390/w14010025

Academic Editor: Francesco De Paola

Received: 7 November 2021
Accepted: 20 December 2021
Published: 23 December 2021

Publisher's Note: MDPI stays neutral with regard to jurisdictional claims in published maps and institutional affiliations.

Copyright: © 2021 by the authors. Licensee MDPI, Basel, Switzerland. This article is an open access article distributed under the terms and conditions of the Creative Commons Attribution (CC BY) license (https:// creativecommons.org/licenses/by/ 4.0/).

1. Introduction

Urban drainage systems (UDS) are used to transport and treat urban runoff and domestic sewage. These systems play an important role in controlling urban floods and improving the quality of the ecological environment. The urban drainage process consists of the collection and discharge of rainwater and sewage, including evaporation, infiltration, treatment, and other processes. With the rapid development of urbanization and increasing water requirements, the constituents and operation of UDS have become increasingly complex. Within this context, there is a need for more comprehensive auxiliary analysis tools that cover the entire life cycle of UDS.

As a digital analysis method, models have been used in the comprehensive simulation of UDS for a long time. In the 1970s, the Environmental Protection Agency released the Storm Water Management Model (SWMM) [1], which initiated the development of a hydrodynamic model of urban drainage systems. The SWMM includes the simulation of rainfall–runoff in catchments and hydraulic processes [2]. With the advent and popularization of personal computers in the 1980s, the use of hydrodynamic models of UDS spread widely, with many commercial software packages becoming available. Simultaneously, research on dynamic simulation models of wastewater treatment plants in the 1980s addressed the limitations of static models used for analyzing activated sludge processes in the 1950s–1970s. In the 1990s, research on the theory of sludge transport and water quality simulation promoted the coupling of hydrodynamic and water quality models. This period also saw the establishment of green infrastructure, represented by low impact development (LID) [3] and the development of specialized tools for LID design and performance evaluation. In recent decades, geographic information systems have been integrated into UDS models [4,5].

Currently, integrated models of UDS are widely used, with real-time control (RTC) technology [6–9] applied to these UDS models [10,11]. A few RTC cases have been successfully designed and operated for many years, and early RTC cases were relatively simple and focused on local control. With the application of model technology, RTC technology is gradually mature and complex. For example, a recent RTC study in Norfolk has shown that the model predictive control (MPC) could reduce overall flooding caused by sea level rise with an average effective percentage reduction of 32% [12]. Another study in a Canadian city has shown that RTC technology with the model application could reduce peak flows (73% to 95% reduction) and significantly improve the quality of outflow during rainfall events [13]. Related studies and cases are widely distributed in Europe [6,7,14,15], America [11,16] and Asia [17].

2. Classification of the Models

In general, UDS models can be divided into simulation-oriented and control-oriented models according to their functions [18–20].

2.1. Simulation-Oriented Model

Simulation-oriented models are used to simulate and then evaluate urban drainage processes. Most simulation-oriented models are mechanistic models that describe the primary physical processes between input and output using mathematical equations [21]. Classified by different UDS objects, simulation-oriented models can be further divided into catchment, drainage network, sewage treatment, and surface water models.

Catchment models simulate rainfall and runoff and include a module for the degradation and transformation of point (or non-point) source pollutants from surfaces to rivers. Thus, catchment models represent the "source" of entire drainage systems and are used to simulate the process of runoff and pollutant transportation in catchment areas as well as the effect of LID. In comparison, drainage network models simulate the transmission of urban rainwater, sewage, and related pollutants in drainage systems. Furthermore, drainage network models simulate the hydrological processes of flow from rainfall to surface runoff, and finally, the entrance of the network. Thus, drainage models include the runoff, storage, and infiltration processes of different land use types in specific catchment areas. In addition, hydraulic processes are simulated, including manholes, pipe networks, natural and artificial channels, culverts, reservoirs, and outlets. Flow velocity and depth in the pipe network are primarily calculated using Saint-Venant equations (SVE) [18,20], which follow the relationship between mass and energy conservation.

Sewage treatment models simulate changes in water quantity and quality during the wastewater treatment process. Various simulation modules can be combined, depending on the specific technological processes in use, including, among others, primary sedimentation tanks, different types of biological reaction tanks (for example, A/O, AAO, SBR [22], and oxidation ditches), and secondary sedimentation tanks. Sewage treatment models were developed based on the International Water Association's (IWA) activated sludge mathematical model ASM [23] (ASM1, ASM2, ASM2D and ASM3, ASM2 + TUD), which can accurately predict the effluent effect and is generally used in the auxiliary design process of sewage treatment structures [24].

Surface water models simulate hydrodynamic processes and include a focus on water quality and water ecology in rivers, lakes, reservoirs, wetlands, estuaries, coasts, and oceans. In the simulation of UDS, hydrodynamic and water quality models of rivers, such as HEC-RAS, HSPF, and Mike11 [25] are commonly used. These models can simulate 1D or 2D hydrodynamics and water quality. Hydrodynamics and water quality models of reservoirs, including EFDC, MIKE 21, and CE-QUAL-W2 [26], are commonly used and simulate 2D or 3D hydrodynamics and water quality.

2.2. Control-Oriented Models

Control-oriented UDS models refer to models that assist in management decision-making by simulating the impact of controllable facilities on the entire system [27]. To improve the computational efficiency of models and assist in real-time decision-making, control-oriented models generally have a low-complexity UDS and can be divided into linear SVE models, data-driven models, and conceptual models [18,20].

Although SVEs generalize the relationship between flow and water level under steady-state conditions in a linear manner, thereby improving computational efficiency [20], linear SVEs [28] do not reflect the dynamic conditions of a system, such as sudden inflow [18]. In response, a simple linear model can be obtained from the Hayami equation by using the moment matching method [29].

Conceptual models can accurately simulate the state of a UDS and, by simulating drainage networks and river channels and adjusting conceptualized facility parameters to monitoring data, can be used for prediction and control decisions [20,30]. According to different conceptualization methods, conceptual models can be divided into mixed logic dynamic models [31], virtual tank-based models, Nash models, Muskingum models [32,33], and integrator-delay models [34,35].

3. Problems and Development of the Model

With the continuous expansion of simulation principles and application scenarios, whether with simulation- or control-oriented models, there is an increasing need for applications to satisfy a larger number of requirements. However, traditional models have a number of problems that have gradually become prominent.

The mechanism of the traditional model is complex. For example, the SWMM includes such a large number of parameters for the catchment, pipes, and other modules that it can be difficult for technical engineers to collect sufficient data to support the modeling. In addition, the large number of parameters can lead to equifinality, requiring a high degree of experience to find the most suitable data fitting method for model calibration.

Although the classic mechanism models represented by SWMM and InfoWorks [36] have high simulation accuracy and industry recognition [37], their complicated mechanisms lead to longer calculation times, conflicting with the actual needs of simulation timeliness. While control-oriented models have a low level of complexity, it is difficult to fine-tune problematic pipelines or pipe networks with complex structures, resulting in a decrease in simulation accuracy and insufficient reliability of results. Therefore, it is necessary to consider coupling the conceptual model and mechanism model to greatly improve the calculation efficiency, while also considering the accuracy of the simulation. In addition, although traditional models generally focus on simulating specific parts of source–process–end facilities in the UDS (such as MIKE11 for rivers and SWMM for pipe networks), the simulation of the integrated water environment requires the simulation of the catchment, LID, pipe network, pumping stations, storage facilities, wastewater treatment plants, river channels, and throttling facilities on a single platform. The lack of comprehensiveness of the simulation objects is a more prominent problem.

In recent years, there has been a change in the use and application of models. For example, while traditionally, models are used to evaluate the effect of the design scheme in the planning and design stages, new applications include the input of required design and operation targets to automatically calculate the appropriate design plan through optimization algorithms. The operation targets usually include the scale of the facility and basic operation rules. Furthermore, the online application of the model requires it to have strong optimization computing capabilities, which are able to meet the calculation requirements of real-time control for optimization schemes. Therefore, models should be able to formulate timely and accurate operation schemes for online applications through a mature optimization algorithm.

4. Approach for UDS Model Improvement

To meet the requirements of a UDS model and to solve the problems associated with current UDS modeling tools, new integrated simulation software should have the following characteristics:

- Ability to integrate multiple processes within the UDS;
- Ability to couple multiple simulation theories;
- Ability to optimize the setting values during the simulation process;
- Ability to support secondary development;
- High reliability and accuracy.

Integrated UDS modeling software should simulate hydrology, hydraulics, and water quality to better understand the continuous dynamic simulation of a UDS, including catchment areas, LID, and urban rainwater, as well as sewage pipelines, water bodies, pump stations, and storage tanks. It should be able to simulate rainfall–runoff, wash-off, hydraulic transmission, and pollutant degradation and should be able to set control rules.

With experience in sewage treatment design and automatic control of infrastructure in the UDS, the North China Municipal Engineering Design & Research Institute Co., Ltd. (NCME) developed a new platform for UDS modeling—namely, the "Smart and Integrated Model of Urban Water (Simuwater, developed by NCME, Tianjin, China)". The framework of Simuwater model is shown in Figure 1.

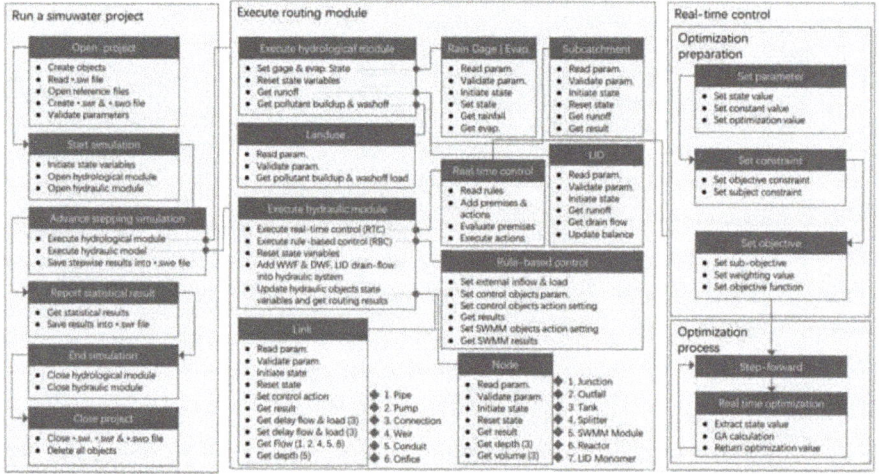

Figure 1. Framework diagram of the developed Smart and Integrated Model of Urban Water (Simuwater).

4.1. Coupling of Different Simulation Theories

To achieve a balance between simulation speed and accuracy, integrated modeling software should contain an assortment of simulation methods for the modeling of various types of objects [25,34,35,38], as shown in Figure 2. Taking the conceptual models as the main part of the project, models with different simulation methods can be connected into a single project, without data interaction problems in the simulation process. The customer hopes to choose a suitable method for improving the modeling accuracy of the local area. For example, in a non-negligible area with various reverse slope pipelines, the SVE can be selected for this area, rather than the Manning equation [25] or the Muskingum method [34]. However, for other regions with insufficient basic data or large data ranges, it will be more appropriate to use the Muskingum method to generalize the simulation pipeline.

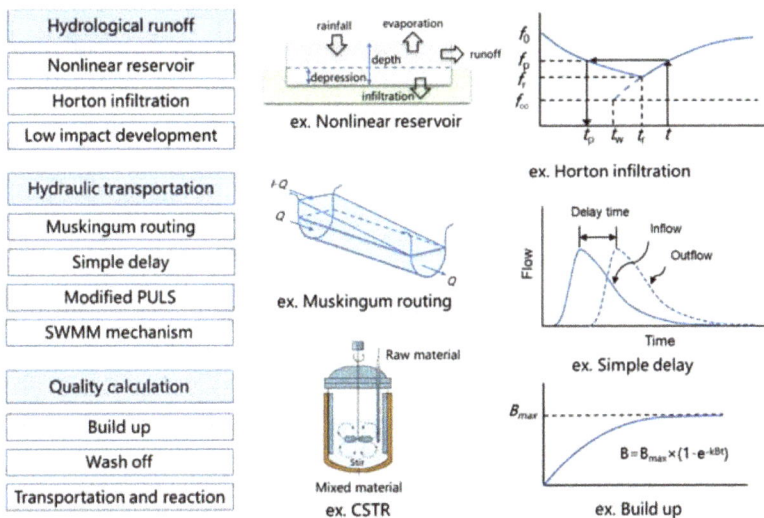

Figure 2. The coupling of various mechanisms in the developed Smart and Integrated Model of Urban Water, Simuwater.

4.2. Integration of Multiple Objects

Ideally, integrated modeling software should have the following four primary modeling objects: source, network, plant, and river objects. Each object can have extensive sub-objects to simulate specific structures (Table 1).

Table 1. List of simulation facilities of the developed Smart and Integrated Model of Urban Water (Simuwater).

Objects	Sub-objects	Diagrams	Functions
Source	Catchment		
	Green Roof		Rainfall–runoff
	Bio-retention cell		Dry weather flow (DWF) Inflow and Infiltration (II)
	Infiltration trench		Source runoff reduction
	Permeable pavement		Source pollution control
	Rain garden		
	Vegetative swale		
	Rain barrel		
Network	Junction		
	Outfall		
	Splitter		Process collection, storage, and treatment of water process transmission and control of water
	Tank		
	Pipe		
	Conduit		
	Connection		

Table 1. Cont.

Objects	Sub-objects	Diagrams	Functions
Water treatment	Pump		
	Weir		
	Orifice		
	Storage		Centralized or decentralized water storage and water quality control
	Filter		
	Wetland		
River	Reach		Terminal flow transmission Terminal pollution decay
	Reservoir		Water allocation and storage

In response to specific objects needing different simulation time-steps (depending on size and functions), an integrated modeling software allows for the application of a unified, complete, and dynamic time-step size in the simulation of different objects.

4.3. Optimization Control Applying

In integrated modeling software, optimization control should ideally be achieved by analyzing the current problems of the system, arranging the system facility state values and constraint conditions, formulating reasonable real-time control objectives, and establishing related functions, as well as building a complete and feasible optimization system [14,17,39]. This kind of method, based on model construction, is widely used in the aspect of CSO control [40], water quality protection [41], and flood reduction [12] and even in integrated environment targets control [10,42]. In the real-time operation/simulation process, data required for the optimization system must be extracted, the strategy formulation completed under real-time multi-control objectives, and the strategy returned to the operation/simulation system to complete the real-time optimization control within the control time-step, as shown in Figure 3.

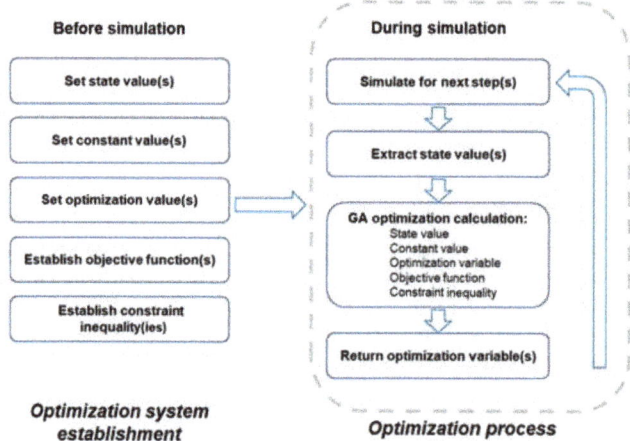

Figure 3. The establishment process of optimization system before and during simulation.

Generally, a genetic algorithm (GA) is coupled in the integrated modeling software as the optimization algorithm and adjusts to suit the UDS optimization calculation [43,44].

With a client-specific objective function and constraint conditions, the optimization algorithm can determine the most suitable operation scheme. The calculation process usually takes less than 1 min, depending on the complexity of the optimization system, thereby meeting the time 5-min limitation of the RTC routing step.

Objective function usually contains several control objectives, and weight values are used to distinguish the importance of every objective.

$$Objective(X) = \sum_{i=1}^{m}((w_i \times ob_i(X)))$$

where X is the collection of controlled variable, $ob_i(X)$ is the ith optimization objective, w_i is the weight value of the ith objective, and m is the number of optimization objectives of the system.

To achieve multi-period predictive control of the UDS, an integer multiple of the control step length (or simulation step length), such as N, can be selected as the predictive period, and the controlled variables and optimized variables can be expanded to N × n accordingly. According to the predicted state of the system in the future period, the algorithm is used to calculate N × n variable values simultaneously, but only the n variable values in the current step are executed—that is, under the premise of considering the system state in the future N time periods, we formulate an optimized operation plan for the current period, as shown in Figure 4. Although this method greatly increases the simulation and optimization calculation time, the formulated plan has a full-time (multi-period) optimality.

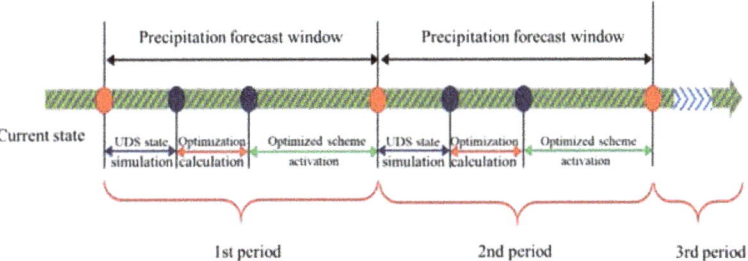

Figure 4. Optimization of the system before and during simulation.

Within this context, we developed Simuwater, an integrated model for UDS, which is based on the rolling optimization method for calculating optimization operation rules according to rainfall prediction and system state simulation data, thereby actualizing the dynamic simulation of optimization control.

4.4. Development of Open-Interface

To meet specific demands, our integrated modeling software provides a large number of interfaces for secondary development (Table 2). Researchers and developers can use common programming languages, such as Python, C, and C++, to call the interfaces to expand the function. In addition, operators can set complicated operation rules in Python and can easily insert them into the software.

Table 2. Interface name and description of the integrated urban drainage software, Simuwater.

Interface Name	Description
Simuwater_run	Run the simulation process.
Simuwater_open	Open the simulation project.
Simuwater_start	Start the simulation process.
Simuwater_step	Advance one routing step of the simulation and update the elapsed time.
Simuwater_end	End the simulation process.
Simuwater_report	Report the simulation process.
Simuwater_close	Close the simulation process.
Simuwater_getMassBalErr	Obtain the continuity error of the simulation process.
Simuwater_setExtInflow	Set the external inflows of Simuwater node during the simulation process (if SWMM module is used).
Simuwater_setExtLoad	Set external inflow loads of Simuwater nodes during the simulation process.
Simuwater_setSplittedValue	Set the splitting values of splitters during the simulation process.
Simuwater_setPumpFlow	Set the pump flow values during the simulation process.
Simuwater_setReactorEmptyingValue	Set the reactor emptying values during the simulation process.
Simuwater_setWeirParams	Set the parameters of weir during the simulation process.
Simuwater_setSetting	Set action values of Simuwater objects during the simulation process (if SWMM module is used).
Simuwater_getResult	Return results according to object type, name, and variable type during the simulation process.
Simuwater_setSwmmExtInflow	Set the external inflows of SWMM node during the simulation process (if SWMM module is used).
Simuwater_setSwmmExtLoad	Set external inflow loads of SWMM nodes during the simulation process (if SWMM module is used).
Simuwater_setSwmmSetting	Set action values of SWMM control objects during the simulation process (if SWMM module is used).
Simuwater_getSwmmResult	Return SWMM results according to object type, name, and variable type during the simulation process (if SWMM module is used).
Simuwater_findObject	Obtain the object index according to the object type and object name.
Simuwater_isInEvent	Judge whether the current calculation time is within the rainfall event of the rain gauge.
Simuwater_init	Set folder path for the temporary swmm5ex.dll file.

5. Case Study of RTC Simulation by Simuwater

A Chinese city's WWTP service area covers an area of approximately 900 ha. It is a combined drainage system, including one WWTP, three primary storage tanks (SUA, SUB, and SUC), and three key distributing wells (DF1#, DF2#, and DF3#) (Figure 5). To control the combined sewer overflow (CSO), storage tank SUA (capacity 15,000 m^3), SUB (capacity 15,000 m^3), and SUC (capacity 5500 m^3) were established in subareas A, B, and C, respectively. Under the operating rules of the original design, the dry weather and combined sewage that do not exceed the discharge capacity of the drainage network in wet weather are directly transported to the WWTP. The combined sewage that exceeds the processing capacity of the pipe network or the WWTP in sub-areas A and B then enters storage tanks SUA and SUB, respectively, and is moved from the storage tank to the WWTP after the rainfall. The sewage from sub-area C enters storage tank SUC through the integrated pumping station and is then discharged into the WWTP.

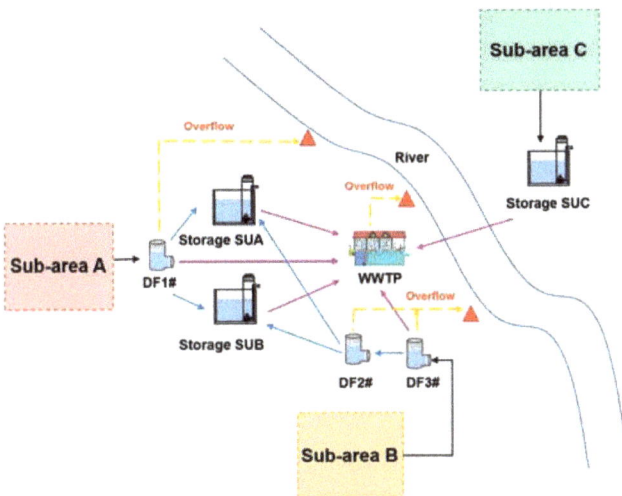

Figure 5. The distribution of primary wastewater facilities in a typical Chinese city, including storage units (SUA, SUB, SUC), distributing wells (DF1#, DF2#, DF3#), and the wastewater treatment plant (WWTP).

According to the topological structure and basic data of the drainage system in this area, an integrated model of the system was built on the Simuwater platform.

5.1. Current Problems and Controlled Areas

During the selected rainfall period, the UDS had clear sewer overflow problems, specifically in distribution well DF2# and the WWTP (Figure 6). Owing to restrictions in the original design of the flow to SUA and SUB, the overload water could not be transported to the tanks in time. In addition, the emptying rules of the pumps in storage tanks were not appropriate and sent a "start" signal to the pumps immediately after the end of precipitation. However, owing to the long delay in sewage transport from sub-area C to the WWTP, the WWTP was not able to function at full capacity immediately after the rain stopped, with the emptying flow to the WWTP increasing the risk of CSO.

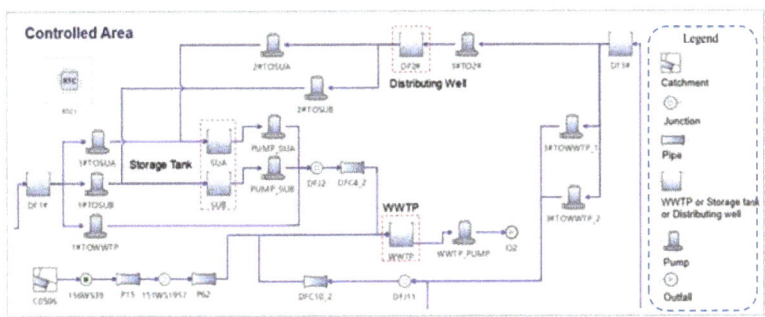

Figure 6. The controlled area of the UDS in a city of China on Simuwater platform.

Considering this information, it became clear that the controlled area should cover the WWTP, storage tank SUA, SUB, and key distribution wells. The flow from storage tank SUC and other components was set as the boundary condition for the controlled area. Notably, the aim of systematic optimization control is to reduce the risk of CSO by making full use of the storage capacities of the storage tanks and optimizing the use of related pumps.

5.2. Reliability Analysis

To ensure the reliability of the simulation results using Simuwater, monitoring and SWMM simulation data were used for comparison with the Simuwater results. Notably, the corresponding NSEs (to monitoring data: 0.84, 0.85; to SWMM data: 0.52, 0.73) were within the confidence interval, allowing the results to be verified, indicating the high reliability and stability of Simuwater simulation results, as shown in Figure 7. In addition, the SWMM model mentioned was calibrated. By comparing the monitoring data with the SWMM simulation results, the NSEs were calculated, and the sensitive parameters of the model were continuously adjusted until the NSEs meet the error requirements. The NSEs of the SWMM model were all above 0.6.

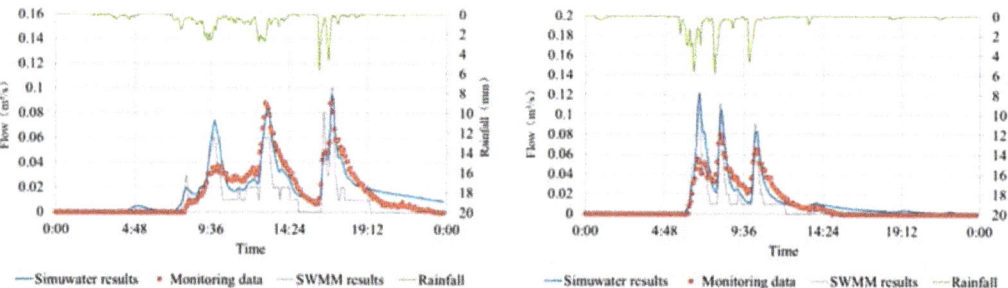

Figure 7. Comparison of results from the storm water management model (SWMM), the coupled Simuwater model, and a monitoring sensor (first NSE: Simuwater data to monitoring data; second NSE: Simuwater data to SWMM data).

5.3. Controlled Variables, Constraints, and Objectives

A real-time value is the state value calculated during a simulation process. This type of value is essential for the optimization of the system. In our case, several real-time values of different components needed to be extracted. These values included the total inflow of distribution well DF2# and the total inflow of SUA, SUB, and the WWTP, as well as the volumes of SUA and SUB and the remaining treatment capacity of the WWTP. The controlled variables of the components included the outflow from DF2# to SUA and SUB and the emptying flows of SUA and SUB.

Constraint conditions describe the basic physical relationships that the entire system must satisfy and abide by in the optimal control calculation. Inequalities are typically used to transform constraint conditions into mathematical forms. In our case study, the constraint conditions were the maximum volumes of SUA and SUB and the maximum power of pumps in SUA and SUB.

In UDS models, the objective function is to transform the optimal objectives of the UDS (such as minimum CSO and stable storage tank volume) into a solvable mathematical function. In our case study, the objective function included the following three parts: minimum CSO in DF2# and the WWTP, target volume of SUA and SUB, and stable operation of pumps.

5.4. Optimization Results

By comparing the simulation results of the original design and the optimal control, we demonstrated the effectiveness of Simuwater as an optimization strategy. Taking 9.4 mm rainfall as an example, the CSO of key facilities was reduced and managed by controlling the outflows of DF2#. Under the original operation rules, the CSO values of DF2# and WWTP were 7426 and 2397 m^3, respectively. In contrast, the CSO under the optimization strategy was reduced to 0 and 1913 m^3, respectively, with a reduction efficiency of 100% and 20.2%, respectively, as shown in Figure 8.

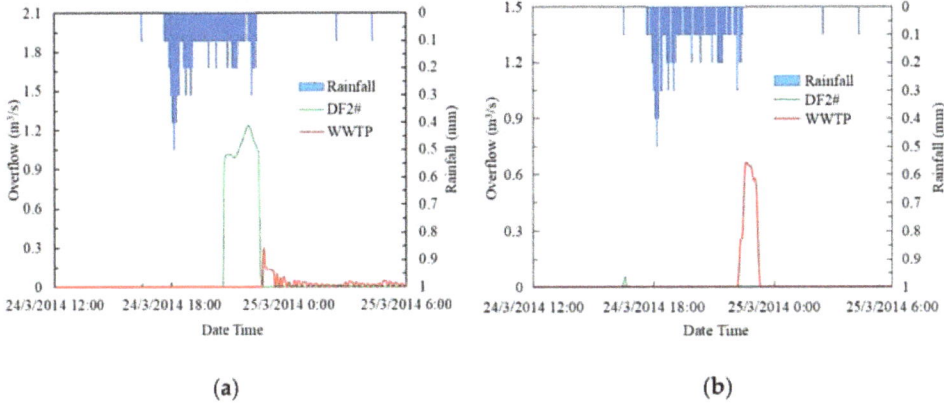

Figure 8. The combined sewer overflow of distribution tank (DF2#) and the wastewater treatment plant (WWTP) in (**a**) the original design and (**b**) optimal control with 9.4 mm rainfall on the developed Simuwater platform.

Simultaneously, in the multi-objective optimization system, by adjusting the weight of the balance volume target of the storage tanks, the volume of the storage tanks SUA and SUB could be maintained at approximately 14,000 m^3, achieving the corresponding control effect (Figure 9).

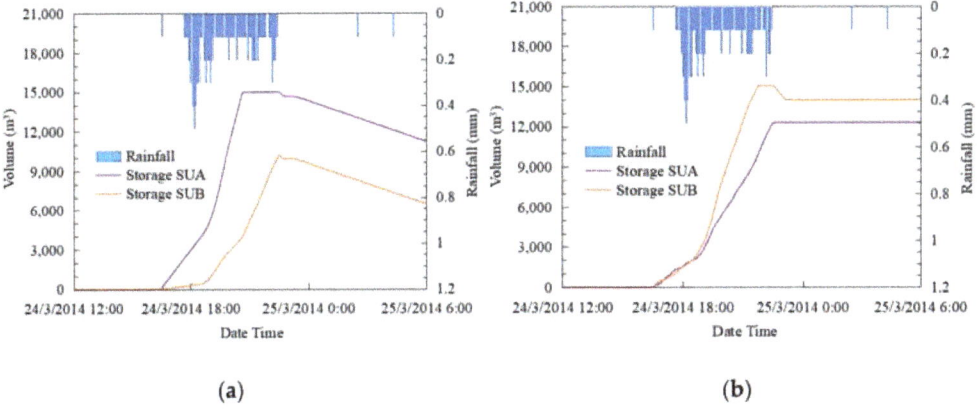

Figure 9. The volume of storage tanks SUA and SUB using (**a**) the original design and (**b**) optimal control with 9.4 mm rainfall on the developed Simuwater platform.

Under the original design, the regulation related to the emptying of storage tanks after rain was implemented (Figure 10). Specifically, after rainfall, the volume of the storage tanks decreased significantly, and the corresponding emptying pump started and stopped repeatedly according to the water-level control principle in the pump. Therefore, the goal of balancing the volume of the storage tank was not only to receive upstream water and reduce the overflow pressure of the upstream DF2# distribution well but also to prevent post-rain emptying from aggravating the load on the WWTP. Furthermore, because of the balanced volume of the storage tanks, the emptying of the pumps was more stable than in the original design, and the start and stop frequencies were significantly reduced, which was conducive to the sustainable operation of the equipment.

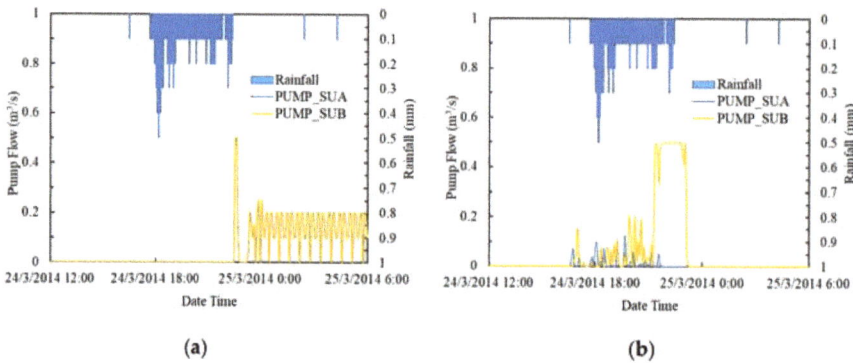

Figure 10. Pump flow of storage units (SUA and SUB) in (**a**) the original design and (**b**) optimal control under 9.4 mm rainfall on the developed Simuwater platform.

6. Conclusions

Within the context of technological development, solving the multiple problems of UDS in China (such as CSO and flooding), together with ensuring the safety of the UDS and improving the quality of the water environment requires the integrated operation of the UDS. This highlights the need for UDS models that include and allow for simulation, evaluation, and decision-making.

As a comprehensive model, Simuwater can simulate the water quantity and quality of all-element objects in the source, network, water treatment, and river modules. Furthermore, the model encompasses a high-precision linearization simulation method and combines the mechanism model with the conceptual model on the same platform to achieve a balance between timesaving and accuracy. Although we indicate the accuracy of the Simuwater model simulation through reliability analysis, future studies should focus on increasing monitoring data support. Notably, the rich control interface and optimization algorithm allows for multi-objective, full-time, and real-time control simulation of the UDS. The developed platform integrates simulation, analysis, optimization, and control, thereby breaking through the single function limitation of traditional models, and the platform encompasses the dual core functions of auxiliary design and optimization control. The application of this model can significantly improve the design, operation, and management of UDS.

To increase the depth and breadth of the Simuwater model, we suggest that future research also focus on simulating the internal process of the WWTP and pipeline deposition. In addition, further research avenues include the integration of machine learning and other related theories with more complete functions, such as sensitivity analyses. Furthermore, we suggest that it is necessary to conduct an in-depth study of the specific hydrology and hydraulic and water quality simulation mechanisms, as well as the coupling of further different methods and models to expand the application scenarios of Simuwater. With the continuous improvement in Simuwater functions and broad application practices, Simuwater can assist in the development of UDS models, from drainage simulation to the simulation of entire water systems.

Author Contributions: Conceptualization, H.W., L.Z. and J.L.; Data curation, Y.Q.; Formal analysis, G.H. and Y.Q.; Funding acquisition, H.W.; Investigation, H.W.; Methodology, H.W., G.H., L.Z. and J.L.; Project administration, H.W.; Resources, H.W.; Software, G.H., L.Z. and J.L.; Supervision, H.W.; Validation, H.W.; Visualization, G.H., Y.Q. and H.J.; Writing—original draft, G.H. and Y.Q.; Writing—review & editing, H.W. and H.J. All authors have read and agreed to the published version of the manuscript.

Funding: This research received no external funding.

Conflicts of Interest: The authors declare no conflict of interest.

References

1. Zeng, Z.; Yuan, X.; Liang, J. Designing and implementing an SWMM-based web service framework to provide decision support for real-time urban stormwater management. *Environ. Model. Softw.* **2021**, *135*, 104887. [CrossRef]
2. Sadler, J.M.; Goodall, J.L.; Behl, M. Leveraging open source software and parallel computing for Model. predictive control of urban drainage systems using EPA-SWMM5. *Environ. Model. Softw.* **2019**, *120*, 104484. [CrossRef]
3. Gironás, J.; Roesner, L.; Rossman, L. A new applications manual for the Storm Water Management Model. (SWMM). *Environ. Model. Softw.* **2010**, *25*, 813–814. [CrossRef]
4. Barnard, T.; Kuch, A.; Thompson, G. Evolution of an Integrated 1D/2D Modeling Package for Urban Drainage. *J. Water Manag. Model.* **2007**, *R227-18*, 343–366. [CrossRef]
5. Alam, M.J. *Two-Dimensional Urban Flood Modelling for Real Time Flood Forecasting for Dhaka City*; Asian Institute of Technology: Khlong Nueng, Thailand, 2002.
6. Langeveld, J.G.; Benedetti, L.; de Klein, J.J.M. Impact-based integrated real-time control for improvement of the Dommel River water quality. *Urban Water J.* **2013**, *10*, 312–329. [CrossRef]
7. Keupers, I.; Wolfs, V.; Kroll, S. Impact analysis of CSOs on the receiving river water quality using an integrated conceptual model. In Proceedings of the 10th International Urban Drainage Modelling Conference, Quebec, QC, Canada, 20–23 September 2015.
8. Butler, D.; Schütze, M. Integrating simulation models with a view to optimal control of urban wastewater systems. *Environ. Model. Softw.* **2005**, *20*, 415–426. [CrossRef]
9. Vanrolleghem, P.; Benedetti, L.; Meirlaen, J. Modelling and real-time control of the integrated urban wastewater system. *Environ. Model. Softw.* **2005**, *20*, 427–442. [CrossRef]
10. Meng, F.; Fu, G.; Butler, D. Regulatory implications of integrated real-time control technology under environmental uncertainty. *Environ. Sci. Technol.* **2020**, *54*, 1314–1325. [CrossRef]
11. Pleau, M.; Pelletier, G.; Colas, H. Global predictive real-time control of Quebec urban community's westerly sewer network. *Water Sci. Technol.* **2001**, *43*, 123–130. [CrossRef] [PubMed]
12. Sadler, J.M.; Goodall, J.L.; Behl, M. Exploring real-time control of stormwater systems for mitigating flood risk due to sea level rise. *J. Hydrol.* **2020**, *583*, 124571. [CrossRef]
13. Shishegar, S.; Duchesne, S.; Pelletier, G. An integrated optimization and rule-based approach for predictive real time control of urban stormwater management systems. *J. Hydrol.* **2019**, *577*, 124000. [CrossRef]
14. Fiorelli, D.; Schutz, G.; Klepiszewski, K. Optimised real time operation of a sewer network using a multi-goal objective function. *Urban Water J.* **2013**, *10*, 342–353. [CrossRef]
15. Kändler, N.; Annus, I.; Vassiljev, A. Smart In-Line Storage Facilities in Urban Drainage Network. *Proceedings* **2018**, *2*, 631. [CrossRef]
16. Schilling, W. A Survey on real time control of combined sewer systems in the United States and Canada. In Instrumentation and Control of water and wastewater treatment and transport systems, Proceedings of the 4th IAWPRC Workshop, 27 April–4 May 1985; Pergamon: Oxford, UK, 1985; pp. 595–600.
17. Xin, D.; Huang, S.; Zeng, S. Design and evaluation of control strategies in urban drainage systems in Kunming city. *Environ. Sci. Eng.* **2017**, *11*, 1–8.
18. Garcia, L.; Barreiro-Gomez, J.; Escobar, E. Modeling and real-time control of urban drainage systems: A review. *Adv. Water Resour.* **2015**, *85*, 120–132. [CrossRef]
19. Bach, P.M.; Rauch, W.; Mikkelsen, P.S. A critical review of integrated urban water modelling—Urban drainage and beyond. *Environ. Model. Softw.* **2014**, *54*, 88–107. [CrossRef]
20. Garcia, L.; Escobar, E.; Barreiro-Gomez, J. On the modeling and real-time control of urban drainage systems: A survey. In Proceedings of the 11th International Conference on Hydroinformatics, New York, NY, USA, 8 January 2014.
21. Meert, P.; Nossent, J.; Vanderkimpen, P. *Development of Conceptual Models for an Integrated Catchment Management: Subreport 1. Literature Review of Conceptual Modelstructures*; Flanders Hydraulics Research: Antwerp, Belgium, 2014.
22. Nasr, M.S.; Moustafa, M.A.E.; Seif, H.A.E. Modelling and simulation of German BIOGEST/EL-AGAMY wastewater treatment plants—Egypt using GPS-X simulator. *Alex. Eng. J.* **2011**, *50*, 351–357. [CrossRef]
23. Saagi, R.; Flores-Alsina, X.; Kroll, S. A Model. library for simulation and benchmarking of integrated urban wastewater systems. *Environ. Model. Softw.* **2017**, *93*, 282–295. [CrossRef]
24. Hauduc, H.; Rieger, L.; Oehmen, A. Critical Review of Activated Sludge Modeling: State of Process Knowledge, Modeling Concepts, and Limitations. *BioTechnol. Bioeng.* **2013**, *110*, 24–46. [CrossRef]
25. Pleau, M.; Colas, H.; Lavallée, P. Global optimal real-time control of the Quebec urban drainage system. *Environ. Model. Softw.* **2005**, *20*, 401–413. [CrossRef]
26. Lindenschmidt, K.-E.; Carr, M.K.; Sadeghian, A. CE-QUAL-W2 Model. of dam outflow elevation impact on temperature, dissolved oxygen and nutrients in a reservoir. *Sci. Data* **2019**, *6*, 312. [CrossRef]
27. Moreno-Rodenas, A.M.; Tscheikner-Gratl, F.; Langeveld, J.G. Uncertainty analysis in a large-scale water quality integrated catchment modelling study. *Water Res.* **2019**, *158*, 46–60. [CrossRef]

28. Ridolfi, L.; Porporato, A.; Revelli, R. Green's Function of the Linearized de Saint-Venant Equations. *J. Eng. Mech.* **2006**, *132*, 125–132. [CrossRef]
29. Litrico, X.; Georges, D. Robust continuous-time and discrete-time flow control of a dam-river system. (I) Modelling. *Appl. Math. Model.* **1999**, *23*, 809–827. [CrossRef]
30. Sun, C.C.; Joseph-Duran, B.; Maruejouls, J. Efficient integrated Model predictive control of urban drainage systems using simplified conceptual quality models. In Proceedings of the 14th IWA/IAHR International Conference on Urban Drainage, Prague, Czech Republic, 10–15 September 2017; pp. 1848–1855.
31. Ocampo-Martinez, C.; Bemporad, A.; Ingimundarson, A. On Hybrid Model Predictive Control of Sewer Networks. In *Identification and Control*; Spinger: London, UK, 2007; pp. 87–114.
32. Achleitner, S.; Möderl, M.; Rauch, W. CITY DRAIN©—An open source approach for simulation of integrated urban drainage systems. *Environ. Model. Softw.* **2007**, *22*, 1184–1195. [CrossRef]
33. Giraldo, J.M.; Leirens, S.; Díaz-Granados, M. Nonlinear optimization for improving the operation of sewer systems: The Bogota Case Study. In Proceedings of the 5th International Congress on Environmental Modelling and Software, Ottawa, ON, Canada, 5–8 July 2010.
34. Bolea, Y.; Puig, V.; Grau, A. Discussion on Muskingum versus Integrator-Delay Models for Control Objectives. *J. Appl. Math.* **2014**, *2014*, 1–11. [CrossRef]
35. Litrico, X.; Fromion, V. Analytical approximation of open-channel flow for controller design. *Appl. Math. Model.* **2004**, *28*, 677–695. [CrossRef]
36. Murla, D.; Gutierrez, O.; Martinez, M. Coordinated management of combined sewer overflows by means of environmental decision support systems. *Sci. Total Environ.* **2016**, *550*, 256–264. [CrossRef] [PubMed]
37. Zoppou, C. Review of Urban Storm Water Models. *Environ. Model. Softw.* **2001**, *16*, 195–231. [CrossRef]
38. Van Daal, P.; Gruber, G.; Langeveld, J. Performance evaluation of real time control in urban wastewater systems in practice: Review and perspective. *Environ. Model. Softw.* **2017**, *95*, 90–101. [CrossRef]
39. Lund, N.; Falk, A.K.; Borup, M. Model. predictive control of urban drainage systems: A review and perspective towards smart real-time water management. *Environ. Sci. Technol.* **2018**, *48*, 279–339. [CrossRef]
40. Svensen, J.L.; Congcong, S.; Cembrano, G. Chance-constrained Stochastic MPC of Astlingen Urban Drainage Benchmark Network. *Control. Eng. Pract.* **2021**, *115*, 104900. [CrossRef]
41. Sun, C.C.; Romero, L.B.; Joseph-Duran, B. Integrated pollution-based real-time control of sanitation systems. *J. Environ. Manage* **2020**, *269*, 110798. [CrossRef] [PubMed]
42. Casal-Campos, A.; Fu, G.; Butler, D. An Integrated Environmental Assessment of Green and Gray Infrastructure Strategies for Robust Decision Making. *Environ. Sci. Technol.* **2015**, *49*, 8307–8314. [CrossRef]
43. Shishegar, S.; Duchesne, S.; Pelletier, G. Optimization methods applied to stormwater management problems: A review. *Urban Water J.* **2018**, *15*, 276–286. [CrossRef]
44. Afshar, M.H.; Afshar, A.; Marino, M.A. Hydrograph-based storm sewer design optimization by genetic algorithm. *Rev. Can. De Génie Civ.* **2011**, *33*, 319–325. [CrossRef]

Article

Optimal Design of Combined Sewer Overflows Interception Facilities Based on the NSGA-III Algorithm

Zhouyang Peng, Xi Jin *, Wenjiao Sang and Xiangling Zhang

School of Civil Engineering and Architecture, Wuhan University of Technology, Wuhan 430070, China; pzy540641398@126.com (Z.P.); whlgdxswj@126.com (W.S.); zxlcl@126.com (X.Z.)
* Correspondence: jinxi@whut.edu.cn; Tel.: +86-136-5985-8356

Abstract: The interception facility is an important and frequently used measure for combined sewer overflow (CSO) control in city-scale drainage systems. The location and capacity of these facilities affects the pollution control efficiency and construction cost. Optimal design of these facilities is always an active research area in environmental engineering, and among candidate optimization methods, the simulation-optimization method is the most attractive method. However, time-consuming simulations of complex drainage system models (e.g., SWMM) make the simulation-optimization approach impractical. This paper proposes a new simulation-optimization method with new features of multithreading individual evaluation and fast data exchange by recoding SWMM with object-oriented programming. These new features extremely accelerate the optimization process. The non-dominated sorting genetic algorithm-III (NSGA-III) is selected as the optimization framework for better performance in dealing with multi-objective optimization. The proposed method is used in the optimal design of a terminal CSO interception facility in Wuhan, China. Compared with empirically designed schemes, the optimized schemes can achieve better pollution control efficiency with less construction cost. Additionally, the time consumption of the optimization process is compressed from days to hours, making the proposed method practical.

Keywords: combined sewer overflows; optimization; SWMM; NSGA-III

Citation: Peng, Z.; Jin, X.; Sang, W.; Zhang, X. Optimal Design of Combined Sewer Overflows Interception Facilities Based on the NSGA-III Algorithm. *Water* **2021**, *13*, 3440. https://doi.org/10.3390/w13233440

Academic Editors: Haifeng Jia, Jiangyong Hu, Tianyin Huang, Albert S. Chen and Yukun Ma

Received: 23 October 2021
Accepted: 1 December 2021
Published: 4 December 2021

Publisher's Note: MDPI stays neutral with regard to jurisdictional claims in published maps and institutional affiliations.

Copyright: © 2021 by the authors. Licensee MDPI, Basel, Switzerland. This article is an open access article distributed under the terms and conditions of the Creative Commons Attribution (CC BY) license (https://creativecommons.org/licenses/by/4.0/).

1. Introduction

Nowadays, many urban areas are still drained by combined sewer systems that collect and transport both municipal wastewater and stormwater/snowmelt runoff with the same pipe network [1]. With rapid economic development and massive population growth, urbanization has become a global trend [2]. Dense urbanization changes the land use of cities and increases surface runoff volume [3–5]. Global climate change has also amplified rainfall intensity in some parts of the world [6,7], which generates huge pressure on the urban drainage system. Therefore, the volume of wastewater can sometimes exceed the capacity of the pipe networks, which lead to combined sewer overflows (CSOs) [1,8].

CSO can be controlled in four ways: operation and maintenance practices; collection system controls, including conventional approaches and green infrastructure; storage facilities; and treatment technologies [9]. However, source control measures are difficult to implement in many older urban areas. Therefore, storage tanks are considered as a cost-effective and straightforward solution to reduce peak runoff and CSOs [10]. Many scholars have attempted to limit the frequency, volume, and/or pollutant load of CSOs by optimizing the design of storage tanks. Lu et al. [11] proposed a two-level optimization (TO) scheme to support the optimal design of storage ponds in urban drainage systems. A new method was proposed to identify optimal rainwater storage locations with the goal of reducing urban inundation damage costs [12]. Wang et al. [13] established a two-stage framework for solving the optimal arrangement of storage tanks using hierarchical analysis and the generalized pattern search method. However, the design of the CSO interception facility is still based on the calculation of empirical formulas, which is not targeted and accurate in specific projects and deviates greatly from the results of actual runoff process. In order to increase the quality of the design scheme and operation efficiency, hydro-hydraulic models and optimization algorithms are introduced in the process of optimal design. Optimization algorithms are used as a framework for generating and selecting better design schemes and hydro-hydraulic models are used to evaluate the quality of each design scheme.

With the increasing complexity of engineering problems and the increase of limiting factors, the traditional optimization algorithms cannot fully meet the needs of engineering practice, so scholars put forward an intelligent optimization method imitating biological evolution theory. Additionally, these algorithms, such as Particle Swarm Optimization (PSO), Genetic Algorithm (GA), Simulation Anneal (SA), Differential Evolution Algorithm (DE), and Ant Colony Optimization Algorithm (ACO), have been tested and proved suitable for solving multi-objective, multi-constraint, nonlinear, and discrete problems [14–18]. Cunha et al. [19] established a rainfall-runoff model to simulate water volume, and obtained the objective function of the volume and location of the storage tank related to peak flow, and obtained the optimal solution of the volume and location of the storage tank based on the SA algorithm. Ryu et al. [20,21] studied the location of the storage tanks based on the storm water management model (SWMM) and PSO algorithm and used the simplified mathematical model, which has a certain guiding significance for engineering design. Tao et al. [22] used the non-dominated sorting genetic algorithm (NSGA-II) to seek the optimal equilibrium for decentralized detention, considering flood control, peak reduction, and investment costs. Oxley and Mays [23] optimized the size and location of a detention pond system based on a simulated annealing approach, including outlet structures in a single detention pond system and multiple detention pond systems.

The aforementioned study cases focused on optimal design scheduling specifically regarding part of CSO interception facilities, for example, only focusing on the storage tank. However, CSO interception facilities are mostly composed of both of a storage tank and pump station, and these two parts play a role together and interact with each other. Partially, optimization cannot consider interactions between storage tanks and pump stations so global optimal solutions cannot be obtained. Therefore, a complete optimization method that considers both the storage tank and pump station as optimal objectives in one optimization process should be used.

When solving the process of optimization models, the hydro-hydraulic model is an important tool for scheme evaluation. Currently, SWMM is the most widely used hydro-hydraulic model in simulation-optimization methods due to its unique features of being open-source and extensible. However, SWMM has no interface functions for parameters setting or result reading, so data exchange between the optimization algorithm and SWMM models has to be implemented by file operations. In addition, SWMM is developed with procedural-oriented programming, so the data structure is organized as global variables. This feature means SWMM cannot be called in a multithreading way during the optimization process. These problems seriously impact the solving efficiency of the optimization method and make the simulation-optimization approach impractical.

Aimed at the outstanding problems of solving efficiency and partial optimization, this study proposes a new optimization model, which considers both the storage tank and pump station as optimal objectives. In order to improve the solving efficiency, SWMM is recoded with object-oriented programming, so that the model data structure is encapsulated in classes and the recoded SWMM model can be called in a multithreading way and fast data exchange without file operations can be achieved.

2. Optimization Model
2.1. Decision Variables

A typical CSO interception facility is formed with an opened or underground storage tank used as the detention volume. Because water stored in the storage tank mostly cannot be drained by gravity, a pump station is needed for emptying the tank so that the facility can be prepared for the next rain event as soon as possible. Therefore, the design parameters of the tank and pump station should both be considered as decision variables and be solved in the optimization process. In the present work, the following parameters are regarded as decision variables:

(1) Tank's cross-sectional area. Most tanks are designed as a columnar shape that has the same cross-section shape and area from top to bottom, and very few tanks will use the sectional area that varied with height. Thus, in this study, only a columnar shape tank is considered and the tank's cross-sectional area is selected as the decision variable.

(2) Tank's effective depth. The total depth of the tank is formed with a sedimentary depth, effective depth, and safe super elevation. The sedimentary depth and safe super elevation can generally be determined by codes and standards, and these two parts only represent a small fraction of the total depth. Thus, only the effective depth is considered as a decision variable in this study.

(3) Pump station's capacity. Generally, the pump station will be formed with several pumps and operated with a scheduling scheme, which describes how pumps start-up or shut off according to pre-specified water depths. In the present work, a simplified scheduling scheme is used, in which all pumps in the pump station are regarded as one pump and start up or shut off together according to pre-specified water depths. Thus, the pump station's capacity is represented by one decision variable.

(4) The pump station start-up water depths. The pump start-up depth, and shutoff depth needed to be determined to control pump operation. However, the pump shutoff depth is generally set to the same as the minimum design water depth of the storage tank. Therefore, for the start-up/shutoff operation control of pumps, only the pump start-up depth needs to be set. This decision variable is represented with the ratio of the water depths to the tank's effective depth.

2.2. Model Formulation

In past studies, the purpose of optimization is to obtain the best cost-benefit solutions through evaluation and comparison of different combinations of decision variables. Thus, the economic objective and ecological objective are the most used objectives. These two objectives are also adopted in the present work. In addition, a new objective of the minimum number of pump start-ups is introduced in the proposed optimization model. Because the storage tank and pump stations are both considered in this optimization process, the interactions between these two parts should take into account and find a feasible combination to achieve a better solution not only with less cost and high interception efficiency but also with a simple and reliable operation scheme. The number of pump station start-ups is used as an indicator to measure the quality of the operation scheme of the considered CSO interception facility. Another reason for adding this objective is that SWMM simulations sometimes will give irrational results caused by computational instability. The new objective can effectively recognize and eliminate individuals that lead to unstable simulations.

2.2.1. Objective Functions

The formulations of the objective functions are as follows:

Economic objective $f_1(x)$. The economic objective minimizes the construction cost of the storage tank and pump station, and can be expressed as the following formula below:

$$\min f_1(x) = \min \left(\sum_{i=1}^{n_1} C_{P_i} + \sum_{i=1}^{n_2} C_{S_i} \right) = \min \left(\sum_{i=1}^{n_1} \alpha \cdot Q_{P_i} + \sum_{i=1}^{n_2} \beta \cdot S_{S_i} \cdot h_{S_i} \right) \quad (1)$$

where C_{P_i} is the construction cost of the i-th pump station, Yuan; C_{S_i} is the construction cost of the i-th storage tank, Yuan; α is the cost of unit drainage capacity of the i-th pump station, Yuan/m^3/s; Q_{P_i} is the drainage capacity of the i-th pump station, m^3/s; β is the unit volume cost of the i-th storage tank, Yuan/m^3; S_{Si} is the bottom area of the i-th storage tank, m^2; and h_{Si} is the depth of the i-th storage tank, m.

Ecological objective $f_2(x)$. In order to reduce the impact of sewage overflow to the receiving water body, the minimum sewage overflow rate is used as the ecological objective, namely:

$$\min f_2(x) = \min C_{of} = \min \frac{V_O}{V_T} \qquad (2)$$

where V_O is the overflow volume of in the study area, m^3; and V_T is the total volume of the combined sewage conveyed to the CSO facility, m^3.

Operational objective $f_3(x)$. The operational objective is to minimize the number of pump startup/shutoff times in pump stations, namely:

$$\min f_3(x) = \min N_{P_{of}} \qquad (3)$$

where $N_{P_{of}}$ is the total number of pump startup/shutoff times.

2.2.2. Constraint Conditions

In the solving process of the optimization model, the calculation of objectives is constrained by several constraints. These constraints can be divided into two categories: general constraints and specific constraints.

General constraints are equations that play roles in the processes of runoff generation and flow conveyance. This kind of constraint follows the same hydro-hydraulic equations for all combined sewer systems and the calculation of the objective values must comply with these general constraints. General constraints in combined sewer systems include the wave surface motion equation to describe the process of runoff generation and St. Venant's equations to describe flow conveyance in a pipe network.

The runoff generation and confluence of the sub-catchment area is controlled by the following wave surface motion equation:

$$\frac{\partial d}{\partial t} = i - e - f - q \qquad (4)$$

where d is the depth of the depression below the surface, m; i is the rate of rainfall and snow melt, mm/s; e is the surface evaporation rate, mm/s; f is the permeability, mm/s; and q is the runoff rate, mm/s.

The movement of the unsteady free surface flow through a channel or pipe is governed by the conservation of mass and momentum equations called St. Venant's equations and can be expressed as:

$$\frac{\partial A}{\partial t} + \frac{\partial Q}{\partial x} = 0$$
$$\frac{\partial Q}{\partial t} + \frac{\partial Q^2/A}{\partial x} + gA\frac{\partial H}{\partial x} + gAS_f = 0 \qquad (5)$$

where t is time, s; x is the distance from a fixed section of the pipeline along the process, m; A is the cross-sectional area of the fixed section, m^2; Q is the flow rate, m^3; g is the acceleration of gravity, m/s^2; H is the water head in the pipeline ($Z + Y$), m; Z is the bottom elevation of the pipeline, m; Y is the pipeline water depth, m; and S_f is the friction slope (head loss per unit length).

In the present work, general constraints are solved by the object-oriented SWMM.

Specific constraints are constraints related to specific optimized facilities and used to limit the value range of decision variables. In this study, the storage tank's sectional area and effective depth and flow capacity of the pump station are constrained by specific constraints and limited in certain ranges.

3. The Solution Method of the Optimization Model

A methodology that combined the use of the object-oriented SWMM and the genetic optimization algorithm NSGA-III [24] is developed and applied in the present work.

3.1. Multithreading Evaluation of Design Schemes

In order to implement multithreading evaluation of design schemes in the evolve process, the SWMM data structure and functions are encapsulated with the object-oriented concept and recoded by C++. A simple diagram about the class definition and relationship is shown in Figure 1.

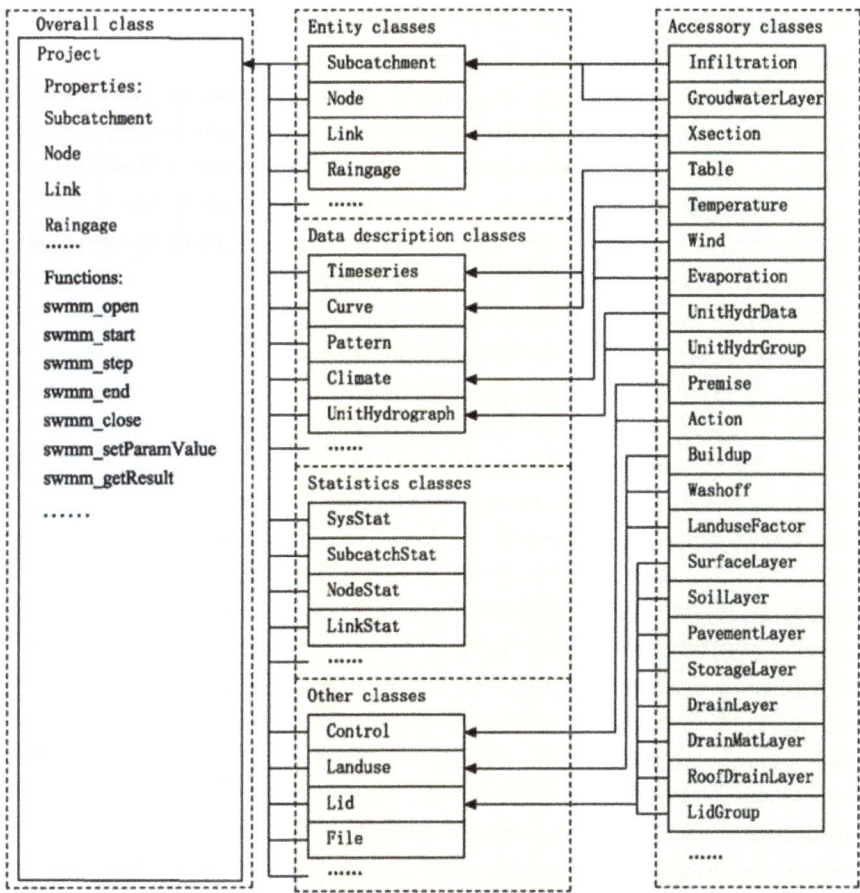

Figure 1. Class definition and relationship of the encapsulated SWMM.

The current data structures in SWMM, such as the sub-catchment, link, node, etc., are encapsulated in the corresponding classes, and embedded into the project class as property members. The object generated from the project class represents an SWMM model, and data accessing, parameter value setting, simulation, and result obtaining can be achieved by calling the function members or visiting variables directly. In the solving process of the optimization model, an SWMM model pool can be generated by declaring an array of project classes, and each element in the array represents an SWMM model of one individual. SWMM models in the model pool can be simulated in parallel.

3.2. Fast Data Exchange

With object-oriented SWMM, the data structure is organized by classes, and parameters and results are declared as public members of the project class. Thus, the data structure of the object-oriented SWMM is transparent to the optimization framework and parameters and results can be obtained at any time in the optimization process. As a result, file operations during the solving process are substantially eliminated and the data exchange efficiency is greatly improved.

Class encapsulation of SWMM and fast data exchange have no effects on the simulation algorithm of SWMM, so the simulation results with these two acceleration measures are totally identical to the results obtained by the original SWMM.

3.3. Simplification of the SWMM Model

The CSO interception facilities are mostly located in the downstream part of sewer systems and their performances are affected by the runoff and flow from the upstream part of the sewer system. In contrast, the performance of these facilities hardly affects the runoff generation and flow conveyance of the upstream part of sewer systems, so that the hydro-hydraulic simulation results of the upstream part of the sewer system can be simulated and saved before solving the optimization model. Additionally, the saved simulation results can be assigned as inflows to nodes located upstream of the CSO interception and storage facilities. With this method, the complexity of the SWMM model is simplified significantly and a lot of simulation time is saved. However, this kind of simplification causes differences between the simulation results of the simplified and original models. According to comparisons between the simulation results, the differences are very little and have almost no effect on the optimization process. However, for the sake of strictness, the practical efficacy of CSO interception facilities should be simulated and evaluated with original models with the optimal schemes obtained.

3.4. Overall Solution Framework

The flow chart of the solving process of the optimization model is shown by Figure 2. The overall solution framework can be divided into two modules: the NSGA-III module and SWMM module. The evolutionary functions are implemented by the NSGA-III module to generate and select better solutions and the SWMM module is in charge of hydro-hydraulic simulation and the objective values' calculation.

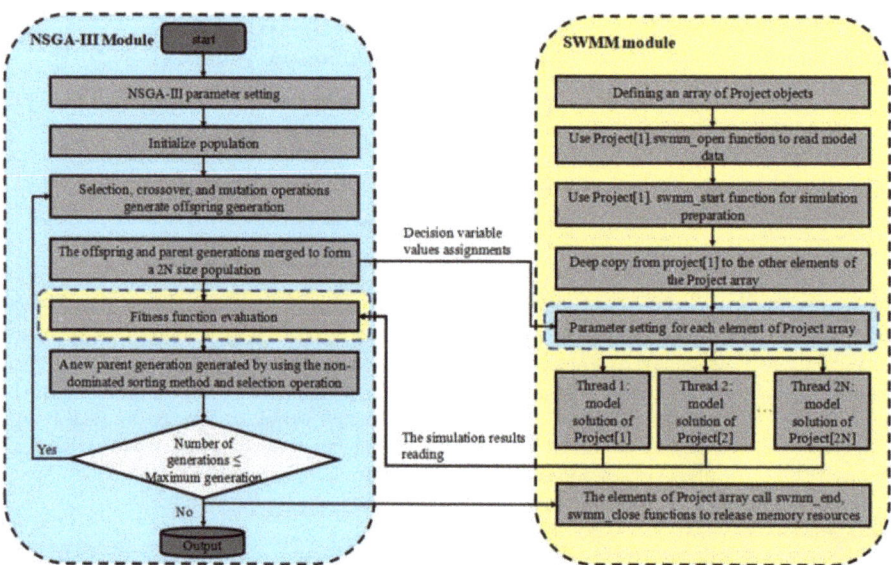

Figure 2. Flow chart of the solving process of the optimization model.

In this study, real number coding, elitist selection strategy, uniform crossover, random mutation, and normal evolving steps are used in the NSGA-III module.

For the SWMM module, at the beginning of the optimization process, an array of project classes is defined, and in order to make the CPU achieve its best performance, the length of the project array should equal the logical core number of CPU. For the sake of minimized file operations, only the first element of the project array is used to read and initialize the model data from the input file by calling the swmm_open and swmm_start functions. Additionally, other elements are initialized by deep copying from the first element. After these works, each element is ready for simulation. In each generation of the evolving process, the SWMM module is called by the following steps:

Step1: Parameters setting. The parameter values that are generated by the NSGA-III module are assigned to the project element. Thus, the schemes of each individual in the NSGA-III module can be represented by an element in the project array.

Step2: Parallel simulation. The project elements are simulated in parallel.

Step3: Simulation result reading. The simulation results used in the individual evaluation are obtained by the NSGA-III module by directly visiting the variable members of the project element.

It should be noted that the population size is generally much larger than the project array length. Thus, the three steps are implemented with a loop way in each generation and a certain number (equal to the length of the project array) of individuals are simulated in parallel in one loop until the population size of individuals is simulated.

4. Case Study

The proposed optimization model was applied to the optimal design of a CSO facility serving a combined sewer system located in Wuhan, China. The service area of the sewer system is about 11.5 hectares.

In the current sewer system, combined flows from the service area are intercepted by two intercepting weirs (IW1 and IW2). The intercepted flow is sent to the dry season wastewater treatment plant. The overflow is directly drained into a nearby river and causes serious pollution to the water body. Therefore, a CSO interception facility is going to be built for interception and storage of overflows from IW1 and IW2. Additionally, the intercepted overflows are finally pumped to the dry season wastewater treatment plant with an acceptable flow rate during and after rain events. Because node 1, node 2, and node 3 are located in the terminal of the sewer system, the invert elevations are very low. If the overflows from IW1 and IW2 flow into the storage tank by gravity, the storage tank must be constructed deep underground. This result in many troubles in construction and maintenance and makes the cost rise sharply. Considering the local land use and distance between IW1 and IW2, a scheme of two storage tanks (SU1 and SU2) with inlet (IP1 and IP2) and outlet (OP1 and OP2) pumps is used as the framework for the CSO interception facility. Because the acceptable extra flow rate of the dry season wastewater treatment plant in wet weather is 3 m^3/s, in order to maximize the interception rate in wet weather, the capacity of OP2 is set as 2.5 m^3/s and is not considered as a decision variable. During rain events when CSOs are generated from IW1 and IW2, the CSOs are pumped into SU1 and SU2 by IP1 and IP2 as much as possible. When CSOs exceed the capacity of IP1 and IP2, the extra part is drained into the nearby river. Additionally, the water stored in SU1 is pumped into SU2 by OP1 and CSOs stored in SU2 are finally pumped by OP2 to the dry season wastewater treatment plant. Due to the complexity of this CSO facility, an optimal design is necessary for obtaining an economical and effective scheme.

The method described in Section 3.3 is used, and the original sewer system model is simplified. The original sewer system model and the simplified model are shown in Figure 3. The simplified model is shown with a sketch map to display the scheme framework clearly.

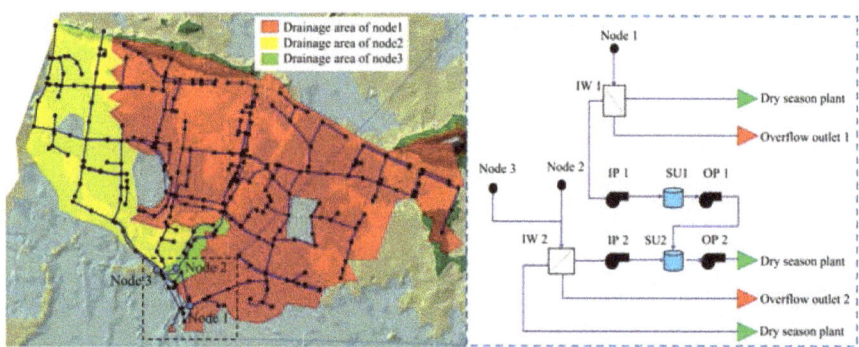

Figure 3. Comparison between the original sewer system model and the simplified model.

In the calculation of the economic objective function, the construction cost of the storage tank is calculated based on the unit volume cost. According to the unit costs of similar projects, the value of 4×10^3 Yuan/m^3 is used in the present work. The construction cost of the pump station is calculated by the formula proposed in the Estimation Index of Wuhan Municipal & Transportation Planning Project (2017 Revised), which as follows:

$$Z_1 = k \cdot \frac{773.6}{q^{0.268}} \qquad (6)$$

where Z_1 is the project investment index of the rainwater pump station (10^4 Yuan/m^3/s), q is the capacity of the rainwater pump station (m^3/s), k is the multiple of project price inflation and 1.40 is used.

In order to evaluate the design schemes comprehensively, a one-year (year of 2013) precipitation is used in the SWMM simulation. The annual rainfall in Wuhan 2013 is near the average annual rainfall value in the recent 20 years in Wuhan. It is a representative year to describe the precipitation condition in the study area. If an annual rainfall larger than the average annual rainfall is used in the optimization process, it will most likely obtain optimal design schemes with a high construction cost and low utilization efficiency, and in contrast, an annual rainfall smaller than the average annual rainfall will lead to optimal design schemes that cause heavy CSOs pollution in the nearby river. Thus, the annual rainfall of 2013 is used in this study as the precipitation data. The rainfall pattern used in this study is shown in Figure 4.

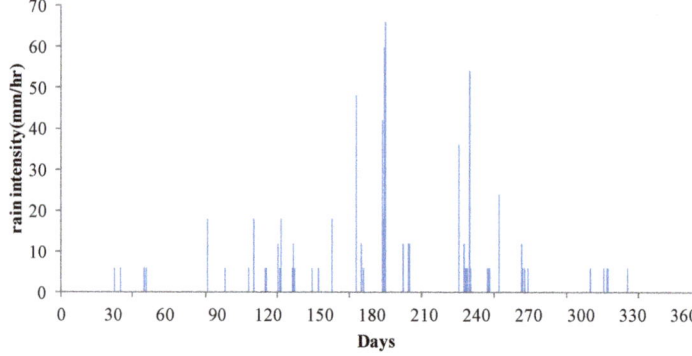

Figure 4. Rainfall pattern of 2013 in Wuhan.

According to the local land use and the capacity of downstream treatment facilities, the values or value ranges of parameters and decision variables of NSGA-III are listed in Table 1. The NSGA-III parameter values are selected according to the most used value range of genetic algorithms. For the decision variables, the value ranges, especially the upper boundaries of the value ranges, are mostly constrained by conditions related to specific study cases. In the present work, the value ranges of the tanks' cross-sectional area and depths are constrained and valued according to the available area and allowable construction depth. The pump stations' capacities of IP1 and IP2 are constrained by the annual allowable overflow times (a value of 5 is used in this work). The pump stations' capacities of OP1 is constrained by an emptying time of 7 h of the storage tank to be emptied because in the present study case, if the interval time between two rains is larger than 7 h, the rainfall process is considered as two independent rain events and the pump station should have the ability to empty the storage tank during the no rainfall period. According to this constraint, the maximum capacity of OP1 is valued as 5 m^3/s. To make the best use of the storage volume and avoid overflow in the storage tank, the start-up water depth should be limited to a certain scope of the full depth of the storage tank. Here, a scope of 0.5~0.9 is used.

Table 1. Values or value ranges of the parameters and decision variables of NSGA-III.

Parameter	Category	Value/Value Range
Population size	NSGA-III parameter	100
Generation size	NSGA-III parameter	100
Crossover probability	NSGA-III parameter	0.6
Mutation probability	NSGA-III parameter	0.1
SU1 tank's cross-sectional area (m^2)	Decision variable	2000~10,000
SU2 tank's cross-sectional area (m^2)	Decision variable	2500~20,000
SU1 and SU2 tank's effective depth (m)	Decision variable	3~6
IP1 pump station's capacity (m^3/s)	Decision variable	1~10
IP2 pump station's capacity (m^3/s)	Decision variable	1~3
OP1 pump station's capacity (m^3/s)	Decision variable	1~5
Pump station start-up water depth (ratio)	Decision variable	0.5~0.9

5. Results and Discussion

The proposed optimization method ran 10 times for the study case. Additionally, for the sake of comparison, the NSGA-II method without the operational objective ran three times for the study case. Although the operational objective value was not used in the evolve process, it was recorded for each individual in the NSGA-II method.

5.1. Effect of Operational Objective

NSGA-III did not always outperform NSGA-II when compared on a variety of multi-objective test problems [25]. To figure out which method has a better performance, comparisons were made between the optimization results from NSGA-III and NSGA-II. The evolve lines of the average objective values are used as representations for the comparison. For the sake comparison more clearly, the average objective values obtained during the evolve process were normalized so that they had an identical range. The comparison is shown in Figure 5.

With the comparison of the economic and ecological objective values, it seems that NSGA-II has a better ability to find schemes with a lower construction cost and lower overflow ratio. However, the comparison of the operational objectives shows that this better performance is achieved at the expense of higher pump start-up times. Generally, the pump start-up times of the NSGA-II schemes are two times higher than the NSGA-III schemes. In the present work, SU1 and OP1 is a combination that is suitable to describe the effect of the operational objective. The optimized results of SU1 and OP1 of the last generation from the runs of NSGA-II and NSGA-III are shown in Figure 6.

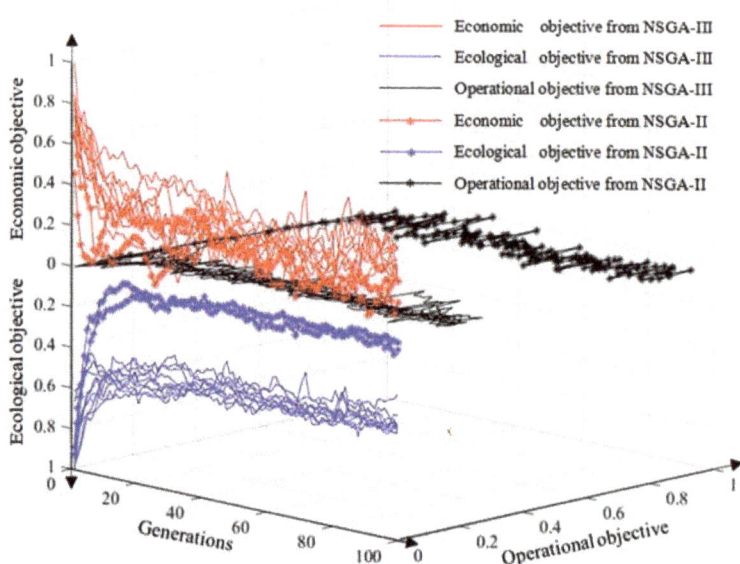

Figure 5. Comparison between the optimization results from NSGA-II and NSGA-III.

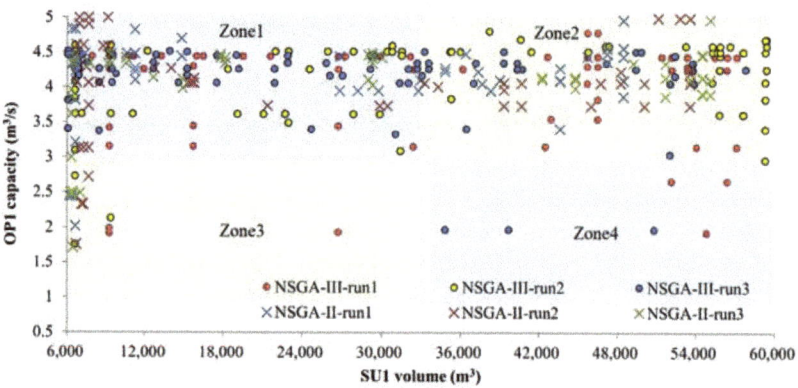

Figure 6. The optimized decision variable values of SU1 and OP1 obtained by NSGA-II and NSGA-III.

In Figure 6, the solution space of SU1 and OP1 is evenly divided into four zones. It can be seen that lots of individuals, regardless of whether they are from NSGA-II or from NSGA-III, are located in zone 1 and zone 2 because schemes with a large pump capacity can achieve a lower overflow ratio, so in order to satisfy the ecological objective, individuals are driven to solution spaces with a larger pump capacity. Similarly, individuals from both NSGA-II and NSGA-III are driven to the solution space with a smaller storage volume, such as zone 1 and zone 3, to satisfy the economic objective. Because there is no operational objective for the NSGA-II method, no individual is located in zone 4, which include schemes with a larger storage volume and smaller pump capacity that can achieve less pump start-up times. In contrast, the NSGA-III method always searches in zone 4 to select schemes with better operational performance. The added operational objective means the optimization model has the ability to search the solution space more thoroughly and obtain schemes that can achieve a balance between the economic, ecological, and operational objectives.

5.2. Verification of Proposed Method

Figure 7 shows the individuals' distribution of the initial and final generations of one NSGA-III run. Figure 7a shows the individuals from a three-dimensional view and Figure 7b shows the individuals with a two-dimensional view of the cost and overflow ratio.

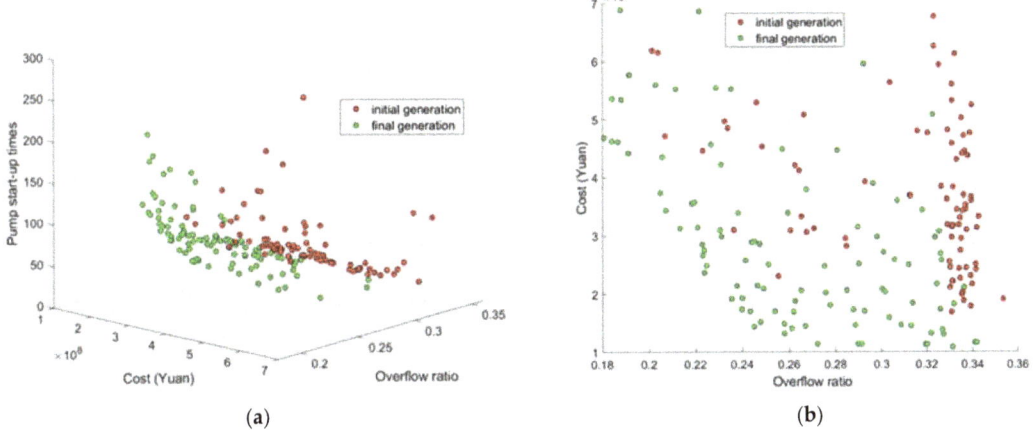

Figure 7. Distribution of individuals of the initial and last generations; (**a**) three-dimensional view; (**b**) two-dimensional view.

It can be seen that with the evolutionary process, the final generation is much closer to the origin of the coordinate system and obtained an obvious improvement compared with the initial generation. A more quantitative measurement for the improvement is the dominant numbers between individuals of the initial and final generations. Table 2 shows the dominant results of 10 NSGA-III runs.

Table 2. Dominant numbers between individuals of the initial and final generations of the NSGA-III runs.

Run	Times of Final Individuals Dominated by Initial Individuals	Times of Initial Individuals Dominated by Final Individuals	Run	Times of Final Individuals Dominated by Initial Individuals	Times of Initial Individuals Dominated by Final Individuals
1	0	1091	6	0	589
2	0	976	7	0	1171
3	0	512	8	0	1040
4	2	539	9	0	822
5	0	959	10	0	929

Table 2 shows that almost all initial individuals are dominated by the final individuals. This indicates that the Pareto front of the final generation was totally separated from the Pareto front of the initial generation, and the solutions of the final generation show an absolute superiority over the solutions of the initial generation.

Another concerned aspect of the optimization result is the detailed values of the construction cost and overflow ratio. Figure 7b shows that much better schemes are generated. In the last generation, a lot of individuals with less construction cost and a lower overflow ratio are generated. These individuals give decision makers more feasible choices for choosing the final design scheme.

5.3. Comparison of Design Schemes

Technique for Order Preference by Similarity to an Ideal Solution (TOPSIS) is a method that is suitable for selecting the best one from solutions with multiple evaluation objectives [26] because it has the advantages of high applicability and low computational effort [27]. Thus, in this study, it is used for selecting the best design schemes from the Pareto solutions of the last generation.

The top-scoring individual was selected and its decision variable values were entered into the original SWMM model of the study case. The simulation results of the empirically designed scheme (Es) and selected optimal designed scheme (Os) are shown in Table 3.

Table 3. Comparison of empirically designed schemes and optimal designed schemes.

Scheme	SU1 (10^3 m^3)	SU2 (10^3 m^3)	IP1 (m^3/s)	IP2 (m^3/s)	OP1 (m^3/s)	OP2 (m^3/s)	CC (10^8 Yuan)	Interception Efficiency (%)			Pump Start-Up Times		
								OM	SM	RE (%)	OM	SM	RE (%)
Es	50.00	100.00	12.00	2.00	2.00	2.50	6.32	78.78	78.98	0.25	475	479	0.84
Os 1	55.00	35.50	8.30	2.90	3.00	2.50	3.93	81.38	81.58	0.25	156	158	1.28
Os 2	55.13	10.00	8.90	2.40	2.60	2.50	3.57	81.11	80.11	−1.23	158	156	−1.27
Os 3	55.85	16.67	8.90	2.70	1.30	2.50	3.80	79.52	79.62	0.13	100	101	1.00
Os 4	57.77	14.27	6.70	2.40	2.50	2.50	3.74	82.11	80.89	−1.49	98	96	−2.04
Os 5	47.00	16.00	7.40	2.90	2.40	2.50	3.44	79.88	79.48	−0.50	145	145	0.00
Os 6	49.41	10.35	7.30	2.70	1.70	2.50	3.25	81.42	79.83	−1.95	155	157	1.29
Os 7	51.40	27.78	8.80	2.70	2.40	2.50	4.13	82.66	81.27	−1.68	182	185	1.65
Os 8	49.74	27.63	6.20	2.50	2.50	2.50	3.94	81.31	80.18	−1.39	163	160	−1.84
Os 9	43.00	17.00	7.80	2.50	2.30	2.50	3.32	81.33	80.05	−1.57	164	165	0.61
Os 10	50.00	10.00	8.50	2.60	2.00	2.50	3.29	79.97	80.30	0.41	190	191	0.53

Note: CC—construction cost, OM—original model, SM—simplified model, RE—relative error.

Table 3 shows that the optimal designed scheme achieves higher interception efficiency with much less construction cost than the empirically designed scheme. In the empirically designed scheme, designers want to reduce the overflow rate with a conservative strategy with a large storage volume, large capacity of the inlet pump, and small capacity of the outlet pump. Thus, the two tanks with large volumes and outlet pumps with a small capacity are used. Due to the expensive unit cost and large volumes of storage tanks, the cost of storage tanks accounts for a large percentage of the total cost of the whole facility and also increased the total cost of the facility.

The simulation results from the original model and simplified model are shown in Table 3. The relative errors between the original model and simplified model are little and the maximum error is smaller than 5%. It proves that the model simplification has little impact on the simulation results and can be used in the optimization process.

Table 4 shows the storage volume utilization of the empirically designed scheme and optimal designed scheme. From Table 4, it can be seen that although storage tanks have large volumes, the volume utilization is relatively low. Especially for SU2, the maximum volume utilization is only 64%. This means that a 36% volume is not used in the whole year. In the optimal designed scheme, smaller volumes are selected. Through the optimal selection of pumps, the volume utilization is increased obviously and the result of the lower overflow ratio and lower construction cost is achieved.

Another concerning aspect of this study is the computational efficiency. With the acceleration measures mentioned in Section 3, the computational efficiency was substantially improved. The proposed optimization model was solved on a computer with 3.6 GHz CPU (4 cores, 8 threads), 16 GB memory, and a thread pool of 8 threads was employed. The mean computation time required for one run was about 8 h. In contrast, the solving process without acceleration measures takes nearly 5 weeks for an optimization run. Thus, the acceleration methods proposed in this study significantly improved the computational efficiency and reduced the computation time to 3.72% compared with the solving method without acceleration. This makes the simulation-optimization approach more practical.

Table 4. Comparison of the storage volume utilization of empirically and optimal designed schemes.

Scheme	Storage Tank	Volume (10^3 m^3)	Average Volume Used (10^3 m^3)	Average Volume Utilization (%)	Maximum Volume Used (10^3 m^3)	Maximum Volume Utilization (%)
Es	SU1	50.00	13.50	27.00	48.25	95.50
	SU2	100.00	5.00	5.00	64.13	64.13
Os 1	SU1	55.00	15.72	28.58	54.51	99.10
	SU2	35.50	3.24	9.13	34.88	98.26
Os 2	SU1	55.13	17.85	32.38	54.15	98.22
	SU2	10.00	1.06	10.60	9.82	98.20
Os 3	SU1	55.85	11.52	20.63	55.29	99.00
	SU2	16.67	0.785	4.71	16.47	98.80
Os 4	SU1	57.77	12.58	21.78	57.19	99.00
	SU2	14.27	0.86	6.03	14.11	98.88
Os 5	SU1	47.00	13.44	28.60	46.60	98.00
	SU2	16.00	1.13	7.06	15.78	98.62
Os 6	SU1	49.97	15.56	31.14	49.59	99.24
	SU2	10.35	0.867	8.36	10.13	97.90
Os 7	SU1	51.40	13.91	27.06	51.26	99.73
	SU2	27.78	1.57	5.65	27.68	99.64
Os 8	SU1	49.74	16.88	21.17	49.57	99.66
	SU2	27.63	1.79	6.48	27.53	99.64
Os 9	SU1	43.00	11.86	27.58	42.73	99.37
	SU2	17.00	1.22	7.18	16.65	97.94
Os 10	SU1	50.00	13.39	26.78	49.79	99.58
	SU2	10.00	0.68	6.80	9.88	98.80

6. Conclusions

In this study, a new optimization model that considered both storage tanks and pump stations as optimized objects was proposed. Additionally, besides the economic and ecological objectives, the operational objective of the minimized the number of pump start-up times was added. The proposed optimization model was solved with the method based on NSGA-III and a new featured SWMM module.

By applying the proposed method to a CSO interception facility in Wuhan, it was found that optimal schemes with a higher CSO interception ratio, less construction cost, and acceptable pump start-up times can be obtained. Compared with schemes obtained by NSGA-II with only economic and ecological objectives, the schemes obtained by the proposed method can achieve a better balance between economic, ecological, and operational objectives.

By using the new featured SWMM module, parallel simulation and fast data exchange can be achieved during the optimization process. Additionally, this makes the solving time compressed from days to hours and makes the proposed method more practical. With the advantage of the high solving efficiency, long-term SWMM simulations can be applied for comprehensive evaluation of individuals. This means the optimization process is no longer an optimization oriented toward a certain rain event, but an optimization oriented to years of precipitation conditions and can give more feasible schemes.

The optimization method developed in this study did not consider the pump electricity cost as an objective. Because the pumps dealing with overflows are generally under intermittent operation, compared with operability and reliability, the electricity cost is not the most important aspect. For the sake of simplification of the optimization model, only the objective of the minimized pump start-up times was considered in the present work. The question of how to combine optimization of the facility design and energy consumption needs further study.

Author Contributions: All authors contributed extensively to the work presented in this paper. Z.P. and X.J. contributed to the subject of research, the development, the writing of the paper and the preparation of algorithms. W.S. adjusted the parameters and performed the simulations and contributed to writing of the paper. X.Z. provided the basics for the optimization algorithm, contributed to the writing of the paper and helped in the final revision. All authors have read and agreed to the published version of the manuscript.

Funding: This research was funded by the National Natural Science Foundation of China for supporting this research project, grant number 31670541.

Institutional Review Board Statement: Not applicable.

Informed Consent Statement: Not applicable.

Data Availability Statement: Not applicable.

Acknowledgments: The kind help of Lei Liu during the submission should be acknowledged.

Conflicts of Interest: The authors declare no conflict of interest.

References

1. Marie-Ève, J.; Sophie, D.; Geneviève, P.; Martin, P. Selection of rainfall information as input data for the design of combined sewer overflow solutions. *J. Hydrol.* **2018**, *565*, 559–569.
2. Moore, T.L.; Rodak, C.M.; Ahmed, F.; Vogel, J.R. Urban Stormwater Characterization, Control and Treatment. *Water Environ. Res.* **2018**, *90*, 1821–1871. [CrossRef] [PubMed]
3. O'Sullivan, A.D.; Wicke, D.; Hengen, T.J.; Sieverding, H. Life Cycle Assessment modelling of stormwater treatment systems. *J. Environ. Manag.* **2015**, *149*, 236–244. [CrossRef] [PubMed]
4. Zhou, N.Q.; Zhao, S. Urbanization process and induced environmental geological hazards in China. *Nat. Hazards* **2013**, *67*, 797–810. [CrossRef]
5. Koc, K.; Ekmekciolu, M.; Zger, M. An integrated framework for the comprehensive evaluation of low impact development strategies. *J. Environ. Manag.* **2021**, *294*, 113023. [CrossRef] [PubMed]
6. Chen, Y.; Samuelson, H.W.; Tong, Z. Integrated design workflow and a new tool for urban rainwater management. *J. Environ. Manag.* **2016**, *180*, 45–51. [CrossRef]
7. Ward, P.J.; Jongman, B.; Aerts, J.C.J.H.; Bates, P.D.; Botzen, W.J.W.; Loaiza, M.A.D.; Hallegatte, S.; Kind, J.M.; Kwadijk, J.; Scussolini, P.; et al. A global framework for future costs and benefits of river-flood protection in urban areas. *Nat. Clim. Chang.* **2017**, *7*, 642–646. [CrossRef]
8. Zhang, D.; Martinez, N.; Lindholm, G.; Ratnaweera, H. Manage sewer in-line storage control using hydraulic model and recurrent neural network. *Water Resour. Manag.* **2018**, *32*, 2079–2098. [CrossRef]
9. Tao, W.; Bays, J.S.; Meyer, D.; Smardon, R.C.; Levy, Z.F. Constructed Wetlands for Treatment of Combined Sewer Overflow in the US: A Review of Design Challenges and Application Status. *Water* **2014**, *6*, 3362–3385. [CrossRef]
10. Bellu, A.; Fernandes, L.F.S.; Cortes, R.M.; Pacheco, F.A. A framework model for the dimensioning and allocation of a detention basin system: The case of a flood-prone mountainous watershed. *J. Hydrol.* **2016**, *533*, 567–580. [CrossRef]
11. Lu, W.; Qin, X.; Yu, J. On comparison of two-level and global optimization schemes for layout design of storage ponds. *J. Hydrol.* **2019**, *570*, 544–554. [CrossRef]
12. Choi, H.; Lee, E.H.; Joo, J.G.; Kim, J.H. Determining optimal locations for rainwater storage sites with the goal of reducing urban inundation damage costs. *KSCE J. Civ. Eng.* **2016**, *21*, 2488–2500. [CrossRef]
13. Wang, M.; Sun, Y.; Sweetapple, C. Optimization of storage tank locations in an urban stormwater drainage system using a two-stage approach. *J. Environ. Manag.* **2017**, *204*, 31–38. [CrossRef]
14. Coello, C.C.; Toscano-Pulido, G.; Lechuga, M. Handling multiple objectives with particle swarm optimization. *IEEE Trans. Evol. Comput.* **2004**, *8*, 256–279. [CrossRef]
15. Das, S.; Suganthan, P.N. Differential Evolution: A Survey of the State-of-the-Art. *IEEE Trans. Evol. Comput.* **2011**, *15*, 4–31. [CrossRef]
16. Deb, K.; Pratap, A.; Agarwal, S.; Meyarivan, T. A fast and elitist multiobjective genetic algorithm: NSGA-II. *IEEE Trans. Evol. Comput.* **2002**, *6*, 182–197. [CrossRef]
17. Kirkpatrick, S.; Gelatt, C.D.; Vecchi, M.P. Optimization by Simulated Annealing. *Science* **1983**, *220*, 671–680. [CrossRef]
18. Dorigo, M.; Maniezzo, V.; Colorni, A. Ant system: Optimization by a colony of cooperating agents. *IEEE Trans. Syst. Man Cybern. Part B (Cybernetics)* **1996**, *26*, 29–41. [CrossRef]
19. Cunha, M.C.; Zeferino, J.A.; Simões, N.E.; Saldarriaga, J.G. Optimal location and sizing of storage units in a drainage system. *Environ. Model. Softw.* **2016**, *83*, 155–166. [CrossRef]
20. Ryu, J.; Baek, H.; Lee, G.; Kim, T.-H.; Oh, J. Optimal planning of decentralised storage tanks to reduce combined sewer overflow spills using particle swarm optimisation. *Urban Water J.* **2017**, *14*, 202–211. [CrossRef]
21. Rossman, L.A. *Storm Water Management Model User's Manual*; US Environmental Protection Agency: Washington, DC, USA, 2010.

22. Tao, T.; Wang, J.; Xin, K.; Li, S. Multi-objective optimal layout of distributed storm-water detention. *Int. J. Environ. Sci. Technol.* **2014**, *11*, 1473–1480. [CrossRef]
23. Oxley, R.L.; Mays, L.W. Optimization–simulation model for detention basin system design. *Water Resour. Manag.* **2014**, *28*, 1157–1171. [CrossRef]
24. Deb, K.; Jain, H. An Evolutionary Many-Objective Optimization Algorithm Using Reference-Point-Based Nondominated Sorting Approach, Part I: Solving Problems with Box Constraints. *IEEE Trans. Evol. Comput.* **2013**, *18*, 577–601. [CrossRef]
25. Ishibuchi, H.; Imada, R.; Yu, S.; Nojima, Y. Performance comparison of NSGA-II and NSGA-III on various many-objective test problems. In Proceedings of the 2016 IEEE Congress on Evolutionary Computation (CEC), Vancouver, BC, Canada, 24–29 July 2016.
26. Hwang, C.L.; Yoon, K. *Methods for Multiple Attribute Decision Making*; Springer: Berlin/Heidelberg, Germany, 1981.
27. Zeng, S.Z.; Xiao, Y. A method based on TOPSIS and distance measures for hesitant fuzzy multiple attribute decision making. *Technol. Econ. Dev. Econ.* **2018**, *24*, 969–983. [CrossRef]

Article

Modelling Infiltration Process, Overland Flow and Sewer System Interactions for Urban Flood Mitigation

Carlos Martínez [1,2,*], Zoran Vojinovic [2], Roland Price [3] and Arlex Sanchez [2]

1. Program of Civil Engineering, Universidad del Magdalena, Carrera 32 No. 22-08, Santa Marta D.T.C.H. 470004, Colombia
2. Environmental Engineering and Water Technology Department, IHE-Delft Institute for Water Education, Westvest 7, 2611 AX Delft, The Netherlands; z.vojinovic@un-ihe.org (Z.V.); a.sanchez@un-ihe.org (A.S.)
3. Emeritus Professor IHE-Delft Institute for Water Education, Scientific Advisor, HydroLogic Research, Westvest 41, 2611 AZ Delft, The Netherlands; rolandkprice@gmail.com
* Correspondence: cmartinez@unimagdalena.edu.co

Abstract: Rainfall-runoff transformation on urban catchments involves physical processes governing runoff production in urban areas (e.g., interception, evaporation, depression, infiltration). Some previous 1D/2D coupled models do not include these processes. Adequate representation of rainfall–runoff–infiltration within a dual drainage model is still needed for practical applications. In this paper we propose a new modelling setup which includes the rainfall–runoff–infiltration process on overland flow and its interaction with a sewer network. We first investigated the performance of an outflow hydrograph generator in a 2D model domain. The effect of infiltration losses on the overland flow was evaluated through an infiltration algorithm added in a so-called Surf-2D model. Then, the surface flow from a surcharge sewer was also investigated by coupling the Surf-2D model with the SWMM 5.1 (Storm Water Management Model). An evaluation of two approaches for representing urban floods was carried out based on two 1D/2D model interactions. Two test cases were implemented to validate the model. In general, similar results in terms of peak discharge, water depths and infiltration losses against other 1D/2D models were observed. The results from two 1D/2D model interactions show significant differences in terms of flood extent, maximum flood depths and inundation volume.

Keywords: Green-Ampt method; infiltration; overland flow; urban flood modelling; 1D/2D coupled modelling

1. Introduction

Hydrological water losses are an important issue within the spatial and temporal distribution of the runoff water in urban catchments. An important component of these losses is infiltration. Although much of a typical urban area is paved, there has been a growing concern to restore natural infiltration functions and reduce impacts to the catchment by allowing rainwater to gradually infiltrate into the ground.

In urban flood modelling, not only the influence of the sewer system in the overland flow is of recognized importance [1,2] but also the interaction between surface water and the infiltration losses, in order to better estimate inundation extent and water depths [3–5]. It is necessary to provide infiltration input in overland flow models, as it plays as a water volume loss that can be defined using empirical laws (e.g., the Horton or Green-Ampt equations).

Some of the current included infiltration approaches focus on: (i) hydraulic models for the simulation of flow routing in drainage canals taking into account the infiltration effect with the Green-Ampt method [6]; (ii) estimating the parameters of the Green-Ampt infiltration equation from rainfall simulation data [7]; (iii) flood routing model incorporating intensive streambed infiltration [8]; (iv) rainfall/runoff simulation with 2D full shallow water equations [9]; (v) modelling two-dimensional infiltration with constant and time-variable water depth [10] and (vi) investigation of overland flow by incorporating different infiltration methods into flood routing equations [11]. Although this is beyond this research, it is important to highlight that not only changes in infiltration and in overland flow should be taken into account in urban catchments studies, but also variations in sediment transport dynamics due to the impact of urbanization, as it has been studied in [12,13].

New approaches including the influence of the sewer system in the overland flow (coupled 1D/2D model) have also been proposed and applied. Some of the current selection approaches focus on: (i) influence of sewer network models on urban flood damage assessment based on coupled 1D/2D models [14], (ii) multi-objective evaluation of urban drainage networks using a 1D/2D flood inundation model [15,16], (iii) the influence of modelling parameters in a coupled 1D/2D hydrodynamic inundation model for sewer overflow [17], and (iv) a coupled 1D/2D hydrodynamic model for urban flood inundation [18].

Recent progress in urban flood modelling reveals that the above mentioned coupled models are accurate and efficient in simulating floods for practical applications. However, the rainfall–runoff transformation on urban catchments involves physical processes governing runoff production in urban areas, such as interception (on rooftops and on trees), evaporation, depression storage and infiltration. Rainfall–runoff models for urban catchments do not usually include these processes. Previous 1D/2D coupled models compute rainfall–runoff in the 1D sewer network [19] and, although this is not real world physics, it is a good approximation. Better approaches compute rainfall-runoff into the 2D model domain without considering infiltration losses [14,18,20,21]. Adequate representation of rainfall–runoff–infiltration within dual drainage models is still needed within a surface water assessment.

This paper aims to develop a new modelling setup which includes rainfall–runoff and the infiltration process on the overland flow and its interaction with a sewer network. The key point is to evaluate the proposed model performance when rainfall–runoff and infiltration losses are included in a dual drainage approach, crucial for proper planning and design of urban drainage systems. For this purpose, we first investigated the performance of an outflow hydrograph generator in a so-called Surf-2D model and used it as an inflow boundary condition. Its results were compared with the nonlinear reservoir method computed in SWMM 5.1 (Storm Water Management Model).

The Surf-2D model was then coupled with SWMM in order to analyze the variation in water depths when overland flow originates not only from rainfall–runoff but also from a surcharge sewer. A benchmark test in Greenfield, Glasgow (UK) produced by the UK Environmental Agency [22] was used to examine water depth predictions and flood extents.

The effect of infiltration losses on the overland flow was evaluated through an infiltration algorithm (Green-Ampt method) added to the proposed Surf-2D model. Infiltration is computed in a grid cell using the Green-Ampt method. In order to show the ability to simulate infiltration from a point source direct runoff resulting from a given excess rainfall hyetograph, a validated FullSWOF_2D open source [23] was used to show the performance of the model, computing water depths for different infiltration parameter combinations in a hypothetical case.

Finally, an evaluation of two approaches for representing urban floods was carried out based on two main 1D/2D model interactions (e.g., rainfall–runoff computed in 1D sewer model vs. rainfall–runoff–infiltration computed in a 2D model domain) to study differences in terms of flood extent, water depths and flood volumes. The methodology applied in this paper is described in the next section, the results will be discussed in Section 3 and the last section will conclude the work.

2. Materials and Methods

Previous 1D/2D coupled models are combined to simulate the flow dynamics in sewer networks and on the aboveground surface [14,18–21]. Approaches to representing urban floods are based on two main 1D/2D model interactions as shown in Figure 1.

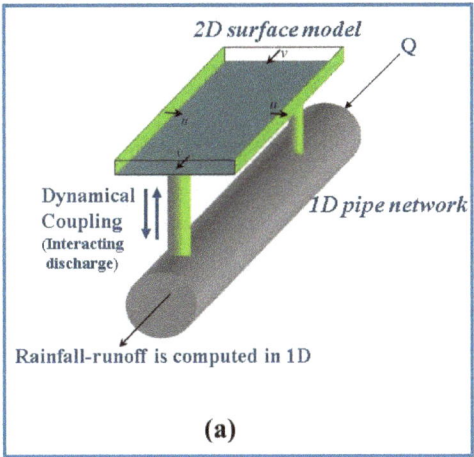
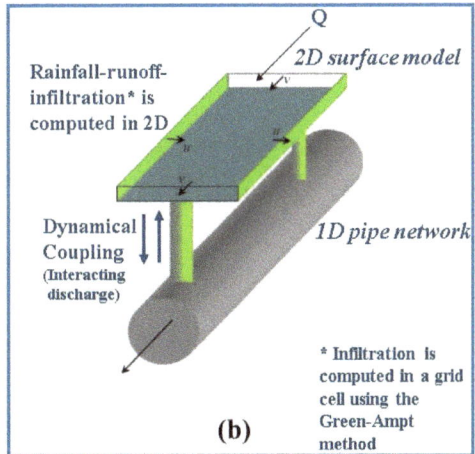

Figure 1. Illustration of the 1D/2D model interactions: (**a**) approach where the rainfall–runoff is computed in a 1D sewer network model (**b**) an approach based on the effect of rainfall–runoff and infiltration process on the overland flow and its interaction with a sewer network. u and v are the fluxes across cell boundaries of the 2D model.

The 1D rainfall–runoff and 1D pipe network coupling are presented in Figure 1a when the hydrological rainfall–runoff process and routing of flows in drainage pipes are performed using the 1D sewer network. When the capacity of the pipe network is exceeded, flow spills into the 2D model domain from manholes and is then routed by a surface 2D model. The surface infiltration 2D and 1D pipe network coupling are presented in Figure 1b. The model uses a rainfall–runoff infiltration dynamically coupled with a sewer network model. In this study a new modelling setup based on this second interaction has been developed. Figure 2 summarizes the whole methodological process.

This framework includes the following main components: (1) an outflow hydrograph generator to estimate the direct runoff within the 2D domain, (2) a proposed Surf-2D model to represent the overland flow, (3) an infiltration module based on the Green-Ampt method, (4) a 1D sewer network simulated using SWMM 5.1, and (5) a coupled SWMM/Surf-2D model for representing urban floods. The details of each component are presented through the following sub-sections:

2.1. Surf-2D Model

In this study, surface water flood simulation builds on the work started in [19]. A non-inertia model system, named subsequently in this work as Surf-2D, was implemented to represent the urban topography, with the ground elevations at the centres and boundaries of cells on a rectangular cartesian grid. This determines the water levels at the cell centres and the discharges (velocities) at the cell boundaries. The system of 2D shallow water equations was obtained by integrating the Navier Stokes equations over depth and replacing the bed stress by a velocity squared resistance term in the two orthogonal directions. The continuity equation for the 2D flood plain flows is expressed as follows:

$$\frac{\partial h}{\partial t} + \frac{\partial (hu)}{\partial x} + \frac{\partial (hv)}{\partial y} = 0 \qquad (1)$$

where h is the water depth and u and v are the velocities in the directions of the two orthogonal axes (the x and y directions), neglecting eddy losses, Coriolis force, variations in atmospheric pressure, the wind shear effect, and lateral inflow; the momentum equation is expressed as in Equation (2) for the x direction and Equation (3) for the y direction:

$$\frac{\partial(hu)}{\partial t} + \frac{\partial(hu^2)}{\partial x} + \frac{\partial(huv)}{\partial y} + gh\frac{\partial H}{\partial x} + gC_f u\sqrt{u^2+v^2} = 0 \quad (2)$$

$$\frac{\partial(hv)}{\partial t} + \frac{\partial(huv)}{\partial x} + \frac{\partial(hv^2)}{\partial y} + gh\frac{\partial H}{\partial y} + gC_f v\sqrt{u^2+v^2} = 0 \quad (3)$$

where H is the water level, g is the acceleration due to gravity and the coefficient C_f appearing in the friction terms is expressed in terms of Chézy roughness. It is known that two-dimensional flow over an inundated urban flood plain is assumed to be a slow, shallow phenomenon [24] and therefore the convective acceleration terms (the second and third terms in Equations (2) and (3)) can be assumed to be small compared to the other terms, and therefore they can be ignored. Expressing the velocities in terms of the discharges and using the Chézy roughness factor, the simplified momentum equations are expressed as Equation (4) (x direction) and Equation (5) (y direction).

$$\frac{\partial}{\partial t}\left(\frac{Q}{Z_Q}\right) + \Delta Y g\frac{\partial H}{\partial x} + g\frac{Q}{C^2 Z_Q^2}\left(\left(\frac{1}{\Delta Y}\frac{Q}{Z_Q}\right)^2 + \left(\frac{1}{\Delta X}\frac{R}{Z_R}\right)^2\right)^{0.5} = 0 \quad (4)$$

$$\frac{\partial}{\partial t}\left(\frac{R}{Z_R}\right) + \Delta X g\frac{\partial H}{\partial y} + g\frac{Q}{C^2 Z_R^2}\left(\left(\frac{1}{\Delta Y}\frac{Q}{Z_Q}\right)^2 + \left(\frac{1}{\Delta X}\frac{R}{Z_R}\right)^2\right)^{0.5} = 0 \quad (5)$$

where H is the water level, Q and R are the discharges in the directions of the two orthogonal axes (the x and y directions), ΔX and ΔY are the grid spacings in the x and y directions, Z_Q and Z_R are the water depths at the cell boundaries, g is the acceleration due to gravity, and C is the Chézy friction factor.

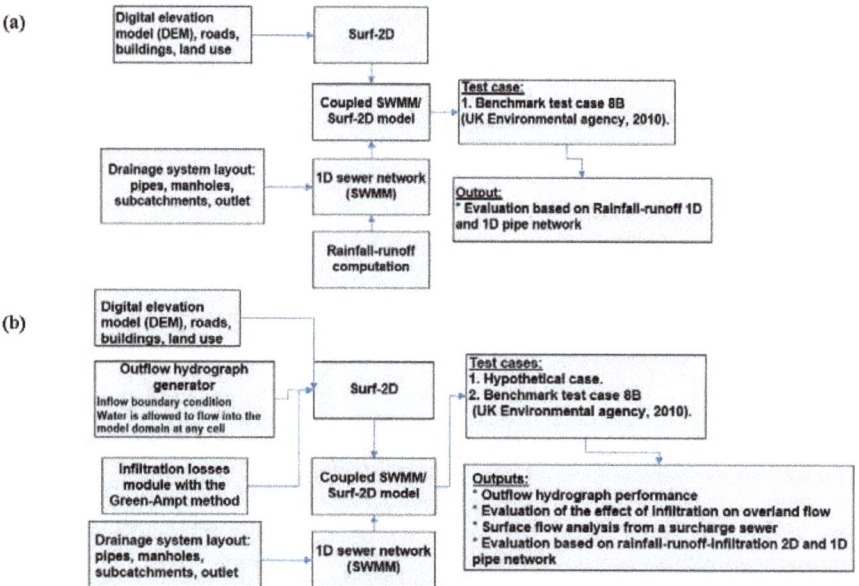

Figure 2. Methodological process: (**a**) approach where the rainfall-runoff is computed in a 1D sewer network model; (**b**) an approach based on rainfall–runoff-infiltration on the overland flow and its interaction with a sewer network.

2.1.1. Numerical Solution

The above conservation of mass and momentum equations (the Saint-Venant equations) given in discretized form were solved using the alternating direction implicit scheme (ADI algorithm). In the ADI algorithm, the solution procedure includes the computation of conservation of mass and conservation of momentum in the corresponding direction and, in the other direction, the conservation of mass is once more computed, but now with the conservation of momentum for that direction (See, [19]). The main features of the Surf-2D model includes a two-point forward spatial and temporal difference scheme adopted on the basis of a uniform time step $\Delta t = t_n + 1 - t_n$, in which n is the time step counter.

2.1.2. Wetting and Drying

The water depth of a grid cell is then calculated as the average depth over the whole cell [19]. When the cell first receives water, the wetting front edge usually lies within the cell. In most cases, only part of the cell will be wetted at that time step. When the flow volume leaving a cell is more than that entering the cell, the cell dries and there is the possibility that the water depth may be reduced to zero or a negative value [19].

In order to avoid negative depth values, the wetting process is controlled by a wetting parameter. When the cell is wetting, the water should not be allowed to flow out of the cell until the wetting front has crossed the cell [25]; each cell has a property called percentage wet when the cell is first wetted, as follows:

$$percentage\ wet = min\left(1, \frac{\sum(v\Delta t)}{\Delta x}\right) \quad (6)$$

where v is the velocity computed from the discharge crossing the cell boundary divided by the cell width and the cell flow depth; Δx is the cell width and Δt is the current time step. Water is not allowed to flow out of the cell until the wetting parameter reaches unity. The wetting parameter is updated at each time step to describe the water traveling across a cell. The whole surface of the cell is used as active infiltration surface, even if rainfall intensity is zero and the cell is only partially wet. In terms of the numerical scheme, the model has the ability to halve or double the time step, halving to meet the convergence criterion, and doubling after a certain number of time steps without halving.

2.2. Infiltration Algorithm

A proposed infiltration algorithm was incorporated into the Surf-2D model code. In this case, infiltration is computed in a grid cell using the Green-Ampt method [26,27]. The cumulative depth of water infiltrated from the soil and the infiltration rate form of the Green-Ampt equation for the one-stage case of initially ponded conditions, and assuming the ponded water depth is shallow, is given in Equations (7) and (8).

$$f(t) = k_e\left[1 + \frac{\Psi \theta_d}{F(t)}\right] \quad (7)$$

$$F(t) = k_e t + \Psi \theta_d ln\left[1 + \frac{F(t)}{\Psi \theta_d}\right] \quad (8)$$

where $f(t)$ is the infiltration rate (mm/h), $F(t)$ is the cumulative infiltration depth (mm), k_e is the effective saturated conductivity (mm/h), θ_d is the moisture deficit (mm/mm), t = time and Ψ depends on the soil and represents the suction head at the wetting front (mm). The ponded water depth h_o computed at the surface of the cell, as described in the previous section and now also available, is assumed to be negligible compared to Ψ as it becomes surface runoff. However, in the case when the ponded depth is not negligible, the value of Ψ-h_o is substituted for Ψ for infiltration computation at time t_n in Equations (7) and (8) [23,28].

Equations (7) and (8) have been solved within the Surf-2D model from a quadratic approximation of the Green-Ampt equation based on the power series expansion presented first by [29]. Stone et al [30] derived their approximation based on two first terms in a Taylor series expansion and it was presented as the modified [29] equation as follows:

$$F_q^*(t_c^*) = 0.5\left(t_c^* + \sqrt{t_c^*(t_c^* + 8)}\right) \quad (9)$$

where $F_q^*(t_c^*)$ is the quadratic approximation of infiltrated depth for the case of the initial ponded conditions and t_c^* is the corrected time (dimensionless). The Taylor series expansion is given as follows [31]:

$$F_{pr}^*(t_c^*) = t_c^* + (2t_c^*)^{1/2}$$
$$-0.2987(t_c^*)^{0.7913}$$
$$t_c^* = \frac{k_e(t+t_s-t_p)}{\Psi\theta_d} \quad (10)$$
$$f_{pr}^*(t_c^*) = 1 + \frac{1}{F_{pr}^*(t_c^*)} = \frac{f(t)}{k_e}$$
$$F_{pr}^*(t_c^*) = \frac{F(t)}{\Psi\theta_d}$$

where $F_{pr}^*(t_c^*)$ is the resulting approximation of the Taylor series for cumulative infiltrated depth; t_s is the time shift, termed as "pseudo time" used as a correction for considering the cumulative infiltrated depth of water at the time of ponding during an unsteady rainfall event; and t_p is the time to ponding. [30], compared to the quadratic approximation of [29], has less error within range values of the ratio of cumulative depth to capillary potential of 0.5 to 150. This range roughly corresponds to coarser textured error soils. Outside this range, both approximations result in small absolute error.

In the Surf-2D model, infiltration is calculated by taking into account the computed velocity at which water enters into the soil (infiltration rate) in the corresponding grid cell (area of the grid) per unit time. This is treated as a discharge point sink within the same time interval. The water infiltration is assumed to be one-dimensional, and thus there is no lateral drainage. To avoid an infinite infiltration rate initially (when the infiltrated volume is still equal to zero), we add a threshold to obtain the infiltration rate *f = min(inf capacity, i$_{max}$)*. Because the infiltrated volume cannot exceed the water depth (*h*) at the surface of the cell that is available for infiltration at time t_n, the volume is updated as shown in Equation (11). Finally, the water depth is updated.

$$V_{inf}^{n+1} = V_{inf}^n + \min(h, f*\Delta t) \quad (11)$$

2.3. Rainfall-Runoff Process

The performance of an outflow hydrograph generator in a 2D model domain was investigated, instead of adding the rainfall rate directly to each cell as a mass source term, as is commonly performed [14,18,23]. In this study an alternative to obtain direct surface runoff resulting from a given excess rainfall hyetograph was obtained by applying the Soil Conservation unit hydrograph known as the SCS-UH method [32]. Figure 3 shows the surface runoff representation in a cell.

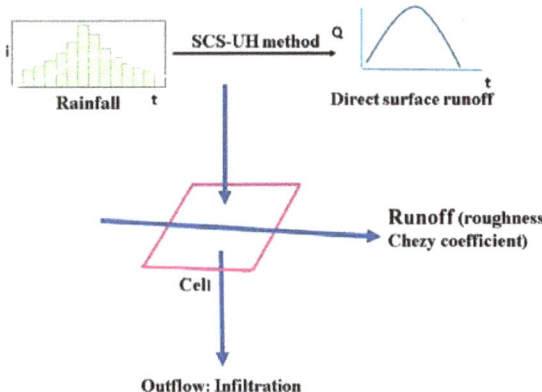

Figure 3. Surface runoff representation in a cell.

The SCS-UH method is a synthetic unit hydrograph, in which the discharge is expressed by the ratio of discharge q to peak discharge qp and the time by the ratio of time t to the time of the unit hydrograph rising, Tp. Given the peak discharge and lag time for the duration of excess rainfall, the unit hydrograph can be estimated from the synthetic dimensionless hydrograph (see for example, [28]). In this case the direct runoff hydrograph accounts for the direct surface runoff, i.e., rainfall minus abstractions or losses, such as initial losses (interception and ponding, considered lower in urban compared to rural areas) and infiltration losses.

Initial losses such as rainfall interception from rooftops, urban trees and depression storage at the start of the design storm are defined as that part of rainfall retained by aboveground objects until it returns to the atmosphere through evaporation. These initial losses have been taken into account in the unit hydrograph computation following the observations found in the literature and recapitulated in [33], expressed as water depth by unit of surface during a frequent rain-event. Infiltration contributes to runoff losses during and after the rainfall event and has been considered thus, as presented in Section 2.2.

The corresponding hydrograph obtained as a result of this process was used as an inflow boundary condition in the Surf-2D model. This means that water is allowed to flow into the model domain at any cell. This can occur either as a source or sink in a cell, with the flow having no horizontal momentum contribution (such as rainfall or flow at surcharged manholes of a drainage network model based on, say, SWMM) or at a cell boundary, in which, in this case, the contribution of the momentum of the inflow is included.

2.4. 1D Sewer Model

SWMM is a dynamic sewer network model that solves the conservation of mass and momentum equations (1D Saint-Venant equations). The model governs the unsteady flow of water through a drainage network of channels and pipes by converting the equations into an explicit set of finite-difference equations. The network system is presented as a set of links which are connected at nodes. Links transport flow from node to node and these nodes are modelled as storage elements in the system. It is assumed that the runoff surface area of a node is equivalent to the surface area of the node itself plus the surface area that is contributed by half of each conduit connected to the node [34]. Continuity and momentum equations are used in the dynamic wave routine at the links, and the continuity equation is used at the nodes. This routing method can account for channel storage, backwater, entrance/exit losses, flow reversal and pressurized flow. The Saint Venant equations and their solution method as implemented in SWMM 5.1 are described in [34].

SWMM defines a node as being in a surcharged condition when all the conduits connected to it are full or when the node's water level exceeds the crown of the highest conduit connected to it [34]. During the surcharge and in order to prevent the surface area at the node from becoming zero (0), a limit on the full conduit width is set, equal to the width when the conduit is 96% full, the so-called minimum full conduit width parameter [20]. To guarantee the mass conservation, a perturbation equation is:

$$\Delta H = \frac{-\sum Q}{-\sum \partial Q / \partial H} \qquad (12)$$

The gradient of flow in a conduit with respect to the head at either end node can be evaluated by differentiating the flow updating the link momentum equation [34], resulting in:

$$\frac{\partial Q}{\partial H} = \frac{-g\overline{A}\Delta t / L}{1 + \Delta Q_{friction}} \qquad (13)$$

where Q is flow rate and H is the hydraulic head of water in the conduit. $\overline{A}$ is the average flow area, Δt is the time step (sec), L is the conduit length (m), ΔH is the adjustment to the node's head that must be made to achieve flow continuity, and $\sum Q$ is the net flow into the node (inflow–outflow) contributed by all conduits connected to the node as well as any externally imposed inflows. $\partial Q / \partial H$ has a negative sign in front of it because, when evaluating $\sum Q$, flow directed out of a node is considered negative while flow into the node is positive. Equation (12) is used whenever heads need to be computed in the successive approximation scheme developed for surcharge flow [34].

2.5. 1D/2D Coupling

In order to simulate the interaction between the sewer network and surface flow, the Surf-2D model was coupled with the SWMM 5.1 open source code through a series of calls built inside a dynamic link library—DLL [34]. The models' linkage follows the work of [20]. To avoid modifications in the original SWMM code, additional functions inside its DLL file were added to feed the interface communication with the Surf-2D model.

The coupling includes two extra functions for exchanging of information between the two models. The first function extracts the node water levels and node depth during every SWMM simulation time, and this function also takes each node ID as inputs to deal with flows. The second function exchanges discharges (node inflow and outflow) between both models. The discharge values can be either positive or negative depending on whether water is being transferred from or to the Surf-2D model. As was stated above, direct surface runoff resulting from a given excess rainfall hyetograph is added directly into the Surf-2D model, thus SWMM computes the dynamic sewer network flow, and its hydrological runoff module was not used.

The Surf-2D model includes two subroutines suggested in previous publications [2,20]. These subroutines calculate the bidirectional discharge between the two models. Discharge between the two models is assumed to take place at the manholes. However, in reality the catchment plays the major role in this. As such, the sewer drainage efficiency depends on the instantaneous and available sewer drainage capacity and the overland flow paths overlying the location of the manholes. Manhole discharges are treated as point sinks or sources in the 2D model within the same time interval as follows.

2.5.1. Drainage Condition

When the water level on the ground surface (h_{2D}) is higher than the hydraulic head at the manhole (h_{1D}) and the ground surface elevation (Z_{2D}), the runoff from the surface flowing into the manhole is determined by either a weir equation if the pressure head in the manhole is below ground surface elevation $h_{1D} < Z_{2D}$ (Equation (14)), or an orifice equation if pressure head in the manhole is above the ground elevation $h_{1D} > Z_{2D}$ (Equation (15)).

$$Q = c_w w h_{2d} \sqrt{2g h_{2D}} \tag{14}$$

$$Q = c_o A_{mh} \sqrt{2g(h_{2D} + Z_{2D} - h_{1D})} \tag{15}$$

where Q is the interacting discharge (m^3/s), and c_w is the weir discharge coefficient. With this form of equation, c_w has a value between 0.6 and 0.7 (0.6), w is the weir crest width (m), A_{mh} is the manhole area (m^2), and c_o is the orifice discharge coefficient with values between 0.6 and 0.62 (0.62). The numbers in parentheses were used as initial values.

2.5.2. Surcharge Condition

Surcharge is determined based on an orifice equation if $h_{2D} < h_{1D}$ (Equation (16)).

$$Q = -c_o A_{mh} \sqrt{2g(h_{1D} - Z_{2D} - h_{2D})} \tag{16}$$

The timing synchronization also becomes an important issue for connecting both models appropriately. Because the sewer network model SWMM and the Surf-2D model use different time steps, the 2D model time step was restricted just before the synchronization time to the value given by applying Equation (17). This time-synchronization technique can be found in detail in [21].

$$\Delta t_{2Dm+1} = \min \left\{ \left(T_{sync} + \Delta t_{1D} - \sum_{i=1}^{m} \Delta t_{2D} \right), \Delta t^*_{2Dm+1} \right\} \tag{17}$$

where Δt_{2Dm+1} is the time step size [s] used for the $m+1$th step, T_{sync} is the time of the previous synchronization [s], $\sum_{i=1}^{m} \Delta t_{2D}$ is the total duration of the time step [s] after m step of the surface water model computation following the last synchronization, Δt_{1D} is the time step used in SWMM, and Δt^*_{2Dm+1} is the time step duration determined by the Surf-2D model for the $m + 1$th step.

2.6. Test Cases

The formerly 2D model was previously tested in [19] with a benchmark case where a wave propagation down a river valley was simulated. To assess the model performance of the proposed Surf-2Dmodel, two tests were selected which enable the studying of specific urban flood aspects and verification of model accuracy.

The first test is a hypothetical case to evaluate the performance of the unit hydrograph. To this purpose, the unit hydrograph (SCS-UH) computed was initially compared with the nonlinear reservoir routing method implemented in SWMM. Figure 4a presents this case; it is a 40 m by 32 m grid plane (8 km^2) with a cell size of 2.5 by 2.5 m and slope of this area equal to 0.007 m.

The second test was taken from the Benchmark test "case 8B" carried out by the UK Environmental Agency (EA) [22]. This corresponds to a hypothetical event happening in the region of Greenfield, Glasgow, UK (see Figure 4b). This test has the objective of assessing the performance of the proposed SWMM/Surf-2D coupled models in terms of water depth predictions and flood volume.

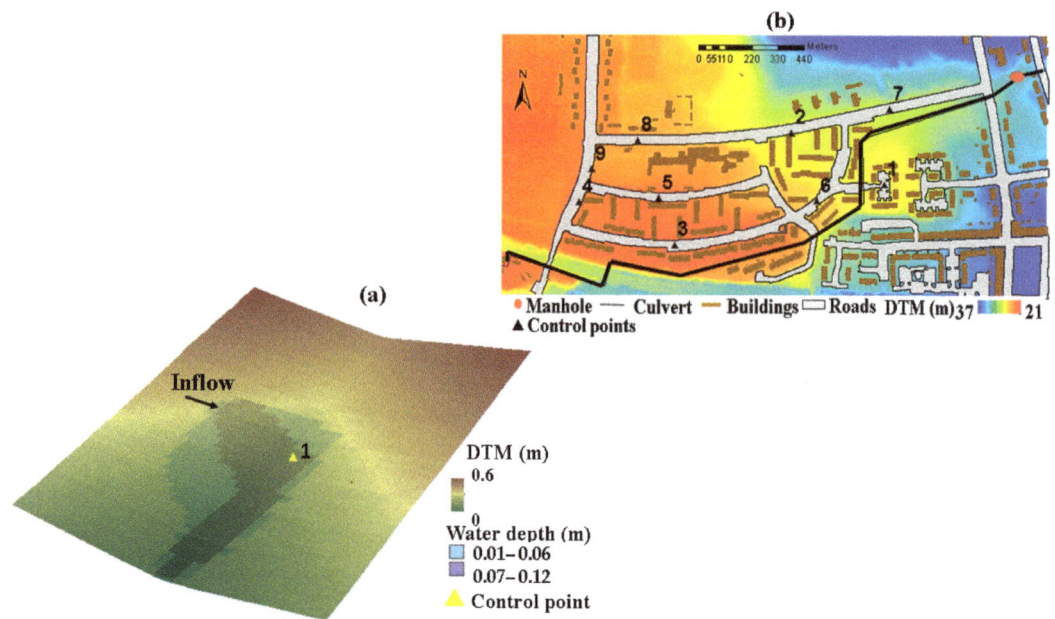

Figure 4. Test cases (**a**) Hypothetical case; (**b**) Benchmark test "case 8B": Greenfield, Glasgow (UK).

A range of software packages, referred to as shallow water equations, "Full models" (i.e., InfoWorks, ISIS, TUFLOW, MIKE FLOOD and SOBEK) and a "simplified model" (some terms of the equations are neglected and simplified equations are solved) known as a UIM model are included in this benchmark test [22]. The infiltration process was not considered in any of these models, so that the infiltration module implemented into the proposed Surf-2D model was not used in this test.

The characteristics of overland flow and the variation in water depths are examined when the overland flow originates not only from rainfall–runoff but also from a surcharging underground pipe. An inflow boundary condition is applied at the upstream end of the pipe. A surcharge is expected to occur at a vertical manhole of 1 m^2 cross-section located 467 m from the top end of the culvert. The flow from the above surcharge spreads across the surface of a 2m resolution DTM created from LIDAR data.

A land cover dependent roughness value was applied with two categories: (1) roads and pavement; (2) any other land cover type. A uniform rainfall of 400 mm/h with 4 min duration and starting at minute 1 was applied with a total simulation time of 5 h to produce direct surface runoff. Similarly, an inflow boundary condition was applied at the upstream end of the pipe (1D model), with a surcharge expected to occur at the manhole.

As was stated in Section 2.6, the benchmark "Test 8B" model packages do not include infiltration processes. For this reason, in order to assess the Surf-2D model's ability to simulate infiltration from a point source (direct runoff resulting from a given excess rainfall hyetograph), the validated open source Full Shallow Water equations for Overland Flow in two dimensions of space (FullSWOF_2D, [23]) was used for comparison purposes. Several features make FullSWOF_2D particularly suitable for applications in hydrology. Small water depths and wet–dry transitions are robustly addressed, rainfall and infiltration (Green-Ampt method) are incorporated, and data from grid-based digital topographies can be used directly. In this software, the shallow water (or Saint-Venant) equations are solved using finite volumes and numerical methods, especially chosen for hydrodynamic purposes (transitions between wet and dry areas, small water depths, and steady-state preservation).

3. Results and Discussion

3.1. Performance of the Outflow Hydrograph Generator

The performance of the outflow hydrograph generator was assessed by the hypothetical case presented in Section 2.6. The initial losses were set to 0.62 mm for both methods following the reviewed values in [33]. This value corresponds to rainfall interception from rooftops, urban trees and depression storage at the start of the design storm. For the nonlinear reservoir, infiltration losses were computed assuming a silt soil class with the following Green-Ampt (GA) parameter values: $k_e = \{5 \text{ mm/h}\}$; $\theta_d = \{0.5\}$; $\Psi = \{190 \text{ mm}\}$. The rainfall intensity was assumed as 70 mm/h at 1-h duration. The result of this process is a hydrograph comparison between the two surface runoff methods (e.g., SCS-UH vs. nonlinear reservoir) as presented in Figure 5.

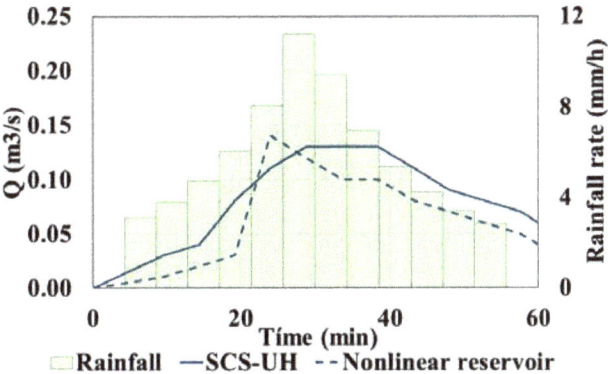

Figure 5. Surface runoff methods comparison for the hypothetical case.

The overland flow rate obtained by applying the SCS-UH method differs slightly around the peak of the hydrograph from the nonlinear reservoir method computed in SWMM. Discharge values in the unit hydrograph method are ±7% higher in comparison to those obtained with the nonlinear reservoir. This can be associated to the different hydrological considerations taken for runoff generation such as the assumed initial losses value. It could also be due to the fact that the infiltration losses were not considered in the unit hydrograph method, as infiltration is computed with the 2D algorithm in the Surf-2D model. However, in general both methods are in good agreement. The corresponding generated hydrograph was then used as an inflow boundary condition in the Surf-2D model for the following analysis.

3.2. Evaluation of the Effect of Infiltration

This test aims to assess the Surf-2D model's ability to simulate infiltration from a point source, direct runoff resulting from a given excess rainfall hyetograph. The hypothetical case (Figure 4a) was applied with an assumed 70 mm/h rainfall intensity of 1 h duration.

As a sensitivity analysis, Figure 6a–c show the comparison of computed water depths for different GA infiltration parameters combinations (k_e, θ_d, Ψ) according to [28,35]. The values of each parameter correspond to sand, silt and clay soil classes, respectively. The obtained water depths were compared to those obtained with FullSWOF_2D software. No groundwater component (neither physically nor parametrized) has been included in both models.

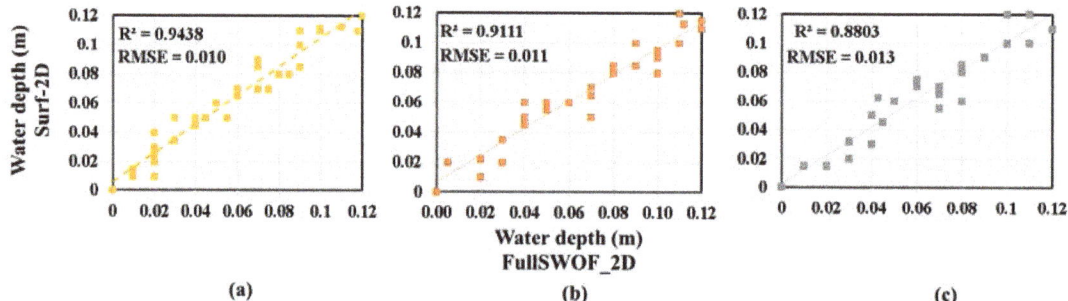

Figure 6. Water depths comparison between Surf-2D and FullSWOF_2D for different GA infiltration parameters (**a**) $k_e = \{117.8 \text{ mm/h}\}$; $\theta_d = \{4.17\}$; $\Psi = \{49.5 \text{ mm}\}$; (**b**) $k_e = \{6.5 \text{ mm/h}\}$; $\theta_d = \{4.86\}$; $\Psi = \{166.8 \text{ mm}\}$; (**c**) $k_e = \{0.3 \text{ mm/h}\}$, $\theta_d = \{3.85\}$; $\Psi = \{316 \text{ mm}\}$.

The water depth results consist of those predicted by FullSWOF_2D for different infiltration parameters combinations (Figure 6). This can be inferred by the R^2 grader at 0.88, the RMSE error statistic which exhibits a small error of an average of 0.011 m, and the total volume difference between Surf-2D and FullSWOF_2D. It is in the order of 4% (see Table 1). The good match can be associated to the similar finite volumes and numerical methods applied in both models to solve the shallow water (or Saint-Venant) equations and to the method used to compute infiltration. These results show the importance, when evaluating the performance of the Surf-2D model, of computing infiltration.

Table 1. Comparison of the overland surface volume differences between Surf-2D and FullSWOF_2D.

Infiltration Parameters	Surf 2D Volume at the Surface (m³)	FullSWOF_2D Volume at the Surface (m³)	RMSE—Water Depths (m)
$k_e = \{117.8 \text{ mm/h}\}$; $\theta_d = \{0.417\}$; $\Psi = \{49.5 \text{ mm}\}$	4570	4760	0.010
$k_e = \{6.5 \text{ mm/h}\}$; $\theta_d = \{0.486\}$; $\Psi = \{166.8 \text{ mm}\}$	4619	4810	0.011
$k_e = \{0.3 \text{ mm/h}\}$; $\theta_d = \{0.385\}$; $\Psi = \{316 \text{ mm}\}$	4667	4860	0.013

Figure 7 presents the results comparison between the Surf-2D model and FullSWOF_2D in a single grid (point 1, Figure 4a) according to the different soil types given in Table 2.

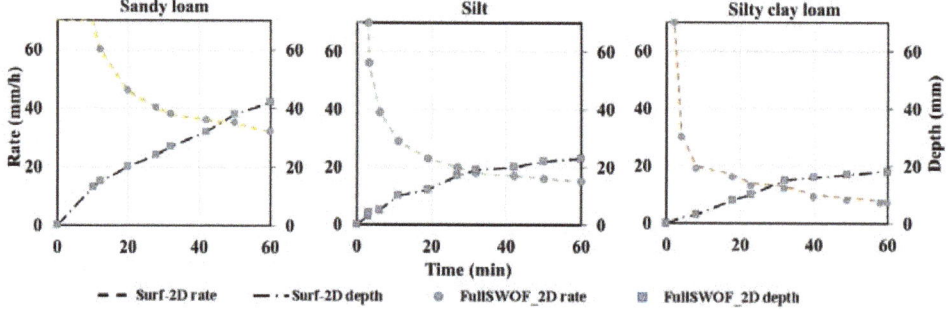

Figure 7. Hypothetical case: infiltration rate and depth results in a single grid (point 1).

Table 2. Green-Ampt parameter values used [28,36].

Soil Texture	k_e (mm/h)	Ψ (mm)	θ_d (mm/mm)
Sandy loam	22	90	0.5
Silt	5	190	0.5
Silt clay loam	1.8	253	0.2

A surface with high infiltration capacity (sandy loam) reduces the surface runoff generated to 36%, and with low infiltration capacity (clay loam) runoff has also been decreased to 27%. The results show the importance of the soil type in determining the overland flow, as this governs the infiltration capacity limits. The hydraulic conductivity is a dominant parameter as it defines the maximum infiltration capacity of the soil, as also presented in [11].

The infiltration rate in a single grid (see control point 1, Figure 4a) was found to be reduced at a decreasing rate at a time up to 20 minutes. It shows an almost steady state after 30 minutes of continuous ponding. The infiltration depth in a single grid was also found to increase at a decreasing rate, which is consistent with previous investigations [30,36]. Figure 7 also shows the logical hydraulic properties of soil from the highest to the lowest: sandy loam and clay loam, respectively (see also [37]).

3.3. Surface Flow from a Surcharge Sewer

This section evaluates the capability of simulating shallow inundation, originating from a surcharging underground pipe. The benchmark test case in the region of Greenfield, Glasgow (UK) and described in Section 2.6 was applied. Figure 8 presents the manhole discharge results using the coupled SWMM/Surf-2D model and its comparison with the mentioned diffusive and dynamic models. Due to the fact that the infiltration process was not considered in any of the benchmark models, the infiltration module implemented in the proposed Surf-2D model was not used for this analysis. Final results have been overlapped with previously published results from the software packages [36].

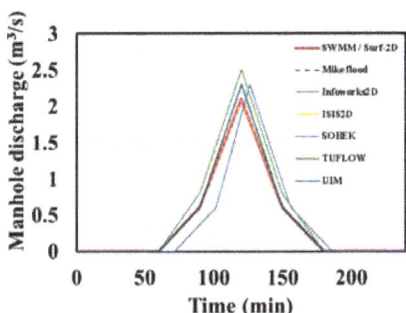

Model	Volume (m³)
SWMM / Surf-2D	5700
Infoworks	5873
ISIS	5864
TUFLOW	5837
UIM	5226
MIKE FLOOD	5024
SOBEK	4987

Figure 8. Manhole discharge predicted by SWMM/Surf-2D superimposed with the results from the models published in the EA benchmark 8B [22].

The SWMM/Surf-2D model predicts similar results in terms of peak discharge at the manhole, as can be seen in Figure 8, although volumes differ within a 12% range (e.g., 5700 m³ for SWMM/Surf-2D model and 5024 m³ for Mike flood). Figure 9a,b present water depths at points 7 and 9, which correspond to a green area (See Figure 4b).

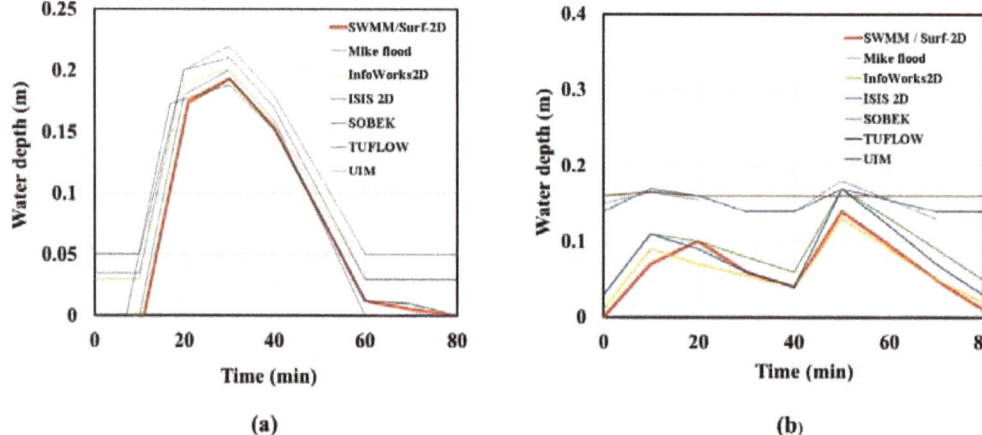

Figure 9. Water depths predicted by SWMM/Surf-2D superimposed with the results from the models published in the EA benchmark 8B [36]. (**a**) Water depths at point 7; (**b**) water depths at point 9.

All models agreed in the prediction of the peak levels (red line). Maximum depths did not exceed 0.23 m at point 7. The coupled model also agrees in the prediction of water depths compared to the other models, and water depths did not exceed 0.20m at point 9. However, the UIM model results do not predict peak levels showing a quasi-constant water depth. This could be a consequence of the scale of the test used here, over smaller domains than one would typically apply a simplified model to (see, [36]). UIM as a simplified model solves the 2D diffusion wave equation which is obtained by neglecting the acceleration terms in the 2D shallow water equations.

In terms of run times, Table 3 presents the efficiency of the proposed SWMM/Surf-2D coupled model compared to the model results reported in [36].

Table 3. Summary of runtimes.

Model	Time-Stepping	Runtime (min)
SWMM/Surf-2D	1 s	18.0
InfoWorks	Adaptive	6.0
ISIS	0.05 s	734.30
TUFLOW	1 s	9.20
UIM	Adaptive	743.30
MIKEFLOOD	1 s	2.08
SOBEK	5 s	18.9

Efficiency obtained herein with SWMM/Surf-2D exhibits similar run times compared to SOBEK and performs better than ISIS and UIM models. However, TUFLOW, InfoWorks and MIKEFLOOD perform better than the method herein mentioned. According to [36], possible explanations for this run times variations include the choice of the time step partly imposed by the numerical approach, the number of iterations performed at each time step, and the efficiency of the numerical algorithm and hardware specification.

3.4. Evaluation of Two Approaches for Representing Urban Floods

The two 1D/2D model interactions illustrated in Figure 1 have been evaluated on the benchmark test "case 8B". Initial losses for the unit hydrograph were set to 0.65mm. The distributed hydraulic conductivity (k_e) in the model for roads–pavement and green areas was set to a very slow infiltration 1.0 mm/h and moderate infiltration 6.5 mm/h, respectively. Figure 10 presents the top view of the flood inundation extent in the region of Greenfield, Glasgow to evaluate two different flood modelling interactions. Flood is condensed across a highway going east to west. The flat slope, especially along the street, facilitates the overland flow's inundation along the street parallel to the main highway.

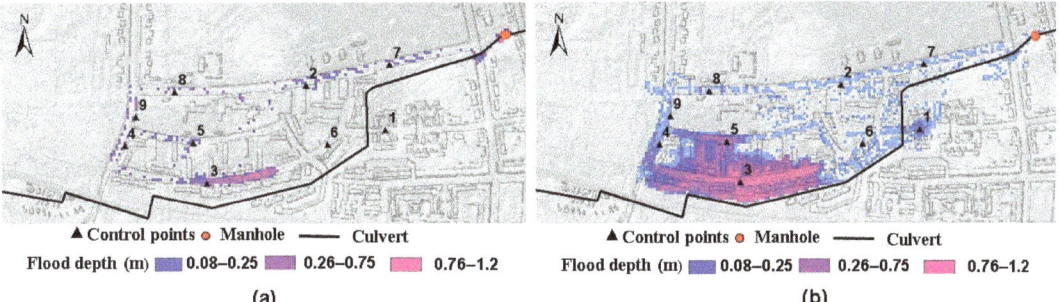

Figure 10. Approaches for representing urban floods: (**a**) rainfall–runoff is computed in 1D culvert, flood extent presented in 2D. (**b**) Rainfall–runoff and infiltration computation accounted in Surf-2D model and its interaction with the culvert.

The first 1D/2D model interaction is presented in Figure 10a. The rainfall–runoff process is performed in the 1D culvert. When the capacity of the pipe network is exceeded flow spills into the 2D model domain from manholes and is then routed by a surface 2D model. In this case, the areal extent, maximum flood depth and volume of the inundation region is about 624 m^2, 0.38 m and 187 m^3, respectively.

The second 1D/2D model interaction, based on rainfall–runoff and infiltration computation. accounted for with the Surf-2D model, is presented in Figure 10b. The culvert initially has no water to simulate overland flow draining back to the system. Here, the impact of rainfall–runoff and infiltration in the 2D domain is presented. An increase in the areal extent (4800 m^2), maximum flood depth (0.9 m) and volume (3600 m^3) around control point 3 is obtained (Figure 10b). Significant differences compared to those obtained in Figure 10a are shown. This approach 2 estimates the areal extent and flood depth caused by the rainfall–runoff, and the infiltration and excess flow from manholes during a flood event (flood inundation). After the event is over, water eventually drains back to the system through the downstream manhole (drainage condition). The inundation region is now about 2820 m^2 of areal extent, 0.8 m of maximum flood depth and 1830 m^3 of volume.

The above results show that differences between the two types of interactions are significant. For instance, differences by around 78% in terms of flood extent, 48% in the maximum flood depths and 90% in inundation volume were found. Integrated modelling approaches are being increasingly promoted as required in order to holistically evaluate urban water systems while facilitating infiltration in urban areas. In terms of modelling, the challenge today is to move from individual considerations of urban drainage system (UDS) performance to integrated applications that include not only the interaction between the sewer network and surface flow (1D/2D coupled models) but also the inclusion of the rainfall–runoff and infiltration process for a better evaluation of the system. Similarly, the modelling of green infrastructure or natural based solutions—NBS—within the evaluation of an UDS needs to be included [38]. The above test results demonstrate an acceptable tool as an advance for further analysis of the performance of these type of infrastructures.

4. Conclusions

In this paper an approach to couple rainfall–runoff-infiltration and sewers is presented. To achieve this, an infiltration module algorithm based on the Green and Ampt method was coded into a model called Surf-2D. Infiltration was calculated by taking into account the computed infiltration rate. Direct surface runoff resulting from a given excess rainfall hyetograph was also computed by applying the unit hydrograph method. The corresponding hydrograph was used as an inflow boundary condition in the 2D domain for each test. The model was then coupled with SWMM 5.1 open source code through a series of calls built inside a dynamic link library—DLL. The presented modelling setup was validated with two cases: a hypothetical case and the real case of Greenfield, Glasgow (UK). A surcharging case came from an EA benchmark report, with the latest validated free software FullSWOF_2D and with two different approaches to representing urban floods.

The following conclusions are reached: the unit hydrograph (SCS-method) implemented was indeed effective for the purpose of producing direct runoff in the 2D model domain compared with the non-linear reservoir method. Despite having different hydrological considerations for runoff generation compared to the nonlinear reservoir method, both methods' results were similar. The inclusion of the Green-Ampt method in the 2D domain had a direct impact on the overland flood-depths. Although determining soil properties may sometimes be difficult for the application of the method, the presented model was capable of reproducing the influence of the infiltration capacity of the soil on the overland flow. As was observed in Figure 7 for a grid cell, the model follows a reasonable range of soil hydraulic properties from the highest to the lowest, sandy loam and clay loam, respectively (i.e., from high infiltration capacity with a sandy loam to low infiltration capacity with a clay loam soil type).

In the benchmark test "case 8B" in the region of Greenfield, Glasgow, the 1D sewer model contributed more reliable analyses of flooding processes due to their impact on the overland flood-depths. The presented model predicted similar results of the software packages (EA benchmark report), compared in terms of peak water depths within a range of a few centimeters. Two approaches for representing urban floods were tested in this work, leading to different flood evolution results. The direct impact of rainfall–runoff and infiltration using benchmark Test 8B allowed the provision of realistic flood volume and gradual recession after the flood peak occurs.

Finally, the presented coupled SWMM/Surf-2D model with the incorporation of rainfall–runoff and infiltration process showed a basis for addressing a better evaluation of urban floods and, in turn, holistically evaluate an urban drainage system.

Author Contributions: Conceptualization, C.M., Z.V. and A.S.; methodology, C.M., Z.V. and A.S.; software, R.P. and C.M.; validation, C.M. and A.S.; formal analysis, C.M., Z.V. and A.S.; investigation, C.M. and A.S.; resources, C.M., Z.V. and A.S.; data curation, C.M. and A.S.; writing—original draft preparation, C.M. and A.S.; writing—review and editing, A.S. and Z.V.; visualization, C.M.; supervision, A.S. and Z.V. All authors have read and agreed to the published version of the manuscript.

Funding: This work received partial funding from the European Union Seventh Framework Programme under Grant agreement No. 603663 for the research project PEARL. The research leading to these results was supported by the Administrative Department of Science, Technology and Innovation, COLCIENCIAS under Grant N.568 of 2012 and the Advanced Training Program for Teaching and Research of the Universidad del Magdalena—Colombia awarded to the first author.

Institutional Review Board Statement: Not applicable.

Informed Consent Statement: Not applicable.

Data Availability Statement: Data sharing not applicable.

Acknowledgments: The authors would also like to thank the UK Environmental Agency for the EA benchmark dataset.

Conflicts of Interest: The authors declare no conflict of interest. The funders had no role in the design of the study; in the collection, analyses, or interpretation of data; in the writing of the manuscript, or in the decision to publish the results.

References

1. Mignot, P.; Nakagawa, H.; Kawaike, K.; Paquier, A.; Mignot, E. Modeling flow exchanges between a street and an underground drainage pipe during urban floods. *J. Hydraul. Eng.* **2014**, *40*. [CrossRef]

2. Chen, A.; Leandro, J.; Djordjevic, S.; Schumann, A. Modelling sewer discharge via displacement of manhole covers during flood events using 1D/2D SIPSON/PDWave dual drainage simulations. *Urban Water J.* **2015**, *13*, 830–840. [CrossRef]
3. Mallari, K.; Arguelles, A.; Kim, H.; Aksoy, H.; Kavvas, M.; Yoon, J. Comparative analysis of two infiltration models for application in a physically based overland flow model. *Environ. Earth Sci.* **2015**, *74*, 1579–1587. [CrossRef]
4. Park, S.; Kim, B.; Kim, D. 2D GPU-accelerated high resolution numerical scheme for solving diffusive wave equation. *Water* **2019**, *11*, 1447. [CrossRef]
5. Martínez, C.; Sanchez, A.; Vojinovic, Z. Surface water infiltration based approach for urban flood simulations. In Proceedings of the 38th IAHR World Congress, Panama City, Panama, 1–6 September 2019. [CrossRef]
6. Pantelakis, D.; Thomas, Z.; Partheniou, E.; Baltas, E. Hydraulic models for the simulation of flow routing in drainage canals. *Glob. Nest J.* **2013**, *15*, 315–323.
7. Van den Putte, A.; Govers, G.; Leys, A.; Langhans, C.; Clymans, W.; Diels, J. Estimating the parameters of the Green–Ampt infiltration equation from rainfall simulation data: Why simpler is better. *J. Hydrol.* **2013**, *476*, 332–344. [CrossRef]
8. Cheng, L.; Wang, Z.; Hu, S.; Wang, Y.; Jin, J.; Zhou, Y. Flood routing model incorporating intensive streambed infiltration. *Sci. China-Earth Sci.* **2015**, *58*, 718–726. [CrossRef]
9. Fernández-Pato, J.; Caviedes-Voullieme, D.; García-Navarro, P. Rainfall/runoff simulation with 2D full shallow water equations: Sensitivity analysis and calibration of infiltration parameters. *J. Hydrol.* **2016**, *536*, 496–513. [CrossRef]
10. Castanedo, V.; Saucedo, H.; Fuentes, C. Modeling two-dimensional infiltration with constant and time-variable water depth. *Water* **2019**, *11*, 371. [CrossRef]
11. Gülbaz, S.; Boyraz, U.; Kazezyilmaz-Alhan, C. Investigation of overland flow by incorporating different infiltration methods into flood routing equations. *Urban Water J.* **2020**, *17*, 109–121. [CrossRef]
12. Martin, R. Evaluating Sediment Dynamics in an Urban Stream with Mobility Frequencies. *Southeast. Geogr.* **2016**, *56*, 409–427. [CrossRef]
13. Plumb, B.; Juez, C.; Annable, K.; McKie, C.; Franca, M. The impact of hydrograph variability and frequency on sediment transport dynamics in a gravel bed flume. *Earth Surf. Process. Landf.* **2019**. [CrossRef]
14. Martins, R.; Leandro, J.; Djordjevic, S. Influence of sewer network models on urban flood damage assessment based on coupled 1D/2D models. *J. Flood Risk Manag.* **2018**, *11*, S717–S728. [CrossRef]
15. Martínez-Cano, C.; Toloh, B.; Sanchez-Torres, A.; Vojinović, Z.; Brdjanovic, D. Flood Resilience Assessment in Urban Drainage Systems through Multi-Objective Optimisation. CUNY Academic Works. 2014. Available online: http://academicworks.cuny.edu/cc_conf_hic/236 (accessed on 12 July 2021).
16. Martínez, C.; Sanchez, A.; Toloh, B.; Vojinovic, Z. Multi-objective evaluation of urban drainage networks using a 1D/2D flood inundation model. *Water Resour. Manag.* **2018**, *32*, 4329–4343. [CrossRef]
17. Ganiyu, A.; Olawale, M.; Pahtirana, A. Coupled 1D-2D hydrodynamic inundation model for sewer overflow: Influence of modelling parameters. *Water Sci. J.* **2015**, *29*, 146–155. [CrossRef]
18. Fan, Y.; Ao, Y.; Yu, H.; Huang, G.; Li, X. A coupled 1D-2D hydrodynamic model for urban flood inundation. *Adv. Meteorol.* **2017**, 1–12. [CrossRef]
19. Seyoum, S.; Vojinovic, Z.; Price, R.; Weesakul, S. A coupled 1D and non-inertia 2D flood inundation model for simulation of urban flooding. *ASCE J. Hydraul. Eng.* **2012**, *138*, 23–34. [CrossRef]
20. Leandro, J.; Martins, R. A methodology for linking 2D overland flow models with the sewer network model SWMM 5.1 based on dynamic link libraries. *Water Sci. Eng.* **2016**, *73*, 3017–3026. [CrossRef] [PubMed]
21. Chen, A.; Djordjevic, S.; Leandro, J.; Savic, D. The urban inundation model with bidirectional flow interaction between 2D overland surface and 1D sewer networks. In Proceedings of the NOVATECH 6th International Conference on Sustainable Techniques and Strategies in Urban Water Management, Lyon, France, 25–28 June 2007; pp. 465–472.
22. Néelz, S.; Pender, G. *Report: Delivering Benefits through Evidence: Benchmarking of 2D Hydraulic Modelling Packages*; Report; Environmental Agency: Bristol, UK, 2010; ISBN 978-1-84911-190-4.
23. Delestre, O.; Darboux, F.; James, F.; Lucas, C.; Laguerre, C.; Cordier, S. FullSWOF: Full Shallow-Water equations for overland flow. *J. Open Source Softw.* **2018**, *2*, 448–486. [CrossRef]
24. Hunter, N.; Bates, P.; Horritt, M.; Wilson, M. Simplified spatially-distributed models for predicting flood inundation: A review. *Geomorphology* **2007**, *90*, 208–225. [CrossRef]
25. Yu, D.; Lane, S. Urban fluvial flood modelling using a two-dimensional diffusion wave treatment, part 1: Mesh resolution effects. *Hydrol. Process.* **2006**, *20*, 1541–1565. [CrossRef]
26. Green, W.; Ampt, G. Studies on soil physics: 1. Flow of air and water through soils. *J. Agric. Sci.* **1911**, *4*, 1–24.
27. Mein, R.; Larson, C. Modelling infiltration during a steady rain. *Water Resour. Res.* **1913**, *9*, 384–394. [CrossRef]
28. Chow, V.T. *Applied Hydrology*; McGraw-Hill Inc.: New York, NY, USA, 1988.
29. Li, R.; Stevens, M.; Simons, D. Solutions to Green—Ampt infiltration equation. *J. Irrig. Drain. Div.* **1976**, *102*, 239–248. [CrossRef]
30. Stone, J.; Hawkins, R.; Shirley, E. Approximate form of Green-Ampt infiltration equation. *J. Irrig. Drain. Eng.* **1994**, *120*, 128–137. [CrossRef]
31. Kale, R.; Sahoo, B. Green-Ampt infiltration models for varied field conditions: A revisit. *Water Resour. Manag.* **2011**, *25*, 3505–3536. [CrossRef]
32. SCS, Soil Conservation Service. *Design of Hydrograph*; US Department of Agriculture: Washington, DC, USA, 2012.

33. Rammal, M.; Berthier, E. Runoff losses on urban surfaces during frequent rainfall events: A review of observations and modelling attempts. *Water* **2020**, *12*, 2777. [CrossRef]
34. Rossman, L. *Storm Water Management Model Reference Manual Volume II—Hydraulics*; EPA/600/R-17/111; U.S. Environmental Protection Agency: Washington, DC, USA, 2017.
35. Rawls, W.; Brakensiek, D.; Miller, N. Green-Ampt infiltration parameters from soils data. *J. Hydraul. Eng.* **1983**, *109*, 62–70. [CrossRef]
36. Zhan, T.; Ng, C.; Fredlund, D. Field study of rainfall infiltration into a grassed unsaturated expansive soil slope. *Can. Geotech. J.* **2007**, *44*, 392–408. [CrossRef]
37. Hossain, S.; Lu, M. Application of constrained interpolation profile method to solve the Richards equation. *J. Jpn. Soc. Civ. Eng.* **2014**, *70*, 247–252. [CrossRef]
38. Martínez, C.; Sanchez, A.; Galindo, R.; Mulugeta, A.; Vojinovic, Z.; Galvis, A. Configuring green infrastructure for urban runoff and pollutant reduction using an optimal number of units. *Water* **2018**, *10*, 1528. [CrossRef]

Article

Designing for People's Safety on Flooded Streets: Uncertainties and the Influence of the Cross-Section Shape, Roughness and Slopes on Hazard Criteria

Luís Mesquita David [1,*] and Rita Fernandes de Carvalho [2]

[1] LNEC—National Laboratory for Civil Engineering, 1700-066 Lisboa, Portugal
[2] MARE, Department of Civil Engineering, University of Coimbra, 3030-788 Coimbra, Portugal; ritalmfc@dec.uc.pt
* Correspondence: ldavid@lnec.pt

Abstract: Designing for exceedance events consists in designing a continuous route for overland flow to deal with flows exceeding the sewer system's capacity and to mitigate flooding risk. A review is carried out here on flood safety/hazard criteria, which generally establish thresholds for the water depth and flood velocity, or a relationship between them. The effects of the cross-section shape, roughness and slope of streets in meeting the criteria are evaluated based on equations, graphical results and one case study. An expedited method for the verification of safety criteria based solely on flow is presented, saving efforts in detailing models and increasing confidence in the results from simplified models. The method is valid for $0.1 \text{ m}^2/\text{s} \leq h.V \leq 0.5 \text{ m}^2/\text{s}$. The results showed that a street with a 1.8% slope, $K \approx 75 \text{ m}^{1/3}\text{s}^{-1}$ and a rectangular cross-section complies with the threshold $h.V = 0.3 \text{ m}^2/\text{s}$ for twice the flow of a street with the same width but with a conventional cross-section shape. The flow will be four times greater for a 15% street slope. The results also highlighted that the flood flows can vary significantly along the streets depending on the sewers' roughness and the flow transfers between the major and minor systems, such that the effort detailing a street's cross-section must be balanced with all of the other sources of uncertainty.

Keywords: dual drainage modelling; extreme rainfall; flooding; safety criteria; urban drainage; uncertainty

1. Introduction

Most of the existing stormwater sewer networks were designed for uniform and steady flows and 5- to 25-year return periods. With urban expansion, the aging of infrastructures, and increasing environmental and quality of life requirements in cities, many sewer networks became undersized. In order to mitigate floods and combined sewer overflows, storage structures and real-time management systems have been implemented in the large sewer networks of some cities, the design and operation of which requires the use of generally complex mathematical models [1–3]. Over the past few decades, many countries have carried out great efforts to make drainage systems more decentralized, integrating nature-based solutions and promoting synergies with other urban infrastructures, such as the green infrastructure and the road and pedestrian infrastructure [4–6]. In this context, the so-called water-sensitive cities have sought to include a chain of components for the retention, infiltration, treatment and use of stormwater in catchments, seeking to replicate hydrological losses and improve urban ecosystems [7,8]. In addition, the redrawing of the urban space to accommodate floods is an adaptation and remediation strategy to deal with climate change, which is also being applied to new developments worldwide [9–11].

The rehabilitation of consolidated urban areas depends on a variety of local constraints and intervention opportunities, the planning of which is complex and requires cross-sectoral, risk-based approaches [12–14]. In new developments, surface flow paths and detention areas (the major system) are often planned for 30- to 100-year return periods.

Several expressions have been derived from laboratory experiments for the thresholds of people's stability against the action of flows. Summaries of empirical expressions, derived by both the original experimental investigators and by third parties, are presented in Shand et al. [15] and Russo et al. [16]. These expressions usually establish thresholds for the velocity and depth of the flood, or for a relationship between these variables, and are increasingly included in design guidelines, manuals of good practices, standards and municipal specifications [9–11]. In most cases, the product of the flood depths and velocities is limited to values between 0.4 and 0.5 m^2/s, although recent studies have proposed lower thresholds for urban floods, such as the thresholds below 0.3 m^2/s proposed by Chanson and Brown [17], and the threshold of 0.22 m^2/s proposed by Martinez et al. [18].

For example, the Australian Guidelines (1987) established that the product of flood depths and velocities in streets should not exceed 0.4 m^2/s. Based on the results of six previous studies from 1973 to 2008, Shand et al. [15] concluded that this criterion ensures a low hazard for children, providing that the maximum depth is limited to 0.5 m and the maximum velocity to 3.0 m/s. However, the authors highlighted that the loss of stability could occur in lower flows when adverse conditions are encountered, including uneven or slippery bottom conditions, unsteady flow, floating debris, poor visibility or human factors such as physical attributes, psychological factors, clothing and footwear. The risk of the instability and buoyancy of vehicles also increases substantially for water depths above 0.2–0.3 m. Melbourne Water [11] proposed more detailed criteria and recommendations according to the type of street, and that allotments are at least 0.3 m above the 100-year flood level. According to some Canadian and USA city manuals, the maximum depth of the overland flow or ponding in streets should be limited to 0.3 m deep at the gutter for the 100-year return period event.

The following three thresholds were proposed by Nanía et al. [19] and Balmforth et al. [9]: the water depth should be limited to 0.3 m or 0.2 m where a highway forms part of the flood channel; the product of the depth and velocity should be limited to 0.5 m^2/s; and the product of the depth by the square of the velocity should be limited to 1.23 m^3/s^2, in order to prevent the risk of pedestrian slipping. Nanía et al. [19] justified a higher threshold for the $h.V$ ratio (0.5 m^2/s) because the 0.4 m^2/s threshold was proposed based on experiments with water depths between 0.5 and 1.2 m, which are excessive for densely occupied areas. However, they introduce the slip criterion, which corresponds to a much more conservative condition for the flows with higher velocities than the $h.V$ condition. Because the higher velocities occur at reduced flow depths, generally lower than 20 cm, the slope and shape of the street cross-sections and the height of the sidewalks can have a significant effect.

Based on laboratory tests, Xia and Falconer [20] obtained graphs for the toppling stability thresholds for adults and for children, of which the products of flow depths and velocities across the range of values are greater than 0.6 m^2/s for adults and greater than 0.4 m^2/s for children. These researchers obtained significantly lower sliding stability thresholds, but decided not to include them in the suggested stability thresholds "because the mode of sliding instability seldom occurs in practice due to the rare occurrence of low depth and high velocity".

However, Russo et al. [16] emphasized that most of the previous relationships were obtained from channels reproducing natural streams, with flows that were generally deep and slow, which is not the case for many urban floods. From the results of hundreds of laboratory tests, in which several subjects of different ages, heights and weights crossed or walked along the flow on smooth concrete surfaces (reproducing the roughness of urban roads), they proposed a new threshold for the maximum allowed flow velocity of $V = 1.88$ m/s. Based on new experiments using the protocol described in [16], and also testing a variety of footwear and situations for both free and busy hands, Martinez et al. [18] established a more restrictive stability threshold for pedestrians of $h.V = 0.22$ m^2/s [17] highlighted the role of hydrodynamic instabilities induced by local topographic effects and large debris (e.g., trees, branches, logs, plastic containers and rubbish) in the real-world hazards. The confluence of flows at street crossings and the transitions from supercritical to subcritical flows also adversely impact pedestrian stability.

Thresholds from 0.3 m^2/s to more than 0.7 m^2/s have been proposed for vehicle stability, depending on the characteristics of the vehicles, such as their weight, length, width, friction coefficient, and others [15,21–23]. In order to avoid the buoyancy of small passenger vehicles and large 4WD, ref. [21] recommended flood depth thresholds of 0.3 m and 0.5 m, respectively. For any type of vehicle, the flood velocity should not exceed 3 m/s.

Salinas-Rodriguez et al. [14] pointed out that setting rigid thresholds can be difficult and even not feasible when applied to existing urban areas or to large catchments as a whole. The safety threshold to be adopted may also take into account the return period considered.

For the analysis and design of large and complex urban systems, two-dimensional (2D) surface models have been increasingly coupled to the sewer network models, which requires high resolution topographic information and a high computational capacity. Their coupling still poses several challenges both in research and in practical applications [24,25]. In the design of new developments, an adequate landscape reshaping should be carried out to prevent ponding and to convey the overland flow along pathways and streets, which can be easily represented in a 1D surface model. 1D/1D models have provided affordable and satisfactory responses to a number of applications [26–29]. Simplified methods have also been employed in the design of small to medium developments [9,11].

This work evaluates and discusses the effect of the cross-section shape, slope and roughness of the streets in compliance with the safety criteria described above. The evaluation is carried out based on analytical expressions, graphical results and a case study.

2. Methodology

The cross-section shape of most streets is similar to the left-hand profile of Figure 1. This composite section and other variants can be represented, in a simpler way, by an equivalent triangular–rectangular cross-section, as shown in the profile at the center of Figure 1.

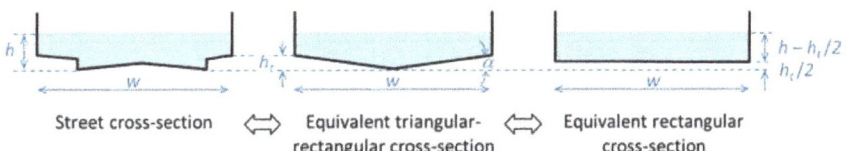

Figure 1. Street cross-section and equivalent cross-sections.

In this work, we will consider the flow in a triangular–rectangular cross-section, admitting a wide range of transversal slopes of the pavement on the triangular base (α): from the null slope, which corresponds to a rectangular section, to a sufficiently high transversal slope, where the cross-section becomes triangular. This cross-section will be considered generic and representative of most streets.

The Manning–Strickler formula, valid for fully rough turbulent water flows, will be used:

$$V/\left(K.\sqrt{S}\right) = Q/\left(K.\sqrt{S}\right) = R^{2/3} \tag{1}$$

where V is the cross-section's average velocity, Q is the discharge, K is the Strickler roughness value (the inverse of the Manning's roughness, n), S is the slope of the street bed, and R is the hydraulic radius.

Only those situations where the width of the cross-section is significantly greater than the flow depth ($W \gg h$) will be considered. From a practical point of view, this happens for virtually all streets. This means that if the cross-section is rectangular, the hydraulic radius corresponds to the flow depth ($R \approx h$), and if the cross-section is triangular, the hydraulic radius corresponds to half the flow depth ($R \approx h/2$).

Consider a triangular–rectangular cross-section with triangular section depth h_t and water depth h above h_t (the central profile of Figure 1). For $W \gg h$, it can be demonstrated that the flow rate and average flow velocity in that cross-section are equal to the flow rate and average velocity in a rectangular cross-section of the same width (W) and a depth equal to $h_{eq} = h - h_t/2$ (right-hand profile of Figure 1).

The concept of the equivalent rectangular cross-section will be used to quantify the effect of the variation of h_t (i.e., the cross-section shape), together with the effects of the variation of the roughness and of the longitudinal slope of the street, in the compliance with the safety criteria. Condition $h.V \leq limA$ will be designated by criterion A, and condition $h.V^2 \leq 1.23 \text{ m}^3\text{s}^{-2}$ by criterion B. We should also consider the variable X, given by $X = K.\sqrt{S}$.

In the next section, the variation of h, V and Q/W for the thresholds of criteria A and B (with $limA$ ranging between 0.22 and 0.5 m^2/s) are evaluated as a function of h_t and $X = K.\sqrt{S}$. The discussion is carried out based on the equations that satisfy each criterion, and on the graphical results of those variations, which are valid for the supercritical flow. As the results for the triangular–rectangular cross-section are iterative, regression equations are obtained to easily compute Q/W as a function of h_t and $K.\sqrt{S}$. Thus, if the flood flow on a street is known, these equations allow us to determine expeditiously the minimum width of the street that fulfills the safety criteria.

In Section 4, this procedure is applied and tested in a case study tailored to cover a diverse set of situations, including subcritical and unsteady flows. Sensitivity analyses are also carried out for the shape and width of the cross-section of the streets, and for the roughness of both the street surfaces and the sewers.

3. Variation of h, V and Q/W Meeting the Thresholds of Criteria A and B as a Function of h_t and $K.\sqrt{S}$

3.1. Equations for the Thresholds of Criteria A and B

Considering $W \gg h$, Table 1 presents the formulae for h, V and Q/W that verify each criterion for the rectangular, triangular and triangular–rectangular cross-sections (Equations (2)–(23)).

Criterion A ($h.V \leq limA$) has the particularity that, for both the rectangular cross-section and the triangular one, the maximum flow per unit of width does not depend on $K.\sqrt{S}$. According to Equations (4) and (7), the threshold for Q/W corresponds to the value of $limA$ for the rectangular cross-section, and to that of $limA/2$ for the triangular cross-section. For example, if $limA = 0.5$ m^2/s, then we have $Q/W \leq 0.5$ m^3/s/m for the rectangular cross-section and $Q/W \leq 0.25$ m^3/s/m for the triangular cross-section.

For both criteria, the calculation of the variables for the triangular–rectangular cross-section is iterative and only valid if one obtains $h \geq h_t$ (otherwise the cross-section is triangular). This check must be carried out before any iterative calculation, using Equation (8) for criterion A and Equation (19) for criterion B.

If the water height is higher than h_t, according to Equation (11) for criterion A, the threshold for Q/W varies between $limA$ and $limA/2$ (in m^3/s/m) depending on the quotient $(h - h_t/2)/h$. This quotient corresponds to the ratio of the hydraulic radius of the equivalent rectangular cross-section with the calculated value of h. If the check shows that $h < h_t$, the values of h and V must be calculated considering the triangular cross-section, using Equations (5) and (6) for criterion A, and Equations (16) and (17) for criterion B. Because the width of the flow at the triangular base will be shorter than the width of the composite cross-section (W) in the ratio h/h_t, the maximum flow rate provided by Equations (7) and (18) must be multiplied by h/h_t, resulting the Equation (12) for criterion A and Equation (23) for criterion B.

Table 1. Equations (adapted from [30]).

	Cross-section	Equation	#
Criterion A: $h.V \leq limA$ m²/s	Rectangular cross-section	$h \leq limA.\left(K.\sqrt{S}\right)^{-3/5}$	(2)
		$V \leq limA^{2/5}.\left(K.\sqrt{S}\right)^{3/5}$	(3)
		$Q/W \leq limA$	(4)
	Triangular cross-section	$h \leq limA^{3/5}.2^{2/5}.\left(K.\sqrt{S}\right)^{-3/5}$	(5)
		$V \leq limA^{2/5}.2^{-2/5}.\left(K.\sqrt{S}\right)^{3/5}$	(6)
		$Q/W \leq limA/2$	(7)
	Triangular–rectangular cross-section	Valid if $h_t \leq limA^{3/5}.2^{2/5}.\left(K.\sqrt{S}\right)^{-3/5}$	(8)
		Iterative calculation:	
		$h \leq limA.\left(K.\sqrt{S}\right)^{-1}.(h - h_t/2)^{-2/3}$	(9)
		$V \leq limA/h$	(10)
		$Q/W \leq limA.(h - h_t/2)/h$	(11)
		If Equation (8) is not valid (triangular flow):	
		$Q/W \leq limA^{8/5}.2^{-3/5}.\left(K.\sqrt{S}\right)^{-3/5}.h_t^{-1}$	(12)
Criterion B: $h.V^2 \leq limB$ m³s⁻²	Rectangular cross-section	$h \leq limB^{3/7}.\left(K.\sqrt{S}\right)^{-6/7}$	(13)
		$V \leq limB^{2/7}.\left(K.\sqrt{S}\right)^{3/7}$	(14)
		$Q/W \leq limB^{5/7}.\left(K.\sqrt{S}\right)^{-3/7}$	(15)
	Triangular cross-section	$h \leq limB^{3/7}.2^{4/7}.\left(K.\sqrt{S}\right)^{-6/7}$	(16)
		$V \leq limB^{2/7}.2^{-2/7}.\left(K.\sqrt{S}\right)^{3/7}$	(17)
		$Q/W \leq limB^{5/7}.2^{-5/7}.\left(K.\sqrt{S}\right)^{-3/7}$	(18)
	Triangular–rectangular cross-section	Valid if $h_t \leq limB^{3/7}.2^{4/7}.\left(K.\sqrt{S}\right)^{-6/7}$	(19)
		Iterative calculation:	
		$\left(K.\sqrt{S}\right)^2.h.(h - h_t/2)^{4/3} \leq limB$	(20)
		$V \leq (limB/h)^{1/2}$	(21)
		$Q/W \leq limB^{1/2}.(h - h_t/2).h^{-1/2}$	(22)
		If Equation (19) is not valid (triangular flow):	
		$Q/W \leq limB^{8/7}.2^{-1/7}.\left(K.\sqrt{S}\right)^{-9/7}.h_t^{-1}$	(23)
Simultaneity of criteria thresholds: $h.V = limA = 0.5$ m²/s and $h.V^2 = limB = 1.23$ m³s⁻²	Rectangular cross-section	$Q/W = limA = 0.5$ m³s⁻¹m⁻¹	(24)
		$K.\sqrt{S} = limB^{5/3}.limA^{-7/3} \approx 7.12$ m^{1/3}s⁻¹	(25)
		$h = limA^2/limB \approx 0.203$ m	(26)
		$V = limB/limA = 2.460$ m s⁻¹	(27)
	Triangular cross-section	$Q/W = limA/2$ m³s⁻¹m⁻¹	(28)
		$K.\sqrt{S} = 2^{2/3}.limB^{5/3}.limA^{-7/3} \approx 11.30$ m^{1/3}s⁻¹	(29)
		$h = limA^2/limB \approx 0.203$ m	(30)
		$V \approx limB/limA = 2.460$ m s⁻¹	(31)
	Triangular–rectangular cross-section	$Q/W = limA - limB/limA.h_t/2$	(32)
		$K.\sqrt{S} = limB/limA.(limA^2/limB - h_t/2)^{-2/3} = f(h_t)$	(33)
		$h = limA^2/limB \approx 0.203$ m	(34)
		$V = limB/limA = 2.460$ m s⁻¹	(35)

3.2. Analytical Expressions for the Separation of the Relevant Criterion

Equations (24) to (35) in Table 1 result from solving the system of equations for criteria A and B, and the Manning–Strickler formula, also considering $W \gg h$.

Because neither h_t or $X = K.\sqrt{S}$ are present in the equations for the calculation of h and V that verify both criteria, we conclude that the values of h and V that equate the thresholds of both criteria depend neither on the cross-section shape or on the roughness and slope of the street. According to Equations (26), (30) and (34) for h, and Equations (27), (31) and (35) for V, this happens for $h = limA^2/limB$ and $V = limB/limA$, which correspond to $h \approx 0.203$ m and $V = 0.46$ m/s for $limA = 0.5$ m^2/s and $limB = 1.23$ m^{3}s^{-2}.

However, $K.\sqrt{S}$ associated with this equality depends on the cross-section shape, as can be seen in the formula of Equation (33); it is $K.\sqrt{S} = limB^{5/3}.limA^{-7/3} \approx 7.12$ m$^{1/3}$s^{-1} for the rectangular cross-section (Equation (25)) and $K.\sqrt{S} = 2^{2/3}.limB^{5/3}.limA^{-7/3} \approx 11.30$ m$^{1/3}$s^{-1} for the triangular cross-section (Equation (29)).

3.3. Graphical Results and Discussion

Figure 2 shows the variation of Q/W, h and V as a function of h_t for different values of $K.\sqrt{S}$ and the adoption of criterion $h.V = 0.22$ m^2/s [18]. Figure 3 shows similar graphs, but now for the combination of criteria $h.V = 0.5$ m^2/s and $h.V^2 \leq 1.23$ m^{3}s^{-2} [9,19].

The results on the y-axis of all of the graphs of Figures 2 and 3 correspond to the particular case of the rectangular cross-section ($h_t = 0$ m). In all of the graphs, the dotted line separates the flow in a triangular cross-section from the flow in the full triangular–rectangular cross-section (Equations (8) and (19) for Q/W). In Figure 3, the area where criterion B is determinant is shaded, which, as previously seen, corresponds to $h \geq 0.203$ m, $V \leq 2.46$ m/s and Q/W below the straight line of Equation (32) if $h_t \leq 0.203$ m, or Q/W below the curve corresponding to $X = K.\sqrt{S} \approx 11.30$ m$^{1/3}$s^{-1} (Equation (29)) if $h_t \geq 0.203$ m.

For both Figures 2 and 3, the upper graphs show and allow us to quantify a significant reduction of Q/W with the increase of h_t and $X = K.\sqrt{S}$. For the case of the rectangular cross-section ($h_t = 0$), Q/W does not depend on $X = K.\sqrt{S}$ for condition A (as discussed earlier), but Q/W can be significantly reduced with the increase of $X = K.\sqrt{S}$ due to criterion B (Figure 3). For example, for $h.V = 0.5$ m^2/s and a rectangular cross-section, we have $Q/W = 0.5$ m^3/s/m for all values of $K.\sqrt{S} \leq 7.13$ m$^{1/3}$s^{-1} ($S \leq 0.9$% for $K = 75$ m$^{1/3}$s^{-1}). However, for $K.\sqrt{S} = 15$ m$^{1/3}$s^{-1} ($S = 4$% for $K = 75$ m$^{1/3}$s^{-1}), Q/W reduces to 0.36 m^3/s/m for a rectangular cross-section due to criterion B, and for 0.18 m^3/s/m for a triangular–rectangular cross-section with $h_t = 0.20$ m (72% and 35% of 0.5 m^3/s/m, respectively).

These results show that for steep streets, the maximum safety flows occur for small flow depths, usually below the sidewalks, and therefore are very sensitive to the resolution and quality of the topographic data. Thus, if high-resolution data are available and their treatment allows for the representation of the lower elevations next to the curbs, the results can be significantly more conservative than if this rigor is not considered.

However, streets with relatively constant average characteristics over tens of meters comprise extensive topographic variability in very detailed descriptions, due to various singularities (e.g., crossings, crosswalks, speed bumps, depressions near gutters and sinks). The detailed modelling of this variability will require great efforts in automatic calculation and result analysis, and it is very questionable whether the whole street design should be conditioned by the strict application of criteria A and B to small singularities, such as depressions next to the inlets. Furthermore, the use of high-resolution digital elevation models will introduce a large amount of information that will be treated as noise, aggravating data handling and computational requirements. The scaling effects of topographic variables and the advantages and disadvantages of using high-resolution digital elevation models for different purposes are discussed in more depth by [28,31].

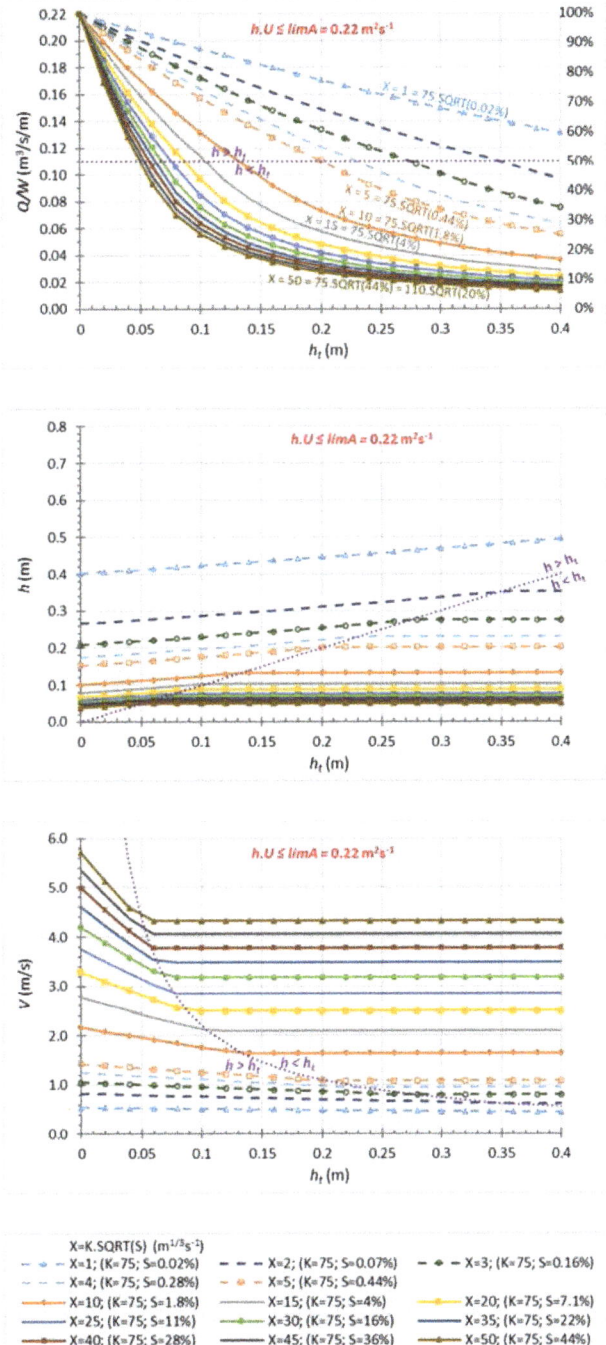

Figure 2. Variations of h, V and Q/W fulfilling the thresholds of criterion $h.V = 0.22$ m^{2}s^{-1}, as a function of h_t and $K.\sqrt{S}$.

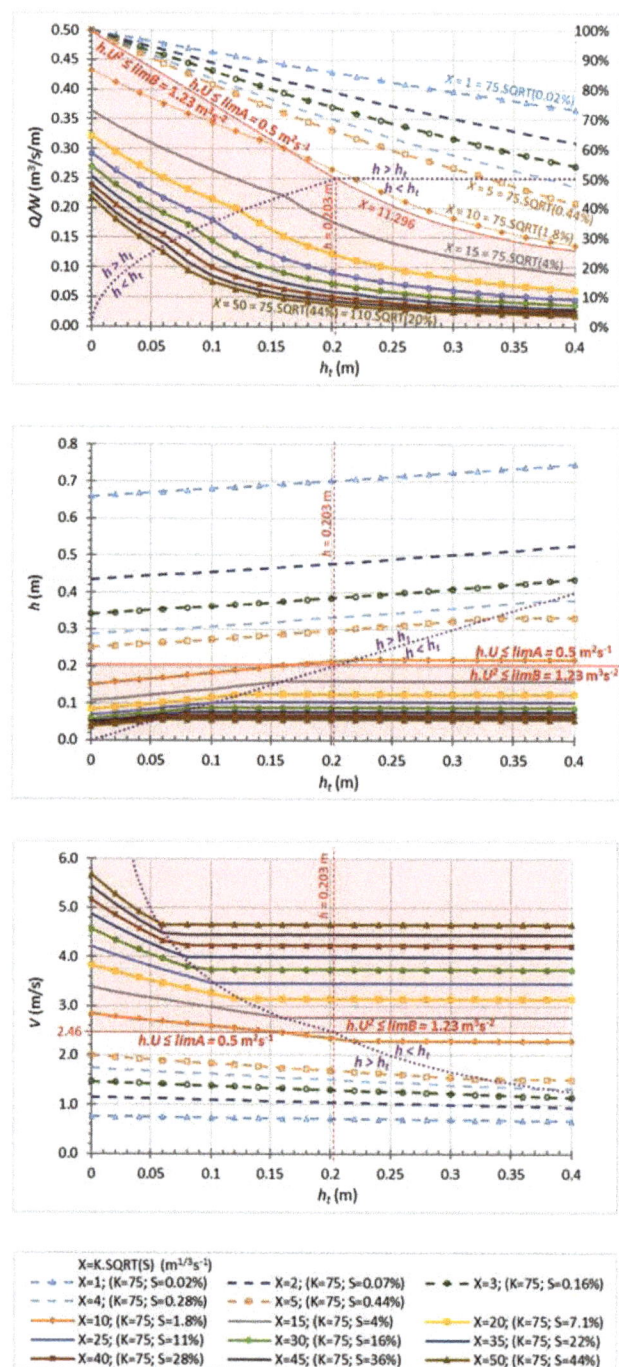

Figure 3. Variations of h, V and Q/W fulfilling the thresholds of both criteria $h.V = 0.5$ m^{2}s^{-1} and $h.V^2 = 1.23$ m^{3}s^{-2}, as a function of h_t and $K.\sqrt{S}$.

3.4. Regression Equations for Iterative Q/W Values and the Proposal of an Expedited Method

The regression equations for Q/W in the triangular–rectangular cross-section were obtained as a function of h_t and $X = K.\sqrt{S}$. Equation (36) is valid for criterion A, considering the coefficients presented in Table 2 (for $limA$ values equal to 0.1, 0.22, 0.3, 0.4 and 0.5 m²/s, or as a function of $limA$).

$$Q/W = 1/\left(a_2.X^2.h_t^3 + (b_1.X + b_0).h_t^2 + \left(c_2.X^2 + c_1.X + c_0\right).h_t + 1/limA\right) \qquad (36)$$

Table 2. Coefficients of the Q/L regression equation for criterion A.

$limA =$	0.10 m²/s	0.22 m²/s	0.3 m²/s	0.4 m²/s	0.5 m²/s	0.22 m²/s ≤ $limA$ ≤ 0.5 m²/s
$a_2 =$	16.93	1.77	0.73	0.32	0.17	$a_2 = 0.023\, limA^{-2.867}$
$b_1 =$	31.26	5.46	2.75	1.45	0.89	$b_1 = 0.191\, limA^{-2.214}$
$b_0 =$	−8.28	−2.28	−1.37	−0.86	−0.59	$b_0 = -0.191\, limA^{-1.637}$
$c_2 =$	−0.0763	−0.0207	−0.0124	−0.0077	−0.0053	$c_2 = -0.0017 \cdot limA^{-1.652}$
$c_1 =$	7.3	2.10	0.81	0.81	0.57	$c_1 = 0.191 \cdot limA^{-1.582}$
$c_0 =$	15.38	4.61	1.85	1.85	1.32	$c_0 = 0.457\, limA^{-1.527}$

Equation (37) presents the regression equation for criterion B.

$$Q/W = 1/\left(0.07.X^2.h_t^2 + \left(0.011.X^2 + 0.36.X - 0.07\right).h_t + 0.86.X^{0.43}\right) \qquad (37)$$

Equations (36) and (37) are valid for $h_t \leq 0.5$ m and $1\ \mathrm{m^{1/3}s^{-1}} \leq X = K.\sqrt{S} \leq 50\ \mathrm{m^{1/3}s^{-1}}$.

For the set of points shown in Figures 2 and 3, the minimum and maximum errors obtained for all of the regressions based on Equations (36) and (37) and Table 2 were −3.1% and +3.2%, respectively. Figure 4 illustrates the errors obtained in the Q/W estimates for $limA = 0.22$ m²/s and $limA = 0.5$ m²/s, and for the combination of $limA = 0.5$ m²/s and $limB = 1.23$ m³s⁻². In each graph, the flat area of null errors corresponds to the cases where $h \leq h_t$.

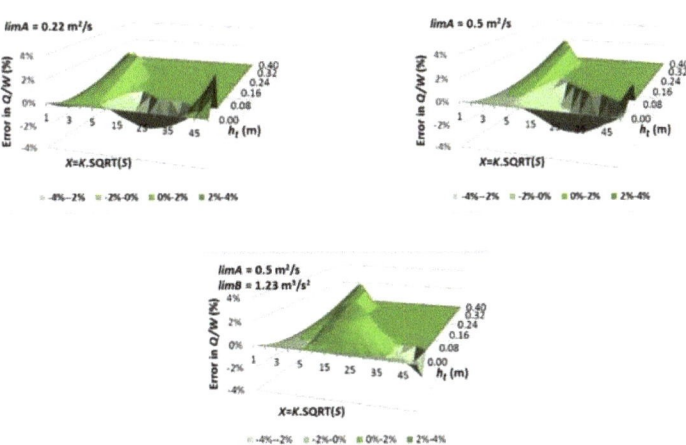

Figure 4. Errors in the estimation of Q/W for $limA = 0.22$ m²/s and $limA = 0.5$ m²/s, and for the combination of $limA = 0.5$ m²/s and $limB = 1.23$ m³s⁻².

Thus, Q/W calculation can be easily implemented in a spreadsheet considering:
- Equation (36) or Equation (12), depending on Equation (8) for criterion A;

- Equation (37) or Equation (23), according to Equation (19) for criterion B.

With an estimate of the maximum flood flow rates obtained from more or less simplified models, these equations will provide the minimum street widths ($Wmin$) to meet safety criteria under supercritical flow conditions.

4. Case Study

The approach proposed was applied to the dual drainage system represented in Figure 5, which was modelled using SWMM 5.1.011 [32]. The case study was built to assess the accumulated effect along a street of simplifications in the street cross-section, covering a diverse set of non-uniform unsteady flows, including subcritical flows. Simulations were run for a 100-year return period Desbordes hyetograph (a hyetograph composed of an increasing triangle, followed by a central peak and a decreasing triangle) and a recording interval of 1 minute. The street has 30 reaches of 50 m each, with different slopes and values of $X = K.\sqrt{S}$, as indicated in the table of Figure 5 (the Manning–Strickler coefficient was considered the same for all of the reaches of streets and sewers, $K = 1/0.013 \approx 77 \text{ m}^{1/3}\text{s}^{-1}$). The changes in the slope of the reaches lead to alternating situations of supercritical and subcritical flows, and included a situation of negative slope in reaches 26 and 27. The cross-section of all of the street reaches was considered irregular, as shown in Figure 6b, with a maximum width of $W = 15$ m. It can be approximated by a triangular–rectangular composite cross-section with $h_t = 0.12$ m.

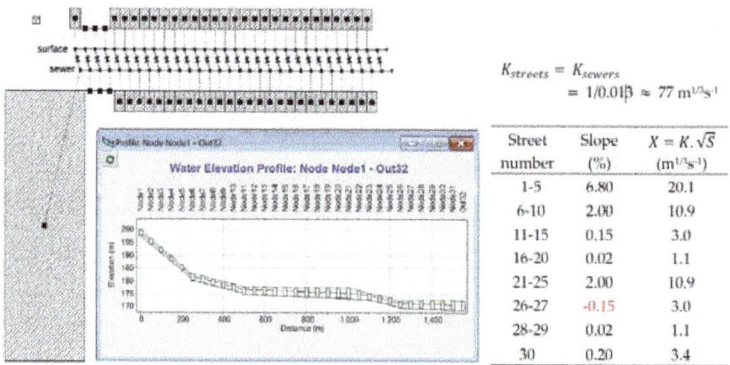

Figure 5. Map and longitudinal profile of the dual drainage system.

The sewer system receives runoff directly from an upstream 52-ha subcatchment, and from 26 subcatchments along the street with 0.5 ha each, representing the inflows from other sewers. The surface system receives runoff from 27 subcatchments along the street with 0.5 ha each. The flow transfers between the major and the minor system are ensured through 30 orifices with a diameter of 0.5 m each, representing the existence of approximately six inlets per reach. The upstream orifice has a larger diameter (2 m) in order to ensure the rapid balance between flows in the major system and in the minor system at the upstream boundary, and thus to guarantee numerical stability.

Figure 6c compares the maximum values of the flood flow (Q) recorded by SWMM in each reach for the following three street cross-sections: the composite section ("Composite"), the triangular–rectangular section ("TriangRect"), and the 15 m wide rectangular section ("Rect15"). The graph shows that the maximum flood flows of the triangular–rectangular cross-section are practically the same as those of the composite cross-section until about reach 19. Downstream of reach 19, the flow rate of the triangular–rectangular cross-section becomes slightly higher than that of the composite cross-section, increasing again a little further downstream of reach 27 (the reach with a negative slope). This very slight and progressive increase in the flow downstream is due to the fact that the area of the base of the triangular–rectangular cross-section is larger than the area of base of the composite cross-section. For the 15 m-wide rectangular cross-section, a similar behavior is observed, albeit with a more pronounced deviation downstream of reach 19. Some flow deviation is also observed in reaches 7 and 8 for the rectangular cross-section, which is recovered downstream.

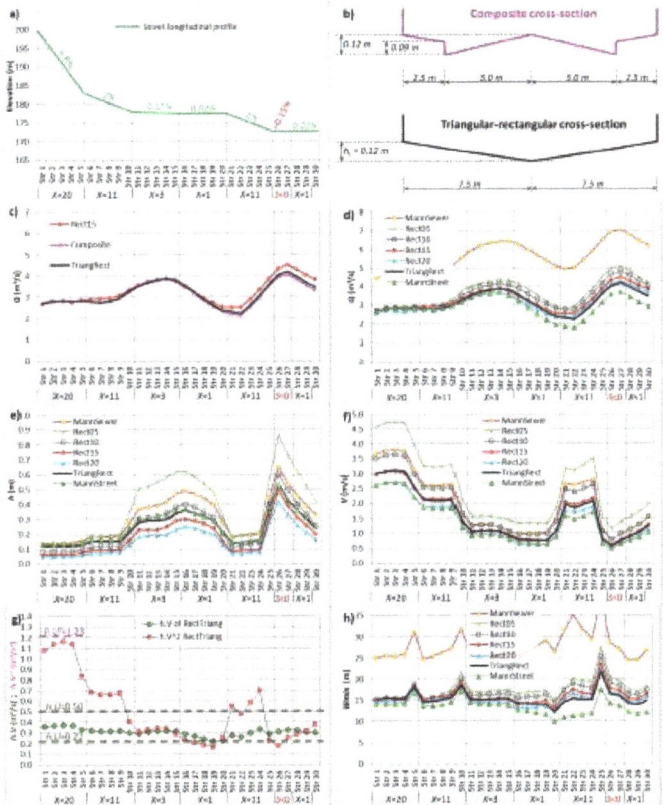

Figure 6. Results from the case study: (**a**) Street longitudinal profile; (**b**) Dimensions of used composite and triangular-rectangular cross-sections; (**c**,**d**) flood maximum flows along the street; (**e**) flood maximum depths along the street; (**f**) flood maximum velocities along the street; (**g**) maximum values of criteria A and B along the street; (**h**) minimum street widths calculated by the expedited method as a function of the flow rates (actual width is 15 m).

Figure 6d–f compare the maximum values of Q, h and V, respectively, for simulations using the triangular–rectangular cross-section ("TriangRect") and the rectangular cross-sections with widths of 5, 10, 15 and 20 m ("Rect05", "Rect10", "Rect15", "Rect20"). These Figures also show the results for the following two additional scenarios, considering the triangular–rectangular cross-section: a 20% decrease of the Strickler coefficient (K) in all streets ("MannStreet"), and a 20% decrease of K in all sewers ("MannSewer") (from $K = 1/0.013 \approx 77 \text{ m}^{1/3}\text{s}^{-1}$ to $K = 1/0.01625 \approx 61.77 \text{ m}^{1/3}\text{s}^{-1}$ for both cases).

For the purpose of this manuscript, it was ensured that $h > h_t = 0.12$ m for the entire length of the street with a triangular–rectangular cross-section, as shown in Figure 6e. In the reaches with a negative slope, the flood depths reach 0.5 m, which would clearly necessitate the resizing of the downstream sewers if this were a real case. As expected, large differences in the h and V values are observed for the different cross-sections.

Figure 6d shows that for rectangular cross-sections with different widths from the triangular–rectangular cross-section (15 m), the results of Q start to differ because of the first reaches, and that this difference is exacerbated downstream. The difference in the results increases with the difference in the width of the rectangular cross-section in relation to that of the triangular–rectangular cross-section.

The results in Figure 6d highlight that the roughness uncertainty has a more significant effect on the calculation of flood flows than the uncertainty in the shape of the cross-section. The 20% reduction in K of the streets ("MannStreet") leads to a significant reduction of the flood flows downstream (Figure 6d) due to the reduction of the flood velocities (Figure 6f). The 20% reduction in K of the sewers ("MannSewer") leads to increases of the flood flows that are much higher than all of the other scenarios. As with roughness, other factors that lead to changes in the transport capacity of the sewers, such as the sewer's slope or diameter, also have a very significant influence on flood flows and, thus, on the fulfillment of the safety criteria.

Figure 6g shows the maximum values of $h.V$ and $h.V^2$ obtained in the successive reaches for the triangular–rectangular cross-section. All of the values of $h.V$ vary between circa 0.22 and 0.4 m^2/s. In the upstream reaches, the $h.V^2$ values are close to but do not reach the 1.23 m^3/s^2 threshold.

Based on the $h.V$ values presented in Figure 6g and the methodology presented in Section 3.4 of this manuscript, the minimum widths of the reaches ($Wmin$) were calculated as a function of the flow rates obtained for each simulation (Figure 6h). Figure 6h shows that, for the triangular–rectangular cross-section, the $Wmin$ results are close to 15 m in most reaches, both in supercritical and subcritical flows. However, in reaches 5, 10 and 25, the $Wmin$ values are significantly greater than 15 m. These three reaches have in common the fact that they are regime transition reaches, being midway between a significant increase of h and a significant reduction of V. In reach 5, there is a transition from a supercritical to a slower supercritical flow, and in reaches 10 and 25 there is the transition from supercritical to subcritical flows. A slight oversizing of $Wmin$ is also observed in reaches 26 and 27, where the slope is contrary to the flow direction.

In practice, flow transitions that lead to hydraulic jumps should be avoided. In cases where there is a significant reduction in the street slope, the smoothing of the street slope reduction must be accompanied by a smoothing of the sewer transport capacity reduction in order to avoid an abrupt increase of floods downstream. At the crossing of a steep street with a flat perpendicular street, designing the street cross-section closer to the rectangular cross-section and capturing the runoff upstream of the crossing may be a solution.

In reach 20, the value of $Wmin$ for the triangular–rectangular cross-section is 12.5 m, about 17% less than 15 m. Reach 20 has a reduced slope, and is immediately upstream of a reach with a high slope. The passage through the critical flow in the transition must be the source of the lower value of $Wmin$. In reaches 16–19 and 28–29, $Wmin$ is slightly less than 15 m (in less than 5%). Part of this difference in the results is probably explained by the rounding to two decimal places of the results of h and V. For the rectangular cross-sections, the $Wmin$ results reflect the flow deviations discussed with respect to Figure 6d.

5. Conclusions

The street's cross-sectional profile plays a relevant role in meeting flood safety criteria, which is quantified and graphically described in this work. For example, a street with a 1.5% slope, $K \approx 75$ m$^{1/3}$s^{-1} and a rectangular cross-section complies with the threshold $h.V = 0.22$ m^2/s for twice the flow of a street with the same width but with a conventional cross-section shape and $h_t = 0.14$ m. The flow will be four times greater for a 15% street slope (Figure 2).

However, the uncertainty associated with the data, models and results is generally significant, and should be taken into account in the compromise between detailing and simplifying the cross-sectional profile. The results of this manuscript confirm that, for various studies and projects of new developments, the consideration of equivalent triangular–

rectangular cross-sections can lead to an adequate compromise between the quality of the data available and the calculation effort.

The results highlight that, on steeper streets, one way to increase the flood flow that meets the safety criteria is to approach the street cross-section to the rectangular section and catch the runoff through grated gutters in the full width of the street.

The approach presented in 3.4 can be used for the expedited verification of a street's compliance with flood safety criteria. This approach allows for the calculation of the minimum width of the street ($Wmin$) based solely on the flood flow (Q). It is valid for most flows, except for rapidly varied flows.

The uncertainty of the street roughness can have a more significant effect on the flood flows, and thus on the fulfillment of the safety criteria, than the uncertainty of the street cross-section.

Finally, the case study highlighted that flood flows can vary exceptionally along the streets depending on the roughness, slope and diameter of the sewer system. During floods, surcharge and increased local head losses can significantly increase the average roughness of the sewer network, with high uncertainty. The existing deposits in the sewers and the dragging effects of mud, stones and debris during floods— which are generally calibrated for frequent storms—also increase the model's uncertainty. The practical difficulty of representing and calibrating all of the flow transfer elements between the surface and the sewer models, and the current knowledge limitations in the parameterization of these devices, also contribute to raise uncertainty. These results emphasize that the effort in detailing the street's cross-section must be framed with the objectives of the work and all of the sources of uncertainty.

Author Contributions: Conceptualization, L.M.D.; methodology, L.M.D. and R.F.d.C.; software, L.M.D.; validation, L.M.D. and R.F.d.C.; formal analysis, L.M.D. and R.F.d.C.; investigation, L.M.D. and R.F.d.C.; writing—original draft preparation, L.M.D.; writing—review and editing, L.M.D. and R.F.d.C. All authors have read and agreed to the published version of the manuscript.

Funding: This work was co-funded by the European Regional Development Fund (FEDER), under programs POR Lisboa2020 and CrescAlgarve2020, through Project SINERGEA (ANI Project n. 33595); and FCT (Portuguese Foundation for Science and Technology) through the Project UIDB /04292/2020, financed by MEC (Portuguese Ministry of Education and Science) and FSE (European Social Fund), under the programs POPH/QREN (Human Potential Operational Programme from National Strategic Reference Framework) and POCH (Human Capital Operational Programme) from Portugal2020.

Institutional Review Board Statement: Not applicable.

Data Availability Statement: Not applicable.

Acknowledgments: Not applicable.

Conflicts of Interest: The authors declare no conflict of interest.

References

1. David, L.M.; Matos, J.S. Combined sewer overflow emissions to bathing waters in Portugal. How to reduce in densely urbanised areas? *Water Sci. Technol.* **2005**, *52*, 183–190. [CrossRef] [PubMed]
2. Campisano, A.; Cabot Ple, J.; Muschalla, D.; Pleau, M.; Vanrolleghem, P.A. Potential and limitations of modern equipment for real time control of urban wastewater systems. *Urban Water J.* **2013**, *10*, 300–311. [CrossRef]
3. Lund, N.S.V.; Falk, A.K.V.; Borup, M.; Madsen, H.; Mikkelsen, P.S. Model predictive control of urban drainage systems: A review and perspective towards smart real-time water management. *Crit. Rev. Environ. Sci. Technol.* **2018**, *48*, 279–339. [CrossRef]
4. Demuzere, M.; Orru, K.; Heidrich, O.; Olazabal, E.; Geneletti, D.; Orru, H.; Bhave, A.G.; Mittal, N.; Feliu, E.; Faehnlej, M. Mitigating and adapting to climate change: Multi-functional and multi-scale assessment of green urban infrastructure. *J. Environ. Manag.* **2014**, *146*, 107–115. [CrossRef]
5. Sörensen, J.; Persson, A.; Sternudd, C.; Aspegren, H.; Nilsson, J.; Nordström, J.; Jönsson, K.; Mottaghi, M.; Becker, P.; Pilesjö, P.; et al. Re-thinking urban flood management-time for a regime shift. *Water* **2016**, *8*, 332. [CrossRef]
6. Stovin, V.R.; Moore, S.L.; Wall, M.; Ashley, R.M. The potential to retrofit sustainable drainage systems to address combined sewer overflow discharges in the Thames Tideway catchment. *Water Environ. J.* **2013**, *27*, 216–228. [CrossRef]

7. Barbosa, A.E.; Fernandes, J.N.; David, L.M. Key issues for sustainable urban stormwater management. *Water Res.* **2012**, *46*, 6787–6798. [CrossRef]
8. Hamel, P.; Daly, E.; Fletcher, T.D. Source-control stormwater management for mitigating the impacts of urbanisation on baseflow: A review. *J. Hydrol.* **2013**, *485*, 201–211. [CrossRef]
9. Balmforth, D.; Digman, C.; Kellagher, R.; Butler, D. *Designing for Exceedance in Urban Drainage–Good Practice*; CIRIA C635: London, UK, 2006.
10. Section 3: Stormwater Management. In *City of Lethbridge Design Standards 2016 Edition*; City of Lethbridge: Lethbridge, AB, Canada, 2016; Available online: https://www.lethbridge.ca/Doing-Business/Planning-Development/Urban-Construction-Right-of-Way-Coordination/Documents/City%20of%20Lethbridge%202016%20Standards.pdf (accessed on 3 July 2019).
11. Melbourne Water. Standards and Specifications. Design. General Guidance, Australia, 2015, Chapter General Approach to Drainage Systems and Chapter Floodway Safety Criteria. Available online: https://www.melbournewater.com.au/planning-and-building/developer-guides-and-resources/standards-and-specifications (accessed on 3 July 2019).
12. Hammond, M.J.; Chen, A.S.; Djordjević, S.; Butler, D.; Mark, O. Urban flood impact assessment: A state-of-the-art review. *Urban Water J.* **2015**, *12*, 14–29. [CrossRef]
13. Nahiduzzaman, K.M.; Aldosary, A.S.; Rahman, M.T. Flood induced vulnerability in strategic plan making process of Riyadh city. *Habitat Int.* **2015**, *49*, 375–385. [CrossRef]
14. Salinas-Rodriguez, C.; Gersonius, B.; Zevenbergen, C.; Serrano, D.; Ashley, R. A Semi Risk-Based Approach for Managing Urban Drainage Systems under Extreme Rainfall. *Water* **2018**, *10*, 384. [CrossRef]
15. Shand, T.D.; Cox, R.J.; Blacka, M.J. Development of Appropriate Criteria for the Safety and Stability of Persons and Vehicles in Floods. In Proceedings of the 34th IAHR Conference, Brisbane, Australia, 26 June–1 July 2011; p. 9.
16. Russo, B.; Gómez, M.; Macchione, F. Pedestrian hazard criteria for flooded urban areas. *Nat. Hazards* **2013**, *69*, 251. [CrossRef]
17. Chanson, H.; Brown, R. Stability of Individuals during Urban Inundations: What Should We Learn from Field Observations? *Geosciences* **2018**, *8*, 341. [CrossRef]
18. Martínez-Gomariz, E.; Gómez, M.; Russo, B. Experimental study of the stability of pedestrians exposed to urban pluvial flooding. *Nat. Hazards* **2016**, *82*, 1259. [CrossRef]
19. Nanía, L.; Gomez, M.; Dolz, J. Analysis of risk associated to the urban runoff. Case study: City of Mendoza, Argentina. In *Global Solutions for Urban Drainage, Proceedings of the 9th ICUD*; Strecker, E.W., Huber, W.C., Eds.; ASCE: Reston, VA, USA, 2002. [CrossRef]
20. Xia, J.; Falconer, R.A.; Wang, Y.; Xiao, X. New criterion for the stability of a human body in floodwaters. *J. Hydraul. Res.* **2014**, *52*, 93–104. [CrossRef]
21. Shand, T.D.; Cox, R.J.; Blacka, M.J.; Smiths, G.P. *Australian Rainfall and Runoff. Revision Projects. Project 10: Appropriate Safety Criteria for Vehicles. Literature Review*; AR&R Report Number P10/S2/020; Water Research Laboratory, The University of New South Wales: Sydney, NSW, Australia, February 2011.
22. Xia, J.; Falconer, R.A.; Xiao, X.; Wang, Y. Criterion of vehicle stability in floodwaters based on theoretical and experimental studies. *Nat. Hazards* **2014**, *70*, 1619–1630. [CrossRef]
23. Martínez-Gomariz, E.; Gómez, M.; Russo, B.; Djordjević, S. A new experiments-based methodology to define the stability threshold for any vehicle exposed to flooding. *Urban Water J.* **2017**, *14*, 930–939. [CrossRef]
24. Krebs, G.; Kokkonen, T.; Valtanen, M.; Setälä, H.; Koivusalo, H. Spatial resolution considerations for urban hydrological modelling. *J. Hydrol.* **2014**, *512*, 482–497. [CrossRef]
25. Carvalho, R.F.; Lopes, P.; Leandro, J.; David, L.M. Numerical Research of Flows into Gullies with Different Outlet Locations. *Water* **2019**, *11*, 794. [CrossRef]
26. David, L.M.; Carvalho, R.F.; Isidro, R.; Sobral, M. Stormwater control of new urban developments–Planning and modelling the Penalva system. In Proceedings of the 12th ICUD, Porto Alegre, Brazil, 11–16 September 2011; p. 8.
27. Jahanbazi, M.; Egger, U. Application and comparison of two different dual drainage models to assess urban flooding. *Urban Water J.* **2014**, *11*, 584–595. [CrossRef]
28. Leandro, J.; Djordjević, S.; Chen, A.S.; Savić, D.A.; Stanić, M. Calibration of a 1D/1D urban flood model using 1D/2D model results in the absence of field data. *Water Sci. Technol.* **2011**, *64*, 1016–1024. [CrossRef] [PubMed]
29. Nanía, L.; León, A.; García, M. Hydrologic-Hydraulic Model for Simulating Dual Drainage and Flooding in Urban Areas: Application to a Catchment in the Metropolitan Area of Chicago. *J. Hydrol. Eng.* **2015**, *20*, 04014071. [CrossRef]
30. David, L.M. Projetar para inundações–efeito do perfil, rugosidade e declive das ruas nos critérios de segurança. *Águas Resíduos* **2019**, *IV.4*, 37–47. (In Portuguese) [CrossRef]
31. Chang, K.T.; Merghadi, A.; Yunus, A.P.; Pham, B.T.; Dou, J. Evaluating scale effects of topographic variables in landslide susceptibility models using GIS-based machine learning techniques. *Sci. Rep.* **2019**, *9*, 12296. [CrossRef] [PubMed]
32. Rossman, L.A. *Storm Water Management Model User's Manual Version 5.1*; U.S. Environmental Protection Agency: Washington, DC, USA, 2015; EPA/600/R-14/413b. Revised September 2015.

Article

A New Strategy for Sponge City Construction of Urban Roads: Combining the Traditional Functions with Landscape and Drainage

Chengyao Wei [1,2], Jin Wang [3], Peirong Li [4], Bingdang Wu [1], Hanhan Liu [4], Yongbo Jiang [1] and Tianyin Huang [1,2,*]

1. School of Environmental Science and Engineering, Suzhou University of Science and Technology, Suzhou 215009, China; wcydxszl@163.com (C.W.); wubingdang@163.com (B.W.); SZjiangyongbo@163.com (Y.J.)
2. Key Laboratory of Suzhou Sponge City Technology, Suzhou 215002, China
3. Housing and Urban-Rural Construction Bureau of Suzhou, Suzhou 215002, China; vae2116@163.com
4. Suzhou Tongke Engineering Consulting Co., Ltd., Suzhou 215000, China; lprkycn@126.com (P.L.); yqdhhqldy@163.com (H.L.)
* Correspondence: huangtianyin111@163.com

Citation: Wei, C.; Wang, J.; Li, P.; Wu, B.; Liu, H.; Jiang, Y.; Huang, T. A New Strategy for Sponge City Construction of Urban Roads: Combining the Traditional Functions with Landscape and Drainage. *Water* **2021**, *13*, 3469. https://doi.org/10.3390/w13233469

Academic Editor: Francesco De Paola

Received: 13 November 2021
Accepted: 1 December 2021
Published: 6 December 2021

Publisher's Note: MDPI stays neutral with regard to jurisdictional claims in published maps and institutional affiliations.

Copyright: © 2021 by the authors. Licensee MDPI, Basel, Switzerland. This article is an open access article distributed under the terms and conditions of the Creative Commons Attribution (CC BY) license (https://creativecommons.org/licenses/by/4.0/).

Abstract: Urban roads play a key role in sponge city construction, especially because of their drainage functions. However, efficient methods to enhance their drainage performance are still lacking. Here, we propose a new strategy to combine roads, green spaces, and the drainage system. Generally, by considering the organization of the runoff and the construction of the drainage system (including sponge city facilities) as the core of the strategy, the drainage and traffic functions were combined. This new strategy was implemented in a pilot study of road reconstruction conducted in Zhangjiagang, Suzhou, China. Steel slag was used in the structural layers to enhance the water permeability of the pavement and the removal of runoff pollutants. The combined effects of this system and of the ribbon biological retention zone, allowed achieving an average removal rate of suspended solids, a chemical oxygen demand, a removal of total nitrogen and total phosphorus of 71.60%, 78.35%, 63.93%, and 49.47%; in contrast, a traditional road could not perform as well. Furthermore, the volume control rate of the annual runoff met the construction requirements (70%). The results of the present study indicate that, combining the traditional basic functions of roads with those of landscape and drainage might be a promising strategy for sponge city construction of urban road.

Keywords: urban water management; drainage function; permeable pavement; biological retention

1. Introduction

With the rapid process of urbanization and the increase of impervious surfaces in urban areas, great changes have taken place in the hydrological environment [1]. In recent years, urban point source pollution has been relatively controlled through continuous treatments and restoration. Urban non-point source pollution has gradually become a major problem for the improvement of the water environment, as it can be transported into flood flows from drainage systems [2,3]. Urban non-point source pollution derives from a complex dynamic process, which mainly refers to the scouring and carrying of surface pollutants by rainfall. Thus, the control of urban non-point source pollution is challenging. In order to control water pollution and improve the quality of the water environment and the utilization of rainwater, many countries have developed different concepts and technical measures according to their own conditions, such as measures based on low-impact development, best management practices, sustainable urban drainage systems, water-sensitive urban designs [4–6]. Considering this background, the sponge city (SC), a concept involving a series of innovative ideas, parameters, and methods, was proposed [7]. The SC emphasizes the usage of engineering and non-engineering measures to realize the accumulation, penetration, and purification of rainwater in urban areas.

Since the proposal of the SC concept, construction methods and models have been researched and actively explored in many cities [8,9]. This has allowed the integration of the concept of SC into many engineering construction fields such as architecture, park and green spaces, water systems, and urban roads [10]. As a special kind of land-use and an important part of the city, urban roads are the main transportation space and the embodiment of the urban landscape. The SC construction of urban roads (SCCUR) can not only achieve the goal of rainwater control, but also build a good platform for technology application and concept publicity [7,8]. However, since priority is given to urban roads' traffic function without considering the relationships between road, landscape, drainage, and other related environmental aspects, the SCCUR concept is rarely taken into account in practical engineering application. Thus, new models and technical strategies are needed for SCCUR.

In the traditional drainage model of urban roads, road runoff rainwater rapidly flows to the gutter inlet and then is discharged by a neighboring rainwater pipe. During this process, the road takes less time to discharge the runoff rainwater. As a result, substances such as suspended solids (SS), total nitrogen (TN), and total phosphorus (TP) may be transferred into the natural water body by runoff, which may cause serious pollution of natural water body and increase the chemical oxygen demand (COD). Thus, the traditional drainage model of urban roads had some disadvantages as reported below.

(1) Serious runoff pollution. As a special area carrying traffic, the runoff pollution of urban roads is serious [11]. With the scouring effect of rainfall, surface pollutants are carried into the water body. Using SC facilities to collect the runoff rainwater and remove the pollutants in rainwater is a critical approach for regional water environment improvement. (2) Large drainage pressure. The density of urban road networks increases rapidly with the development of urbanization, resulting in high watermark and poor soil permeability. After raining, a surface runoff will quickly form and enter rainwater pipes. Areas with an old rainwater system will face a large drainage pressure, which may lead to poor drainage, road ponding, and waterlogging. Using SC facilities is possible to collect the source runoff rainwater and extend its discharge time. This strategy can not only alleviate the drainage pressure on a rainwater drainage system, but also facilitate the natural water replenishment of plants [10]. (3) Poor traffic experience. Urban roads include motor lanes, non-motorized lanes, and sidewalks, which are usually impermeable [8]. In the traditional drainage system, rainwater inlets are used to collect road rainwater, and the interval between the rainwater inlets is generally 30–50 m. After long usage, the pavement surface of the sidewalks becomes loosen, and slight deformations appear in non-motor vehicle lanes, which causes ponding on roads. Travel on sidewalks and non-motorized lanes becomes difficult in rainy days.

With these disadvantages, many difficulties appear for the implementation of SCCUR. (1) Limitation of green space. According to design standards [12] and statistical analyses of the greening rate of multiple existing roads, the current greening rate of urban roads is generally less than 30%. The green space of urban roads is composed of a middle zone, a side zone, and a certain space on its two sides. Due to their particular function, belt-like urban roads are greatly limited in green space. Thus, the effective use of limited green space is a difficulty for SCCUR. (2) Establishing a drainage system. In the process of implementing SCCUR, drainage facilities need to be included in multiple SC facilities to ensure drainage security. For facilities with regulation and storage functions, overflow systems are especially needed. The overflow system generally includes rainwater collection pipes in the drainage layer of the facilities, overflow inlets, and rainwater pipes connected with the conventional drainage system. In actual engineering projects, the drainage safety should be ensured, and two sets of drainage systems should be avoided, which requires the establishment of a complete system including an anterior collection system and a back-end discharge system. (3) Professional coordination. SCCUR mainly uses various technical SC facilities to collect the runoffs and reduce pollutants. These SC facilities are mainly constructed in roads and green space, but roads and greening are not the main scope of drainage. Thus, these three features of road, landscape, and drainage need to be highly integrated during the actual process of SC construction.

Based on the above disadvantages, a new strategy for SCCUR with a drainage system as the core is proposed in the present study. Suitable SC facilities were constructed on a road, which was reconstructed using the proposed strategy. By analyzing the construction process and related achievements, the feasibility and key points of this strategy are discussed.

2. Materials and Methods

2.1. Project Overview

The present work was conducted on Liangfeng Road in Zhangjiagang, Suzhou, as a pilot project (Figure 1). The road was being remodeled for the improvement of streets and lanes in the old district within the construction of SC with the specific aim to improve road, landscape, and drainage. The length and width of the reconstructed site were about 400 m and 27–32 m, respectively, and a typical section is shown in Figure 1. The whole road space includes two-way motor lanes, non-maneuverable lanes, and sidewalks. The green space was scarce and concentrated at the end of the road. The main problems of this road were damage of the pavement, low green coverage, disorder of the drainage system.

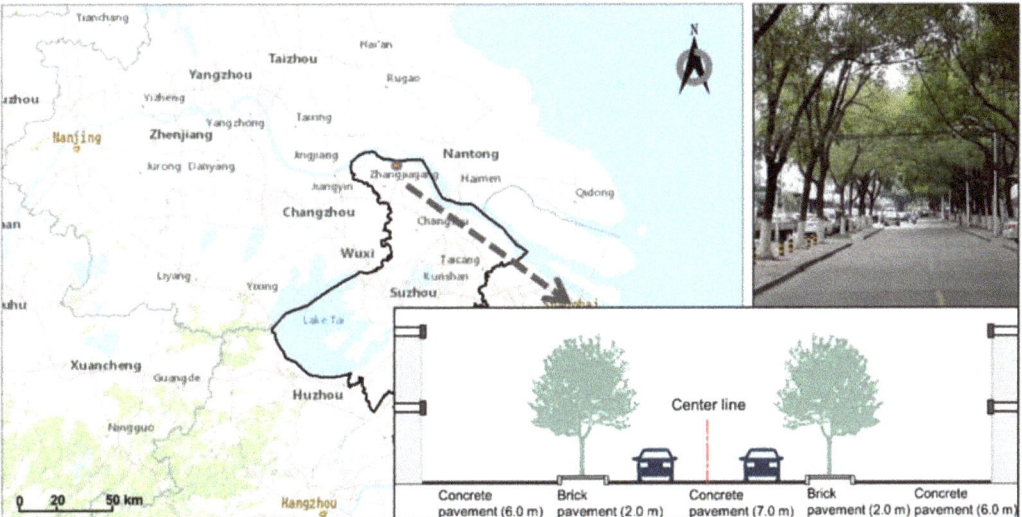

Figure 1. Location and typical section of the pilot road.

Facilities have become diversified with the development of SC. The main functions of SC facilities are infiltration (permeable pavement, sunken green space, biological retention, infiltration well), storage (wet pond, stormwater wetland), transfer (grass filter, infiltration pipe, infiltration channel), and purification (vegetation buffer zone, initial rainwater disposal facility). The selection of SC facilities should be based on the use of each road partition, and an optimization design should be realized according to the site conditions on the basis of combining the actual construction demands. The parameters should be defined pertinently to ensure the efficiency of the SCCUR.

2.2. Design Scheme

The process of SCCUR involves different features, such as road, landscape, and drainage. SCCUR needs to build a drainage system including the collection, treatment, and discharge of the runoff rainwater. Thus, a comprehensive consideration of the slope, elevation, and plant configuration of the urban road is necessary. After analyzing the construction conditions of the urban road and formulating a preliminary SC scheme, the construction of different features should be coordinated, with the drainage system as the core. The main aspects of different features and the used integration methods are presented in Figure 2a.

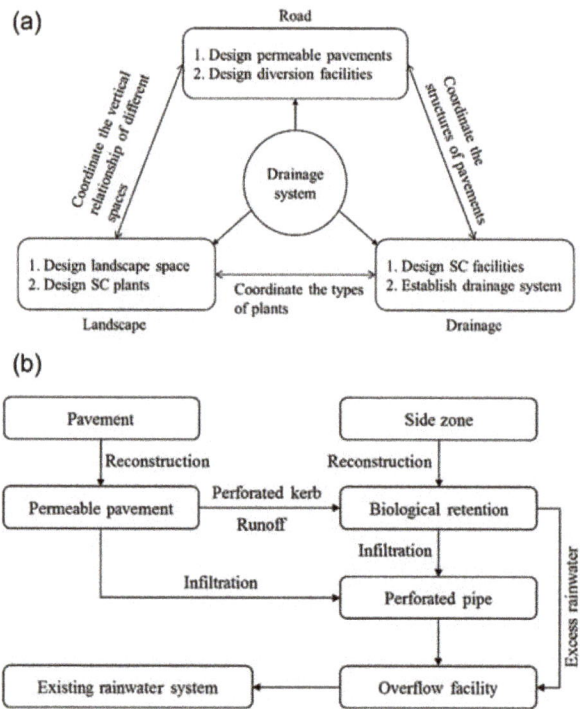

Figure 2. (a) Main aspects of different features in the design of SCCUR; (b) transformation strategies for different surfaces.

The vertical and transverse of the pavement need to be optimized while reconstructing the pavement, which plays a critical role in the collection of runoff rainwater. Due to the construction conditions (schedule, investment, etc.), the renewal of the motor lane involved paving asphalt concrete on the existing concrete pavement. To achieve the goals of this reconstruction, the pavement of the non-motor lane and sidewalk were planned to be demolished and rebuilt from foundation to surface. The street trees of this road were located on the sidewalk, which had a serious impact on the traffic function of the sidewalk. In order to improve the green space and optimize the traffic function of this road, we decided to transform the sidewalk into green space and divide the sidewalk in the outer space of the non-motor lanes. The SC scheme of this road and the transformation strategies of different surfaces are shown in Figure 2b.

2.3. Sampling and Analytical Methods

The reduction of runoff pollution is mainly realized through the comprehensive action of various SC facilities in the construction of SC. Research shows that SC facilities (e.g., permeable pavement, biological retention, etc.) studied in the laboratory can effectively reduce a variety of pollutants. In order to further analyze the effect of SCCUR, the reduction capacity of natural runoff pollution of SC facilities constructed in this case was studied.

According to the actual construction situation under exam, we studied the pollutants removal performance of steel slag–permeable asphalt pavement (fully pervious) and ribbon biological retention zone (RBRZ) by analyzing the pollutant indexes of water samples.

The sampling locations were selected based on the guidelines of different sampling techniques [13]. Generally, the locations were the typical cross sections of the road, and the sampling points were water inlet, overflow outlet, and outlet of porous drainpipes [14]. Multiple samples were collected with the random points method on the typical cross sections of the road. These water samples were collected by self-made devices, and the collection points of effluent water were at the end of the pipes in the drainage layer of steel slag–permeable asphalt pavement and RBRZ. The water samples of the control group were collected at the rainwater inlet on the section of the pilot road not undergoing reconstruction. The pollutants in urban runoff rainwater are generally subjected to the first flush phenomenon, and the collection of surface runoff rainwater in the initial stage of rainfall is particularly important [15]. The sampling interval time increased gradually with the progress of rainfall, and the sampling times were 5, 10, 15, 20, 25, 30, 45, 60, 90, 120 min after the formation of runoff. The water samples were collected in 500 mL polyethylene bottles and labeled. All collected water samples were immediately transported to the laboratory and subjected to analysis within 48 h.

Influent and effluent water samples were tested for water quality parameters (SS, COD, TN, TP). Water quality was monitored by measuring SS, COD, TN, and TP with standard procedures [16]. Generally, SS was measured with the gravimetric method, COD was measured with potassium dichromate titration, TN was measured with potassium persulphate digestion–UV spectrophotometry, and TP was measured with ammonium molybdate spectrophotometry.

3. Results

3.1. Establishment of a New Model of SCCUR

Figure 3a shows the typical runoff organization and drainage model of urban roads. Although this model can rapidly collect and discharge the runoff, it also carries a large amount of pollutants to the water body. The pollutants brought by runoff are important contributors to water pollution. The key of our new strategy is a change in the mode of controlling runoff rainwater. The control of the runoff and the removal of the pollutants can be realized by optimizing urban roads and green spaces and by building suitable SC facilities.

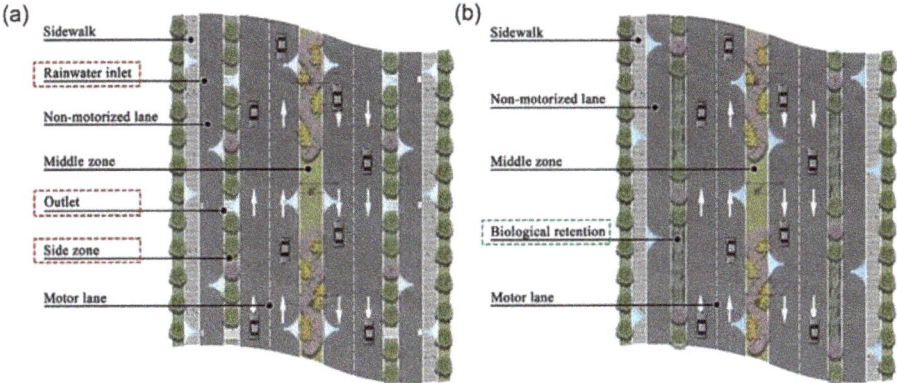

Figure 3. (a) Typical runoff organization and drainage model of urban roads; (b) optimization of runoff rainwater.

The conditions of SC construction in areas of urban roads should be analyzed based on the actual construction requirements. For reconstructed roads, functional requirements (pavement repair, etc.) and non-functional requirements (activity space optimization, etc.) should also be analyzed. A reasonable scheme of the SCCUR should be formulated on the basis of thorough investigations.

In order to realize the runoff control of urban roads, the conventional mode of runoff should be changed (the red dash boxes marked in Figure 3a). A new strategy should not only avoid runoff entering the rainwater pipes rapidly, but also avoid road ponding. This goal can be achieved through two measures: by improving the water permeability of the pavement and delaying the formation of runoff, and by setting up SC facilities with rainwater regulation, storage, and purification capacity to collect and purify rainwater. In the new strategy of SCCUR, the vertical and transverse design of roads and green spaces should be considered comprehensively. The optimized strategy is shown in Figure 3b.

3.2. Pilot Study

3.2.1. Background of the Site

The site was selected at Suzhou (Figure 1). In combination with the construction objectives and engineering documents, the site was surveyed, and a preliminary scheme of the SCCUR was formulated. The construction objectives of the SCCUR generally include several aspects, and the specific objectives need to be comprehensively analyzed according to the requirements and demands of the project location. According to these, the main problems of this project were analyzed, and solutions were proposed. Regarding the road, the concrete pavement was cracked, and the sidewalk could not properly allow walking. Regarding the landscape, the green space was limited, and the green layer was unvarying. Finally, the drainage system of this road had been built for a long time without meeting the standard requirements. Ponding also occurred in this road. With these problems, the following solutions were proposed: realization of a permeable pavement, optimization of the spatial layout to improve the green coverage rate and enrich the green layer, and use of several SC facilities to optimize the organization of the runoff and relieve the pressure on the rainwater pipes. A comprehensive design with the drainage as the core was formulated to eliminate road waterlogging and reduce the pollution of the runoff.

Based on the analysis of the requirements, objectives, and conditions of this road, the major features to integrate in the reconstruction were determined to be the road, the landscape, and the drainage. An SC scheme of this road with the drainage system as the core was formulated to renew the pavement, improve the landscape, and optimize the drainage system.

3.2.2. Design of the SC Facilities

In order to further study the application of steel slag and the performance of the steel slag–permeable asphalt mixture in the construction of SCCUR, two structures of steel slag–permeable asphalt pavement were designed for the renewal of the motor lane and non-motor lane. The structures of the steel slag–permeable asphalt pavement are shown in Figure 4a,b. The structural layers were different between the semi-pervious and the fully pervious structures. Steel slag and more layers were designed in the fully pervious structure for the water permeability of the pavement. For the steel slag–permeable asphalt mixture, the production mixture proportion and the production gradation of steel slag are shown in Tables 1 and 2, respectively. It should be noted that the whetstone ratio (weight ratio between oil and stone) was 4.5%, as shown in the Table 1. The main steel used in the present study was from 3# and 4# hot bin, with steel slag size of 6–15 mm. A larger size (greater than 9.5 mm, Table 2) of the steel slag was favorable for the drainage.

The structural layers of the steel slag–permeable asphalt pavement consisted of steel slag–permeable brick, screed coat, steel slag-pervious concrete, and a gravel drainage layer (Figure 4c). With these layers, the steel slag–permeable asphalt pavement acquired a perfect water permeability, allowing rainwater to infiltrate through the structural layers and finally enter the rainwater system through a pipe in the gravel drainage layer, which had 1% openings for water collection.

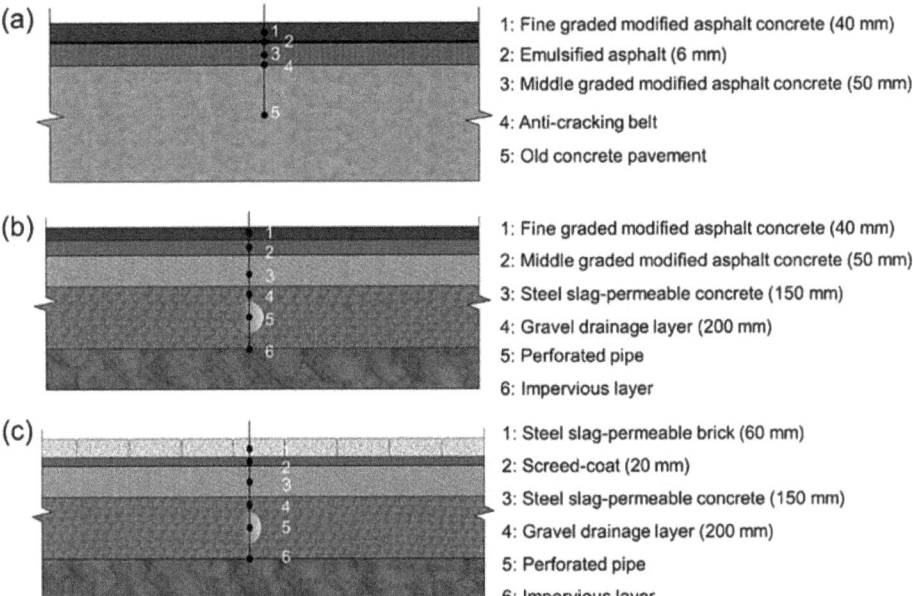

Figure 4. Structures of the steel slag–permeable asphalt pavement: (**a**) semi-pervious; (**b**) fully pervious; (**c**) structures of the steel slag–permeable concrete pavement (fully pervious).

Table 1. Proportion of steel slag in the mixture.

Hot Material Warehouse (Size, mm).	Proportion (%)
1# (0–3)	11.5
2# (3–6)	0
3# (6–11)	49.0
4# (11–15)	34.5
Mineral powder	5.0

Table 2. Production gradation of the steel slag.

Diameter of Square-Opening Sieve (mm)	Mass Proportion of Sifting (%)	
	Design Gradation	Design Standard
16	100	100
13.2	98.1	90–100
9.5	60.8	50–80
4.75	16.8	12–30
2.36	13.7	10–22
1.18	11.3	6–18
0.6	9.2	4–15
0.3	8.0	3–12
0.15	7.1	3–8
0.075	5.1	2–6

Based on the above design, the steel slag–permeable asphalt mixture was prepared, and performance tests were conducted. Leakage loss, dispersion loss, and Marshall residual stability were analyzed, and the results are reported in Table 3. On the basis of the standard technical requirements e, the steel slag–permeable asphalt mixture had a good performance, especially as concerns its stability and permeability (Table 3).

Table 3. Performance test results of the steel slag–permeable asphalt mixture.

Test Items	Detection Value	Technical Requirements
The number of actual hits of the Marshall specimen	50	-
Whetstone ratio (%)	4.5	-
Relative density of gross volume	2.421	Measured
Theoretical relative density	3.036	Calculation
Porosity (%)	20.3	20–22
Stability (KN)	9.86	≥5.0
Binder loss of Schellenberg asphalt leakage test (%)	0.22	≤0.8
Mixture loss of the Fort Kentucky flying test (%)	7.0	≤15
Mixture loss of the water immersion Kentburgh scattering test (%)	14.6	20
Residual stability in immersion Marshall test (%)	92.8	≥85
Permeability coefficient (mL/min)	5433	≥5000

3.2.3. Ribbon Biological Retention Zone

Considering the actual conditions of the site under exam, the structural layers of the RBRZ included an aquifer (150 mm), a planting soil layer (400 mm), a filter layer (300 mm), and a gravel drainage layer with perforated pipes (250 mm). A permeable geotextile was laid between the different structural layers to ensure pollutant removal and stability of RBRZ. The drainage safety was ensured by the perforated drainage pipes in the gravel layer, and a fabric protecting against osmosis was laid under the gravel layer to ensure the safety of the subgrade. The values of the pivotal design parameters of the RBRZ are shown in Table 4. Among these parameters, planting soil plays a key role in pollutant removal, while filter layer and gravel drainage layer favor permeability.

Table 4. Design parameters of the RBRZ.

Structural Layer	Minimum Permeability Coefficient K (m/s)	Minimum Void Ratio (%)	Material Specifications
Planting soil	1.5×10^{-5}	5	Evenly mix 45% medium sand, 10% pine bark, 5% nutrient soil
Filter layer	10×10^{-5}	10	Ceramsite matrix of φ 7–10 mm
Drainage layer	100×10^{-5}	15	φ 20–30 mm gravel with perforated drainage pipe

3.2.4. Diversion Facilities

The key of the SCCUR is to ensure the runoff rainwater can effectively enter the SC facilities with water storage capability, such as the RBRZ. For the purpose of traffic safety, kerbs are generally set between the pavement and the green belts of urban roads (Figure 5a,b). Several kerbs were perforated to ensure that the runoff rainwater formed on the pavements could enter the RBRZ. The kerbs were made of granite, with size of 99 cm × 30 cm × 15 cm.

Based on practical considerations and requirements, two holes with size of 30 cm × 7.5 cm were made in each kerb. The distance between the perforated kerbs was 15 m, and the kerbs were fixed with concrete. The vertical relationship between pavement, perforated kerb, and RBRZ is shown in Figure 5c.

Overflow inlets were set in the RBRZ for drainage safety, so to allow the excess rainwater in the RBRZ to enter the drainage system through the overflow inlets, when rainfall exceeds the water storage capacity of the RBRZ. As shown in Figure 5d, the used overflow inlets in the shape of a platform have five inlet surfaces, which significantly reduces the risk of blockage. The maximum overflow capacity of the overflow inlets with a size of 75 × 45 cm was 30 L/s, and the interval between them was 25–30 m.

Figure 5. The road before (**a**) and after (**b**) reconstruction; (**c**) vertical relationship between different parts; (**d**) schematic diagram of the overflow inlet.

3.2.5. Runoff and Pollution Control Capacity

The runoff volume control capacity of this road was checked based on the reconstruction scheme, using Equations (1)–(4) [17,18].

$$V_w = A_f h_m (1 - f_v) \quad (1)$$

where V_w (m³) is the water storage volume, A_f (m²) is the area of the sampling location, h_m (m) is the maximum storage height (<0.3 m), f_v is the proportion of the plant cross-sectional area to the water storage area (0.15–0.3);

$$G = A_G \left(n_1 d_{f1} + n_2 d_{f2} \right) \quad (2)$$

where G (m³) is the water storage of the internal structure, A_G (m²) is the area of the facility area, n_1 is the average porosity of the planting soil, d_{f1} (m) is the depth of the planting soil, n_2 is the porosity of packing layer, d_{f2} (m) is the depth of the filler layer;

$$V = 10 H \Psi F \quad (3)$$

where V (m³) is the water storage, H (m) is the depth of the design rainfall, Ψ is the rainfall runoff coefficient, F (m²) is the service area;

$$R_w = \frac{V_w + G}{V} \quad (4)$$

where R_w is the water volume control rate.

For the examined location, the values of the A_f, h_m, and f_v were 4, 0.15, and 0.2; thus, V_w resulted to be 0.48. The values of the A_G, n_1, d_{f1}, n_2, and d_{f2} were 4, 0.05, 0.4, 0.12, and 0.55; thus, G was calculated as 0.35. The values of the H, Ψ, and F were 19.4×10^{-3}, 0.65, and 23; thus, V was calculated as 0.29. Based on the above results, R_w resulted to be 2.86. This value was greater than that in a previous report (1.32) [19], which indicated that we obtained a greater runoff volume control capacity.

Event mean concentration analysis is used as the standard to evaluate the runoff quality; it has a high confidence level in the comparison between different catchment areas. The average event means the concentration removal rate of SS, the COD, and the removal rates of TN and TP by the steel slag–permeable asphalt pavement (fully pervious) were

61.8%, 66.84%, 56.33%, and 47.36%; the results for RBRZ were 71.60%, 78.35%, 63.93%, and 49.47% (Figure 6). The removal rate was comparable to that in previous reports [20,21]. For example, for porous asphalt–bio-retention combined roads, the removal rates of SS, TN, and TP were 70. 26%, 46. 29%, and 19. 27% [20].

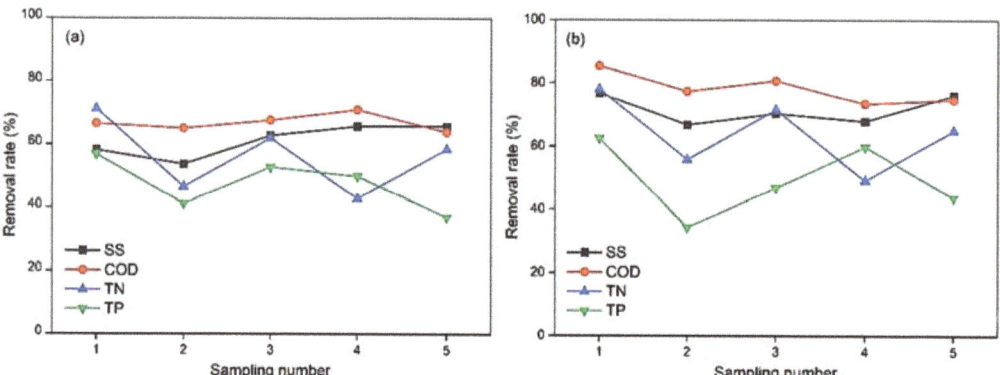

Figure 6. Removal rate of SS, COD, removal rates of TN and TP of the steel slag–permeable asphalt pavement (**a**) and the ribbon biological retention zone (**b**) in different samples.

4. Discussion

As a strategy of stormwater management, SC has many advantages. In terms of urban roads, the construction goals mainly include the following aspects: 1. The volume control rate of annual runoff. In order to meet the requirement of the volume control rate of annual runoff, various suitable facilities can be used to store the runoff rainwater. These SC facilities can meet the requirement that a certain amount of runoff will not be discharged. 2. Removal of runoff pollutants. The interception and filtration functions of these facilities can be used to reduce the pollutants in the runoff rainwater, so as to reduce the total discharge of rainwater pollutants of the site. 3. Integration of different functions. With the development of cities, the functional requirements of urban roads have gradually changed from simplification to diversification. The SCCUR needs not only to control the runoff and remove pollutants, but also to integrate the requirements of road, landscape, and drainage, thus creating a multi-functional road space while realizing road renewal.

In the present SCCUR, several SC facilities were constructed, and the runoff organization was optimized to reduce the total amount of discharged rainwater. In the examined case, the volume control rate of annual runoff needed to approach 70%, and the dependable rainfall was 19.4 mm. The removal of pollutants (SS, TN, TP, etc.) and a proper COD in runoff rainwater is another objective in road reconstruction, and the degree of SS removal is generally used as a typical pollutant index in SCCUR due to its certain correlation with other pollutant indexes [22]. In this case, the comprehensive removal rate of SS in the runoff needed to approach 50% after the reconstruction. Furthermore, with the development of the city, urban roads are endowed with more functions. Therefore, a road needs to be built into an ecological block integrating traffic, recreation, and walking on the basis of runoff control and road renewal.

In particular, a permeable pavement, a typical type of SC facilities, can greatly improve hydrology and water quality. With the development of pavement engineering, the types of permeable pavements suitable for various underlying surfaces are gradually increasing in number. In many occasions, permeable pavements can not only meet their functional requirements, but also improve runoff control and runoff pollution reduction compared to impervious pavement [23]. As an alternative to traditional asphalt pavements, permeable asphalt pavements are generally used in traffic roads, parking areas, etc. Permeable asphalt pavements have excellent performance in drainage, antiskid function, and noise reduction.

When producing permeable asphalt pavements, structural competition, connected air voids, and aggregates should be selected and designed systematically according to the rainfall characteristics, as they have a critical impact on permeability and bearing capacity [24].

However, natural aggregate resources are facing the problem of overexploitation; the incorporation of recycled aggregate in asphalt mixtures is an efficient method to preserve resources [25]. As an industrial solid waste with the largest output, steel slag has the characteristics of large porosity, high hardness, and good particle shape. Compared with ordinary asphalt mixtures, the steel slag–permeable asphalt mixture has better performance as regards repeated fluctuation under low temperatures, the snow melting process on an electrical-thermal pavement system, permeability, and water stability [26,27]. In the present study, steel slag was applied. The results of the runoff volume control capacity showed that the above parameters met the requirements of relevant standards. The appearance of the paved steel slag–permeable asphalt pavement was uniform and flat, without segregation and oil spots, and the performance met the design requirements. These results suggest a new way of steel slag utilization. More in-depth research can be carried out in future practical projects.

After optimizing the spatial layout, the sidewalk was moved to the outside of the non-motor lane, located on the outermost side of this road. In order to improve the travel experience in rainy days and meet the requirements of the SCCUR, the steel slag–permeable concrete pavement was placed on the sidewalk. Although permeable pavements have a positive effect on runoff coefficient reduction and runoff pollutants interception, their storage capacity of rainwater is limited. The slope of the permeable pavements should be designed considering other facilities, so that rainwater can enter those facilities with water storage capacity when it exceeds the infiltration capacity of the permeable pavements.

Biological retention (bioretention) is a kind of SC facility that can be subdivided into many types according to its functions and use. Bioretention has a good removal effect on pollutants in runoff, and the design of parameters (such as the type of filter material, the thickness of the aquifer, etc.) has a certain impact on its removal capacity [28,29]. In the present study, after the demolition of the brick sidewalk (2 m wide), the space was restored to green belt. According to the demands of rainwater storage and purification, these green belts needed to be sunk.

Furthermore, in order to control rainwater of this road, the runoff formed on the pavements was diverted to the RBRZ, and the number of rainwater inlets on this road was reduced markedly for better implementation. For analyzing the SCURR achievements, the volume control rate of annual runoff (not less than 70%) was verified. The SC facility can fully control the discharge of runoff rainfall within the design range when its water volume control rate is ≥ 1. The results of runoff control capacity show that the rainwater in the service of RBRZ could not be discharged directly when the rainfall was 19.4 mm, which met the construction requirements of this road (for which the volume control rate of annual runoff was 70%). Due to the fact that the plants in the RBRZ occupied a certain storage space, a reduction factor of water storage capacity had to be considered during the of verification. The effective space of the aquifer in this case was calculated as 80% based on the different plants.

As runoff is a complex hydrological process, pollutant concentration is easily affected by rainfall, runoff, and other factors. Thus, event mean concentration was used to benchmark the performance of pollutants removal in this study [30]. As shown in Figure 6, both steel slag–permeable asphalt pavement (fully pervious) and RBRZ constructed in this pilot work had a significant removal capacity of conventional pollutants in runoff rainwater. The RBRZ showed a higher removal capacity than the steel slag–permeable asphalt pavement (fully pervious), which was likely due to the specific removal mechanism of the RBRZ.

5. Conclusions

Owing to the limited service area of point facilities, it is difficult that the traditional drainage mode of urban roads can meet the requirements of SC; therefore, in this case, the effect of SCCUR is insignificant. A new construction strategy with the drainage system as the core was proposed in this paper. The key findings are as follows:

1. New design strategy. The construction strategy focused on the organization of runoff rainwater in the road space, which could be achieved through vertical optimization of the structure. Thus, the vertical relationship of different facilities needs to be designed in an integrated way, and the basic functions of the facilities need to be met. While ensuring the drainage safety, the runoff rainwater shall be diverted to the nearest SC facilities with capacity of retention and pollutant removal.
2. Additional advantages of the new strategy. Generally, the road and the green space were used to create a sponge road combining basic road functions with landscape effect and drainage capacity. The reconstruction of a road in Zhangjiagang with this new strategy was successful. Through the construction of permeable pavements, biological retention facilities, and other facilities, optimization of runoff organization, volume control of the runoff, removal of pollutants, and renewal of the road were realized. The construction of SC not only can achieve the goal of stormwater management, but also is an important way to realize urban renewal.

Therefore, the combined usage of SC facilities has a significant effect on the control of runoff rainwater pollution according to the effective pollutant's removal performance of the reconstructed road. The selection and design of facilities shall be based on the characteristics of runoff pollutants, and the construction quality should be strictly controlled. These results support the design and construction of SC urban roads.

However, there are some limitations in the present study. For example, the composition of steel slag was not analyzed, the design of the steel slag–permeable asphalt pavement was not optimized, and the runoff pollution process was not investigated. In a further study, the optimization, decontamination mechanism, effect of climate change, and cost evaluation should be considered. With the above information, the methods in the present study could be largely adopted.

Author Contributions: Conceptualization, C.W. and T.H.; methodology, C.W., J.W. and P.L.; validation, H.L. and T.H.; investigation, C.W., H.L. and Y.J.; data curation, C.W. and B.W.; writing—original draft preparation, C.W.; writing—review and editing, T.H.; visualization, C.W. and B.W.; supervision, T.H.; project administration, T.H.; funding acquisition, T.H. All authors have read and agreed to the published version of the manuscript.

Funding: This research was funded by Key Laboratory of Suzhou Sponge City Technology (Grant No. SZS2021265), the Suzhou Science and Technology Plan Project—Minsheng Science and Technology (Suzhou science and Technology Bureau, Grant No. SS202002), the Water Pollution Control and Treatment, National Science and Technology Major Project (Grant No. 2017ZX07205002).

Acknowledgments: We are grateful to Xiaoyi Xu and Wei Wu of Suzhou University of Science and Technology for insightful discussions.

Conflicts of Interest: The authors declare no conflict of interest.

References

1. Zhao, Y.; Xia, J.; Xu, Z.; Zou, L.; Qiao, Y.; Li, P. Impact of urban expansion on rain island effect in Jinan city, north China. *Remote Sens.* **2021**, *13*, 2989. [CrossRef]
2. Risch, E.; Gasperi, J.; Gromaire, M.; Chebbo, G.; Azimi, S.; Rocher, V.; Roux, P.; Rosenbaum, R.K.; Sinfort, C. Impacts from urban water systems on receiving waters -How to account for severe wet-weather events in LCA? *Water Res.* **2018**, *128*, 412–423. [CrossRef]
3. Beg, M.N.A.; Rubinato, M.; Carvalho, R.; Shucksmith, J. CFD Modelling of the transport of soluble pollutants from sewer networks to surface flows during urban flood events. *Water* **2020**, *12*, 2514. [CrossRef]
4. Bae, C.; Lee, D.K. Effects of low-impact development practices for flood events at the catchment scale in a highly developed urban area. *Int. J. Disaster Risk Reduct* **2020**, *44*, 101412. [CrossRef]

5. Rubinato, M.; Nichols, A.; Peng, Y.; Zhang, J.; Lashford, C.; Cai, Y.; Lin, P.; Tait, S. Urban and river flooding: Comparison of flood risk management approaches in the UK and China and an assessment of future knowledge needs. *Water Sci. Eng.* **2019**, *12*, 274–283. [CrossRef]
6. Kuller, M.; Bach, P.M.; Ramirez-Lovering, D.; Deletic, A. Framing water sensitive urban design as part of the urban form: A critical review of tools for best planning practice. *Environ. Model. Softw.* **2017**, *96*, 265–282. [CrossRef]
7. Zha, X.; Luo, P.; Zhu, W.; Wang, S.; Wang, Z. A bibliometric analysis of the research on sponge city: Current situation and future development direction. *Ecohydrology* **2021**, *14*, 2328. [CrossRef]
8. Qi, Y.; Shun Chan, F.K.; Griffiths, J.; Feng, M.; Sang, Y.; O'Donnell, E.; Hutchins, M.; Thadani, D.R.; Li, G.; Shao, M.; et al. Sponge city program (SCP) and urban flood management (UFM)-the case of Guiyang, SW China. *Water* **2021**, *13*, 2784. [CrossRef]
9. Wang, K.; Zhang, L.; Zhang, L.; Cheng, S. Coupling coordination assessment on sponge city construction and its spatial pattern in Henan province, China. *Water* **2020**, *12*, 3482. [CrossRef]
10. Yin, D.; Chen, Y.; Jia, H.; Wang, Q.; Chen, Z.; Xu, C.; Li, Q.; Wang, W.; Yang, Y.; Fu, G.; et al. Sponge city practice in China: A review of construction, assessment, operational and maintenance. *J. Clean. Prod.* **2021**, *280*, 124963. [CrossRef]
11. Hu, D.; Zhang, C.; Ma, B.; Liu, Z.; Yang, X.; Yang, L. The characteristics of rainfall runoff pollution and its driving factors in Northwest semiarid region of China—A case study of Xi'an. *Sci. Total Environ.* **2020**, *726*, 138384. [CrossRef]
12. *Code for Planting Planning and Design on Urban Road*; Ministry of Construction of the People's Republic of China: Beijing, China, 1997; p. 4.
13. *Water Quality-Guidance on Sampling Techniques*; Ministry of environmental protection of the people's Republic of China: Beijing, China, 2009; pp. 1–16.
14. Li, J.; Sun, Y.; Li, X.; Wang, W. Research on stormwater runoff quality monitoring in sponge city. *Water Wastewater Eng.* **2021**, *6*, 68–74.
15. Perera, T.; McGree, J.; Egodawatta, P.; Jinadasa, K.B.S.N.; Goonetilleke, A. New conceptualisation of first flush phenomena in urban catchments. *J. Environ. Manag.* **2021**, *281*, 111820. [CrossRef]
16. *Analytical Methods for Water and Wastewater Monitoring*, 4th ed.; China Environmental Monitoring Press: Beijing, China, 2002; pp. 107–258.
17. *Technical Guide for Sponge City Construction: Rainwater System Construction for Low Impact Development: Trial Implementation*; China Construction Industry Press: Beijing, China, 2015; p. 58.
18. *Guide for Construction, Operation and Maintenance of Rainwater Garden in Jiangsu Province*; Department of Housing and Urban Rural Development of Jiangsu Province: Nanjing, China, 2018; pp. 32–34.
19. Wang, N. Discussion on design schemes for urban road reconstruction based on sponge city concept in Xiamen. *China Water Wastewater* **2016**, *22*, 112–116.
20. Gong, M.; Zuo, J.; Ren, X.; Zhao, H.; Luo, X.; Liao, Y.; Li, X. Evaluation on effect of urban non-point source pollution control on porous asphalt-bio-retention combined roads. *Environ. Sci.* **2018**, *9*, 4096–4104.
21. Li, P.; Jia, Y.; Lu, J.; Teng, J.; Zhang, H.; Tao, H. Pollution control for storm runoff water quality on elevated road. *China Water Wastewater* **2016**, *19*, 142–146.
22. Paudel, B.; Montagna, P.A.; Adams, L. The relationship between suspended solids and nutrients with variable hydrologic flow regimes. *Reg. Stud. Mar. Sci.* **2019**, *29*, 100657. [CrossRef]
23. Kayhanian, M.; Li, H.; Harvey, J.T.; Liang, X. Application of permeable pavements in highways for stormwater runoff management and pollution prevention: California research experiences. *Int. J. Environ. Sci. Technol.* **2019**, *8*, 358–372. [CrossRef]
24. Li, Z.; Cao, Y.; Zhang, J.; Liu, W. Urban rainfall characteristics and permeable pavement structure optimization for Sponge Road in North China. *Water Sci. Technol.* **2021**, *83*, 1932–1945. [CrossRef]
25. Mikhailenko, P.; Piao, Z.; Kakar, M.R.; Bueno, M.; Poulikakos, L.D. Durability and surface properties of low-noise pavements with recycled concrete aggregates. *J. Clean. Prod.* **2021**, *319*, 128788. [CrossRef]
26. Zhu, B.; Liu, H.; Li, W.; Wu, C.; Chai, C. Fracture behavior of permeable asphalt mixtures with steel slag under low temperature based on acoustic emission technique. *Sensors* **2020**, *20*, 5059. [CrossRef]
27. Liu, W.; Li, H.; Zhu, H.; Xu, P. Properties of a steel sla-permeable asphalt mixture and the reaction of the steel slag-asphalt interface. *Materials* **2019**, *12*, 3603. [CrossRef]
28. Vijayaraghavan, K.; Biswal, B.K.; Adam, M.G.; Soh, S.H.; Tsen-Tieng, D.L.; Davis, A.P.; Chew, S.H.; Tan, P.Y.; Babovic, V.; Balasubramanian, R. Bioretention systems for stormwater management: Recent advances and future prospects. *J. Environ. Manag.* **2021**, *292*, 112766. [CrossRef] [PubMed]
29. Lim, F.Y.; Neo, T.H.; Guo, H.; Goh, S.Z.; Ong, S.L.; Hu, J.; Lee, B.C.Y.; Ong, G.S.; Liou, C.X. Pilot and field studies of modular bioretention tree system with talipariti tiliaceum and engineered soil filter media in the tropics. *Water* **2021**, *13*, 1817. [CrossRef]
30. Yang, L.; Li, J.; Zhou, K.; Feng, P.; Dong, L. The effects of surface pollution on urban river water quality under rainfall events in Wuqing district, Tianjin, China. *J. Clean. Prod.* **2021**, *293*, 126136. [CrossRef]

Article

Study on the Influence of Sponge Road Bioretention Facility on the Stability of Subgrade Slope

Wensheng Tang [1,2], Haiyuan Ma [1,2,*], Xinyue Wang [1,2], Zhiyu Shao [1,2], Qiang He [1,2] and Hongxiang Chai [1,2,*]

[1] Key Laboratory of Three Gorges Reservoir Region's Eco-Environment, Ministry of Education, Chongqing University, Chongqing 400045, China; tangwensheng87@126.com (W.T.); wxinyue@hnu.edu.cn (X.W.); shaozhiyu@cqu.edu.cn (Z.S.); Hq0980@126.com (Q.H.)

[2] National Centre for International Research of Low-Carbon and Green Buildings, Chongqing University, Chongqing 400045, China

* Correspondence: mahy@cqu.edu.cn (H.M.); chaihx@cqu.edu.cn (H.C.)

Abstract: With the large-scale application of sponge city facilities, the bioretention facility in urban roads will be one of the key factors affecting the safety of construction facilities in areas with abundant rainfall. In this study, by establishing a three-dimensional finite element model for numerical analysis and combining it with geotechnical tests, the effects of bioretention facility on water pressure distribution, seepage path, and slope stability under rainwater seepage conditions are proposed. In addition, this study puts forward the relationship between the parameters of the bioretention facility and the stability of the slope in combination with the effect of runoff pollution control, which provides direction and basis for the planning, design, and construction of sponge cities in road construction.

Keywords: sponge city; bioretention facility; rain infiltration; slope stability

Citation: Tang, W.; Ma, H.; Wang, X.; Shao, Z.; He, Q.; Chai, H. Study on the Influence of Sponge Road Bioretention Facility on the Stability of Subgrade Slope. *Water* **2021**, *13*, 3466. https://doi.org/10.3390/w13233466

Academic Editor: Haifeng Jia

Received: 3 November 2021
Accepted: 2 December 2021
Published: 6 December 2021

Publisher's Note: MDPI stays neutral with regard to jurisdictional claims in published maps and institutional affiliations.

Copyright: © 2021 by the authors. Licensee MDPI, Basel, Switzerland. This article is an open access article distributed under the terms and conditions of the Creative Commons Attribution (CC BY) license (https:// creativecommons.org/licenses/by/ 4.0/).

1. Introduction

With the rapid development of cities and the advancement of urbanization, non-point source pollution has begun to become the main factor affecting the urban ecological environment [1]. The gradually developed traffic network leads to a substantial increase in the area of impervious underlying urban surface, and the increase in engineering measures and traffic volume leads to an increase in the emission of pollutants such as dust and tail gas [2,3]. During heavy rains, the surface comprehensive runoff coefficient increases, resulting in increasingly serious surface runoff pollution. This has become one of the important hazards affecting the health of urban water bodies [4–6]. In order to reduce the flood disaster in the city during the rainy season and create a green and livable urban ecological environment, the United States first proposed the low-impact development (LID), including bioretention facilities, green roofs, permeable pavement, and other measures, which can effectively control rainwater runoff pollution that is the dominant source of pollution [7–9]. To play the natural ecological function, restore the urban ecological environment, and realize the natural collection, storage, penetration, and purification of rainwater, China proposed the concept of "Sponge City" in 2014 and issued relevant guidelines [10]. Since then, China has vigorously promoted the construction of sponge cities to reduce or even remove pollutants in rainwater runoff and realize the coordinated development of water resources and cities [11,12].

Bioretention facilities are an important technical measure in the road construction of sponge city. After introducing rainwater runoff formed by road impervious underlying surface into facilities mainly by gravity flow, they utilize the complex physical, chemical, and biological synergies between plants, soil, and filler to purify the rainwater and simultaneously absorb and reduce in situ rainwater runoff of roads to achieve the purpose of controlling the total amount of rainwater runoff and the peak value of runoff, reduce runoff pollution, and restore the natural hydrological cycle of the site [13,14]. In developed countries such as the United States, the research on bioretention technology started earlier, and much research was conducted on the reduction in peak runoff and the removal of pollutants. Therefore, a relatively complete technical system has been formed so far [15–17]. Recently, British scholar Maksimovic and his team put forward the concept of "blue-green dream", which provides innovative methods for urban planning and transformation through the combination of blue-green components and integrated urban design [18]. With the popularization and application of sponge cities, research on bioretention facilities in China is also gradually deepening. Liang et al. (2020) used numerical simulation to analyze the changes in water content and settlement of the foundation in two bioretention facilities with different lengths of anti-seepage membranes [19]. Meng et al. (2013) summarized the effects of design parameters of bioretention facility on the retention effect and purification effect of road rainwater through years of experimental observations [20]. Pan et al. (2012) studied the total amount of rainwater runoff control, peak reduction, and delay effects of different peak flow in bioretention facilities through simulation experiments [21].

The bioretention facility in Sponge City will absorb most of the surface runoff, effectively reducing the probability of urban waterlogging. Road rainwater runoff enters the bioretention facility through the pavement and curb openings. The structural layer of the bioretention facility is usually composed of the water storage layer, planting layer, filter layer, and drainage system, etc. The typical cross-section is shown in Figure 1. The water storage layer is located on the surface of the facility and acts as temporary retention of rainwater runoff. The planting layer plays the role of infiltration and purification of rainwater runoff and preserves part of the rainwater to meet the need for plant growth. The filter layer can store part of the infiltration runoff and enhance the purification effect of the facility. The drainage system at the bottom is usually composed of a gravel layer and perforated drainage pipes wrapped in it, which serves to enhance the hydraulic conductivity and transport properties of the retention zone.

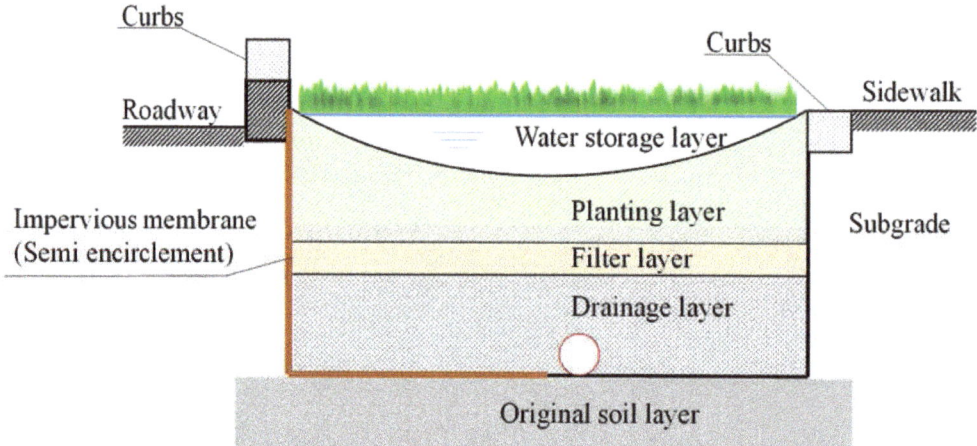

Figure 1. Cross-sectional view of road bioretention.

At present, research on bioretention facilities mainly focuses on the purification mechanism of pollutants and runoff regulation effects. However, few studies pay attention to its influence on the seepage and stability of subgrade slopes. Because most of the carriageways and sidewalks are impervious paving, the rainwater grate on the roadway is easily blocked under heavy rain or continuous rainfall. The biological retention facility will become the main rainwater drainage channel on the road. The influx of a large amount of rainwater caused the biological retention facility to reach saturation. The rainwater seeps into the roadbed and the slope, making the soil of the slope that was originally unsaturated gradually become saturated, the shear strength between the soils reduces. This will further lead to the instability of the road slope [22].

Based on the Green-Ampt model, Pradel et al. (1993) accurately derived the function relationship between rainwater seepage depth and rainfall duration by considering rainfall intensity, rainfall duration, the volume fraction of water in slope soil, and the substrate suction of unsaturated soil [23]. Ng et al. (1998) used the finite element analysis method to analyze the transient seepage changes under different rainfall situations and different initial conditions and analyzed the influence of hydrogeological conditions, slope seepage prevention, and rainfall characteristics on slope stability [24]. Sun et al. (2009) established a finite element model and carried out research with examples, considering slope softening includes damage softening and water-induced softening [25]. The results show that with the increase in water content, the increase in rigidity ratio of soil is the main factor leading to slope instability under rainfall conditions.

When rainwater seepage occurs inside the road slope under the action of rainfall infiltration, the water content of the soil increases, and the shear strength decreases, which has a very negative impact on the stability of the slope. Therefore, it is necessary to consider the influence of pore water pressure when analyzing slope stability. The failure of soil slope manifests as local stability failure and overall stability failure. The former is caused by concentrated seepage or seepage gradient greater than critical gradient, which manifests as internal erosion, leading to piping and flowing soil on slope surface. The latter is caused by pore water pressure prevalent in the seepage field because the pore water pressure, especially the excess pore water pressure, causes the shear strength of the soil to decrease, which leads to instability of the entire slope.

This study focused on the influence of sponge road bioretention facility on the stability of subgrade slope and analyzed the water pressure distribution, seepage path, and slope stability under rainwater seepage conditions by combined use of three-dimensional finite element model and geotechnical tests.

2. Materials and Methods

2.1. Road Runoff Infiltration Analysis

H. Darcy, a French engineer, put forward the famous Darcy's law through long-term observation, experiment, and deduction to describe saturated flow in homogeneous sand columns. According to Darcy's law (Equation (1)), there is a proportional relationship between the water velocity and hydraulic gradient in the process of saturated soil seepage:

$$v = KJ, \quad J = \frac{dh}{dL} \tag{1}$$

where J is the hydraulic gradient (−), h is the hydraulic head (m), and L is the length through which the water flows (m). The proportionality coefficient K quantifies the resistance of the porous medium to the movement of the fluid, and it is called the hydraulic conductivity of the saturated porous medium. K was originally referred to as the permeability coefficient [26]. In the International System of Units (SI), K is expressed in m/s, but in the current practice, the most used unit is m/d.

Rainfall infiltration of soil is a movement of water in the gas of the soil. It is a process of two-phase flow, in which water replaces air in the process of infiltration. After the rain infiltrates the slope, the substrate suction field and pore pressure field of the slope are changed, which causes the redistribution and adjustment of the original stress field of the slope.

Based on Darcy's law and the mass conservation, Richards (1931) formulated the flow equation under transient conditions for a variably saturated porous medium as a function of the water content θ and the pressure head [27]. The hydraulic conductivity is a function of the water content (Equation (2)):

$$K = K(\theta) \tag{2}$$

This is the first time that Darcy's law has been applied to the analysis of water flow in unsaturated soils. The filling of the urban road subgrade generally controls the degree of soil compaction, which ensures the optimal moisture content of the soil and the unsaturated state of subgrade soil. The volume of water and air contained in unsaturated soils is the key factor affecting rain infiltration. Regardless of thermal, chemical reactions, and electric gradients, the driving force of water seepage in unsaturated soils comes from the positional head and the negative pore water pressure in the soil.

2.2. Project Overview

This paper takes the newly-built municipal sponge city road in Chongqing as the research object. The soil layer in the construction site is mainly quaternary Holocene plain fill (Q_4^{ml}) and distributed residual Diluvial silty clay (Q_4^{el+dl}); the underlying bedrock is composed of interbedded sandstone and sandy mudstone of the Middle Jurassic Shaximiao Formation (J_2s). According to the geological prospecting evaluation, there is no adverse geological phenomenon at the site, and the rock and soil are stable. The newly-built road is an urban trunk road with a design speed of 60 km/h. It is a two-way six-lane standard road with a width of 38 m, of which the sidewalk is 7.5 m wide. The side slope of the roadbed is graded every 8 m, of which the bridle-way is 2 m wide, and the slope rate is 1:1.5 and 1:1.75 from top to bottom.

Biological retention facilities are set up between the sidewalk and the curbs of the road and are arranged longitudinally along the road. According to the requirements of the "Sponge City Planning and Design Guidelines of Chongqing", the total annual runoff control rate in the region is ≥70%, and the rainwater runoff pollutants reduction rate is ≥50%. The biological retention facility adopts a typical cross-section form with a design width of 3.5 m, in which the depth of the water storage layer, the thickness of the planting layer, the thickness of the filter layer, and the thickness of the gravel layer are 20 cm, 50 cm, 10 cm, and 30 cm, respectively. The anti-seepage membrane is in the form of a half-wrap, and the structure layer is in contact with the roadbed to prevent rainwater from entering the range of the roadbed during the process of rainwater seepage, which affects the quality of the roadbed.

2.3. Calculation Conditions

The three-dimensional finite element program MIDAS/GTS is used for the numerical simulation calculation. The software can simultaneously conduct three-dimensional spatial model analysis, stress analysis by stratum structure method, and stress-seepage coupling analysis considering the influence of groundwater on the structure.

The experiment selects a typical roadbed cross-section with a roadway width of 19 m and a roadbed height of 15 m. The transverse range, vertical range, and length range of the model calculation are 55 m, 25 m, and 60 m, respectively. The calculation model regards the rock and soil as an isotropic ideal elastoplastic material and uses the Mohr–Coulomb yield criterion for analysis. According to the size of the model, the bridle-way in the middle of the slope is selected as the control point 1, the slope foot as the control point 2, 8 m depth below the midpoint of the bioretention facilities as the control point 3, and 8 m depth below the midpoint of the carriageway as the control point 4 (Figure 2).

According to the engineering characteristics of road slopes, this study uses solid element tetrahedral meshes to divide the three-dimensional model, which can perform not only stress-strain analysis on subgrade slopes but also conduct stress-seepage coupling analysis considering the impact of rainwater seepage. The grid division simulation method has become a commonly used numerical analysis method in the engineering field, and the calculation results were verified by actual projects with high credibility [28]. The three-dimensional finite element meshing is shown in Figure 2. The boundary conditions are as follows: for the boundary conditions of the three-dimensional model, the displacement on the left and right sides of the model is constrained in the X direction, the displacement in the front and back directions is constrained in the Y direction, and the displacement in the lower part is constrained in the X, Y, and Z directions. Considering that the bio-retention facility and the slope surface are permeable layers, the rainfall boundary conditions are set at the surface of the bio-retention facility and the side slope surface of the roadbed. The rainfall intensity is selected according to heavy rainfall in the same area of the road project [29], the rainfall duration is 24 h, and the rainfall intensity duration curve is shown in Figure 3.

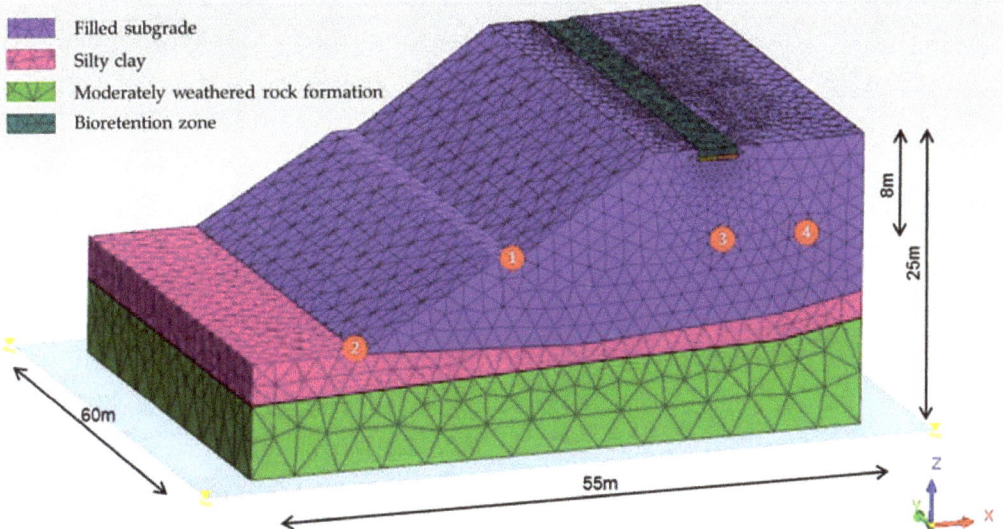

Figure 2. Finite element mesh model used in this study, 1–4 in the figure indicates the control points in the model.

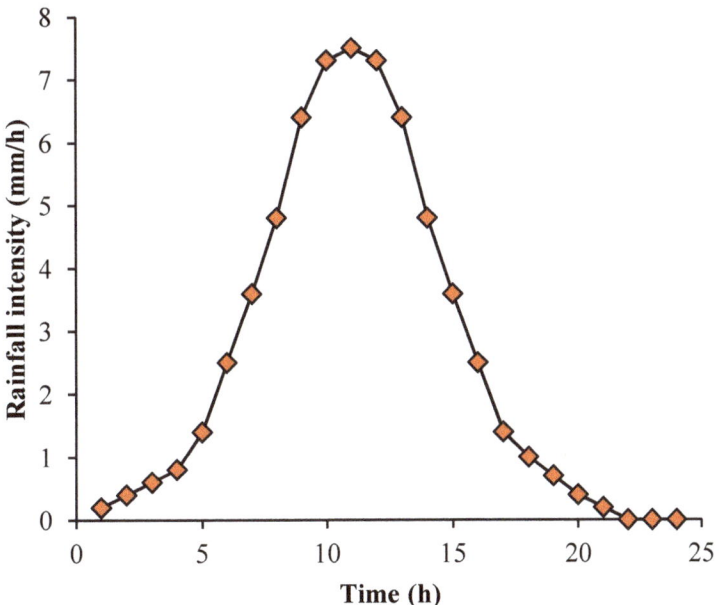

Figure 3. Rainfall intensity over time.

2.4. Model Parameters

In calculation, according to the compactness of compacted fill in the actual project, the value of compactness of subgrade slope is 94%, and the current soil layer is taken as a silty clay layer. Because the clay cover layer has a low hydraulic conductivity compared to the conductivity of the weathered bedrock, the impact of the rainwater infiltration in the rock formation is not considered. The parameters of the planting soil layer in bioretention facility are consistent with those of subgrade fill, in which the parameters of the lower permeable layer are selected according to the gravel material with higher hydraulic conductivity, and the physical and mechanical parameters are selected according to geological engineering investigation and regional experience, as shown in Table 1.

Table 1. Mechanical parameters.

Material	Elastic Modulus E/MPa	Poisson Ratio μ	Density $\gamma/(Kg \cdot m^{-3})$	Cohesion C/(KPa)	Internal Friction Angle $\Phi/(°)$
Subgrade slope	3200	0.32	2000	5	25
Bioretention facility	8000	0.25	2100	0	35
Silty clay	5500	0.3	2050	20	16
Moderately weathered bedrock	21,000	0.2	2500	330	38.5

The hydraulic conductivity curves and soil-water characteristic curves of different materials are shown in Figure 4. The initial pore water pressure distribution and provided in the Supplementary Materials.

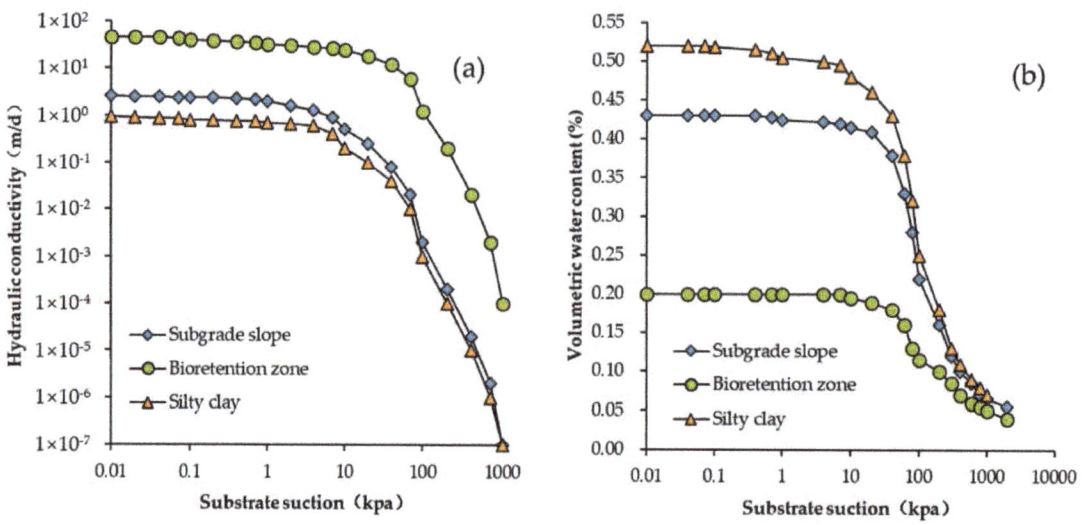

Figure 4. Hydraulic conductivity (a) and soil-water characteristic (b) curves of different materials.

3. Results

3.1. Influence of Bioretention on Pore Water Pressure

In order to analyze the influence of bioretention facility on pore water pressure distribution of subgrade slope under rainfall, it is considered that the layer below the slope toe is the current silty clay layer and the moderately weathered bedrock layer, and the subgrade slope soil is unsaturated being above the groundwater table.

Figure 5 shows a cloud diagram of pore water pressure distribution of subgrade slope within 48 h after rainfall without and with bioretention facility.

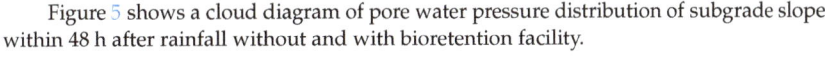

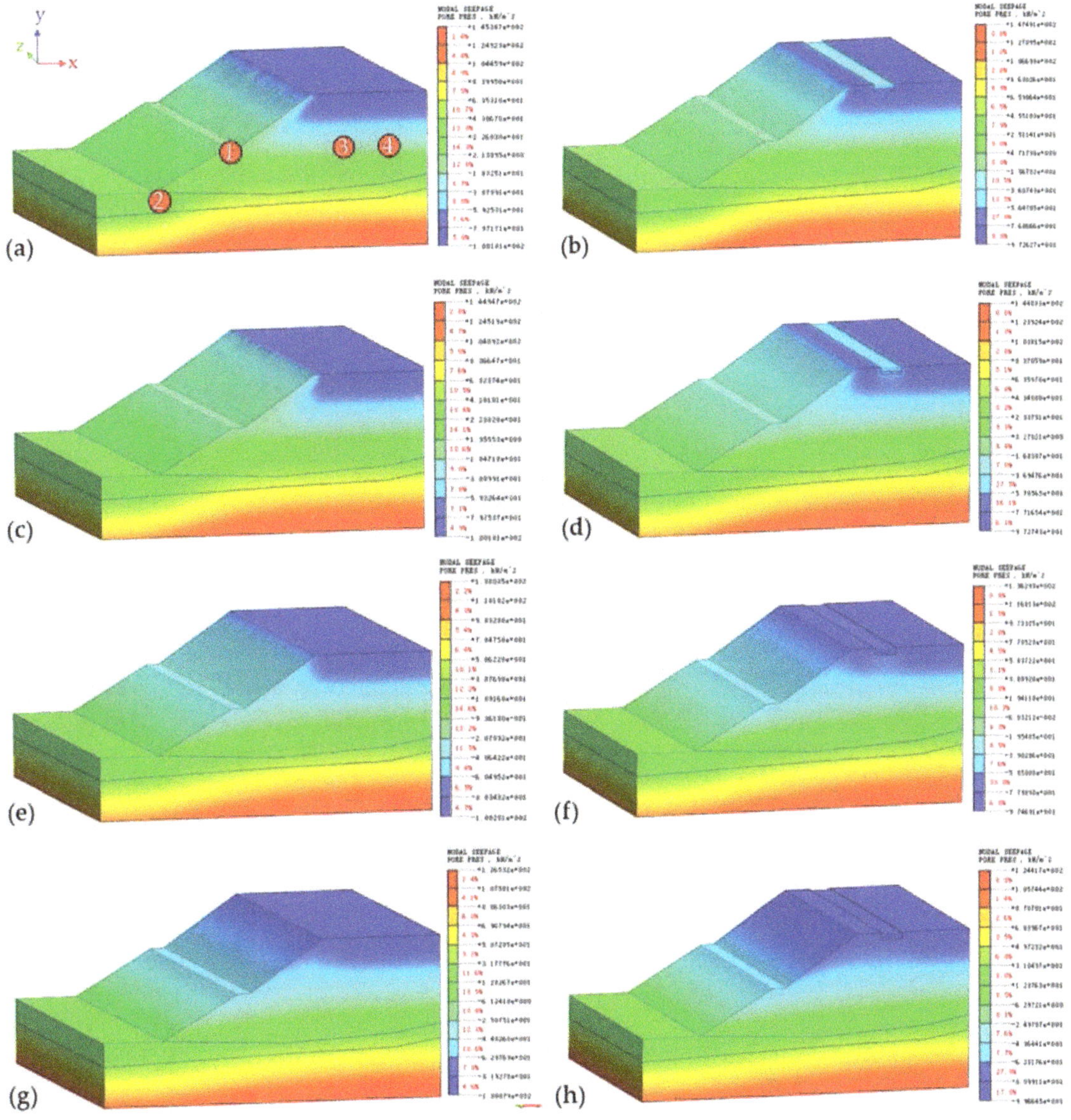

Figure 5. Comparison of water pressure distribution on subgrade slopes without or with biological retention facilities within 48 h after rainfall: (**a**,**b**) 6 h; (**c**,**d**) 12 h; (**e**,**f**) 24 h; (**g**,**h**) 48 h.

Pore water pressure curves (Figure 6) at different control points show that the pore water pressure of rock and soil increases continuously with the rainfall time. When pore water pressure is negative, it indicates that the rock of the slope is unsaturated, and there is substrate suction. The substrate suction gradually decreases as the water content in the rock and soil increases. When the pore water pressure is positive, the inter-particle voids in the soil are filled with water and are in a saturated state. At this time, the substrate suction is zero.

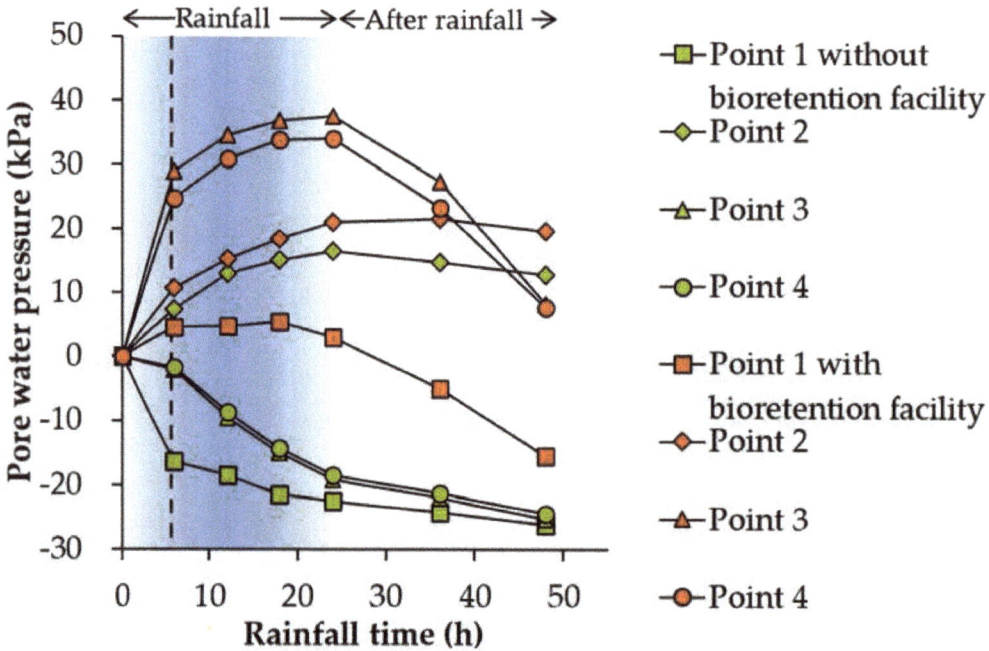

Figure 6. Pore water pressure curves of different control points.

At the beginning of rainfall (0–6 h), the pore water pressures at the four control points showed an increasing trend under the conditions of no biological retention facilities and biological retention facilities. Compared with the roadbed slope without biological retention facilities, the pore water pressure of control point 3 and control point 4 increased significantly from −2.1 kPa to 28.9 kPa and −1.7 kPa to 24.8 kPa, respectively, under the condition of biological retention facilities. It suggested that the rainwater infiltration of biological retention facilities has a greater impact on the pore water pressure of the sidewalk and the soil under the roadway, and the impact is basically the same.

In the middle and late stages of the rainfall (6–24 h), the increasing extent of pore water pressure at each control point under the two conditions decreased. Under the condition of biological retention facilities, the pore water pressure of control points 2, 3, and 4 increased from 10.8 kPa to 21.0 kPa, from 28.9 kPa to 37.5 kPa, and from 24.8 kPa to 34.0 kPa, respectively, showing a trend of slowly increasing with the rainfall time.

During the 24–48 h after the end of rainfall, the pore water pressure of control points 1, 3, and 4 showed different decreases. A decrease from 3.03 kPa to −15.4 kPa, from 37.5 kPa to 8.3 kPa, and from 34.0 kPa to 7.8 kPa were observed for three control points. The pore water pressure of control point 2 only slightly decreases from 21.0 kPa to 19.6 kPa. It indicates that after the rainfall ends, the pore water pressure of the soil under the slope, pavement, and roadway decreases with the infiltration of rainwater. The rainwater in the slope gradually concentrates on the slope foot along the seepage field after the rainfall ends so that the pore water pressure at the slope foot does not show a significant downward trend.

For control points without a bioretention facility, the pore water pressure values at points 1, 3, and 4 are all negative, indicating that the soil under the slope, pavement, and roadway is not saturated under this condition. While the pore water pressure value of point

2 is positive, which indicates that during the rainfall period, the rainwater is concentrated at the foot of the slope and reaches a saturated state. For control points with bioretention facility, the pore water pressure values are all positive values, indicating that the rainwater seepage through bioretention facility causes the saturation condition of slope surface, pavement, and soil under roadway, and has an obvious influence on the distribution of pore water pressure in subgrade slope.

3.2. Influence of Bioretention on Flow Velocity

Figure 7 is the cloud diagram of the flow velocity distribution of subgrade slope within 48 h after rainfall without and with bioretention facility.

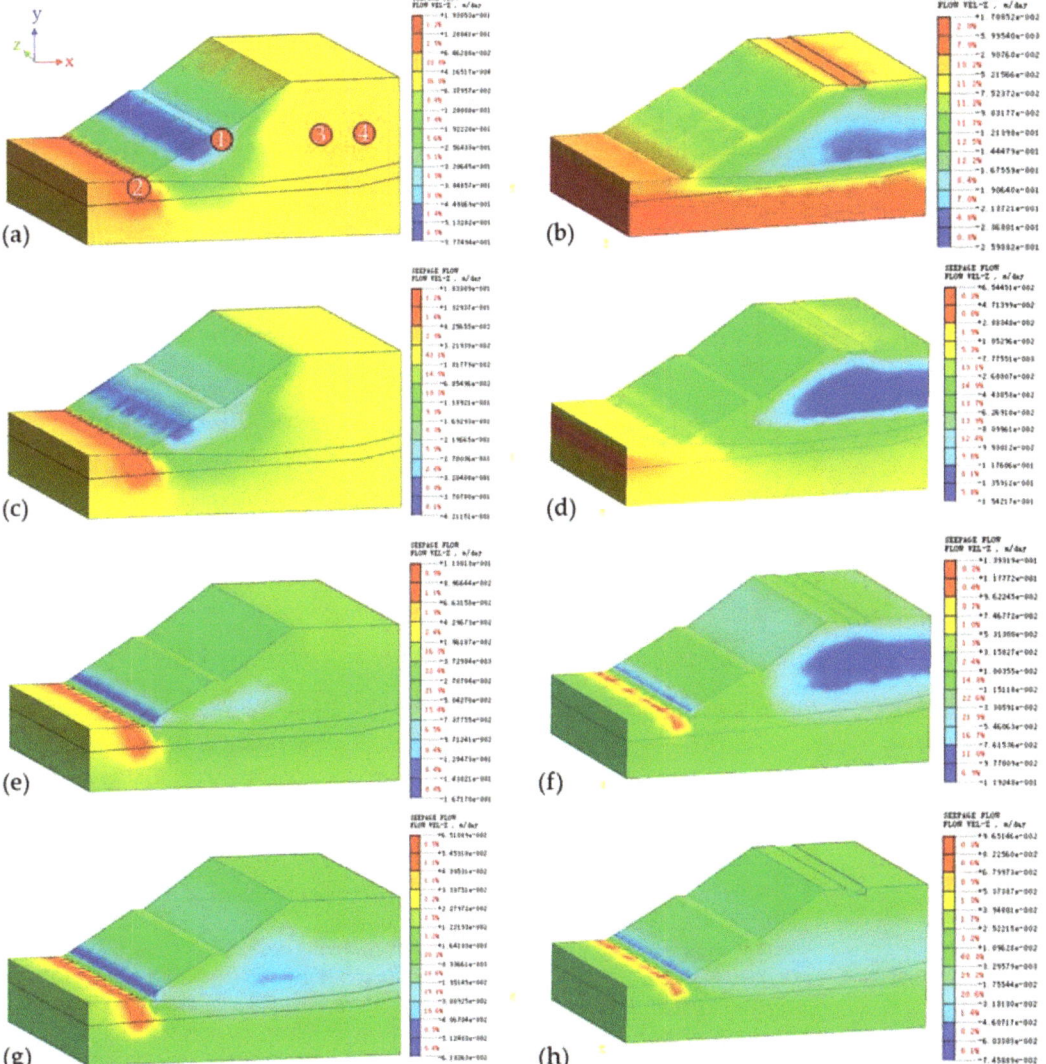

Figure 7. Comparison of water flow rate distribution in subgrade slopes without or with biological retention facilities within 48 h after rainfall: (**a**,**b**) 6 h; (**c**,**d**) 12 h; (**e**,**f**) 24 h; (**g**,**h**) 48 h.

Figure 8 is a graph of the water flow velocity at different control points. In the figure, a positive value means that the direction of water flow is upward along the Z-axis, and a negative value means that the direction of water flow is downward along the Z-axis. The flow velocity curves of different control points were analyzed. In the initial rainfall period of 0–6 h, the flow velocity of control point 1 increased to 9.97 cm/d, control point 2 to 3.52 cm/d, control point 3 to 18.63 cm/d, and control point 4 to 18.74 cm/d. The flow velocity at all four control points increased linearly with the rainfall time, indicating that the slope soil saturated faster in the early rainfall stage, the flow velocity increased greatly, and the water flow at each control point flowed down to the unsaturated soil along the gravity direction. In the middle and late period of rainfall (6–24 h), the flow velocity at control point 1, control point 3, and control point 4 decreased gradually from 9.97 cm/d to 1.45 cm/d, 18.63 cm/d to 10.57 cm/d, and from 18.74 cm/d to 10.19 cm/d under the condition of no biological retention facilities. The flow velocity at control point 2 gradually increased from 3.52 cm/d to 8.21 cm/d, and meanwhile, the flow at control point 2 changed into upward seepage.

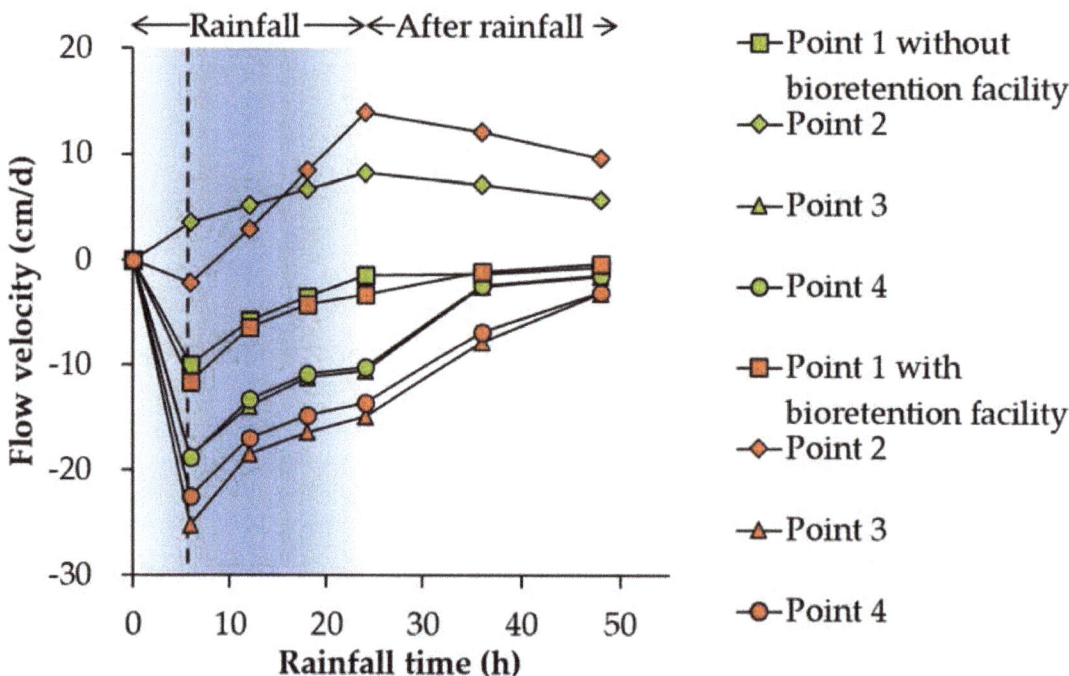

Figure 8. Flow velocity curves of different control points.

Therefore, in the middle and late stages of rainfall, as the slope surface and the interparticle pore of soil are gradually filled with water, and the rainfall intensity decreases, the internal seepage velocity of the saturated soil gradually decreases. Under the effect of the seepage field, the rainwater seepage from the subgrade slope gradually concentrates at the foot of the slope, causing the increase in water flow velocity at the foot of the slope and upward seepage. After the end of the rainfall (24–48 h), the flow velocity of all control points gradually decreases and approaches zero. Among them, the time required for control point 2 to reduce the flow rate to 0 is longer than that of other control points due to the concentration of rainwater.

Compared with the point without bioretention facility, the changing trend of water flow velocity and direction at the point with bioretention facility is the same, and the flow velocity increases in different degrees. At the beginning of rain, the difference between the flow velocity of each control point under the two conditions is small, indicating that the rainwater seepage of the bioretention facility at the initial stage of rainfall has little effect on the water flow velocity inside the subgrade slope soil. In the middle and late periods of rain, when there is a bioretention facility, the flow velocity at point 2 increased from 2.88 cm/d to 13.93 cm/d, much higher than that at point 1, point 3, and point 4. It shows that the rainwater seepage in the bioretention facility is concentrated at the foot of the slope through the seepage field and generates a greater flow velocity. In addition, the increased value of control point 3 is the same as that of control point 4, indicating that the impact of rainwater from the bioretention facility on the flow velocity of the soil under the pavement and the roadway is the same.

3.3. Influence of Bioretention on Slope Displacement

Figures 9 and 10 show the horizontal and the vertical displacement distribution of the subgrade slope within 48 h after rainfall without and with bioretention facilities.

Figure 11 shows the horizontal and vertical displacement curves of different control points. In Figure 11, the positive value represents the upward displacement along the Z-axis, and the negative value represents the downward displacement along the Z-axis. The horizontal displacement of the slope conforms to the dynamic characteristics of the circular slip of the soil slope and is consistent with the displacement trend of the conventional soil slope [30,31]. With the change in rainfall time, the horizontal displacement and vertical displacement at each control point of the subgrade slope gradually increase.

Under the condition without biological retention facilities, the horizontal displacement of control points 1, 2, 3, and 4 is increased to 25.1 mm, 39.5 mm, 10.7 mm, and 6.8 mm, respectively. The increase in control point 1 and control point 2 was higher, indicating that the horizontal direction of the slope and the soil of the slope toe was greatly affected by the rainwater infiltration during the rainfall seepage process, and the horizontal displacement changed greatly. In particular, the slope toe of control point 2 is subjected to stress concentration and gradual rainwater seepage concentration, and the horizontal displacement is greater than that at the slope of control point 1.

At the same time, vertical downward displacement of 10.1 mm, 17.2 mm, and 17.7 mm at control point 1, control point 3, and control point 4 were observed. The vertical displacement of control point 3 and control point 4 is greater than that of control point 1. The results show that under the action of gravity and osmotic force, the pores between soil particles decrease, and the gravity and osmotic force at control point 3 and control point 4 are higher, so the vertical displacement of soil mass changes greatly. The maximum displacement at control point 2 is 12.6 mm, indicating that the slope toe heave occurs under the effect of stress concentration and seepage.

Twenty-four to forty-eight hours after the rain ends, the horizontal and vertical displacements of each control point increased slightly, which suggested that the slope surface and internal soil particles gradually became compacted under the effect of gravity and seepage force, and the variation in soil displacement decreases. The influence of infiltrating rainwater on the displacement of subgrade slope and internal soil gradually weakened.

Compared with the displacement value of the subgrade slope without biological retention facilities, under the condition of biological retention facilities, the changing trend of the horizontal and vertical displacements of each control point is basically the same, and there are different degrees of increase on the basis of the displacement value without biological retention facilities. The maximum horizontal displacement of control point 1 increased from 25.9 mm to 45.6 mm, the maximum vertical displacement increased from 10.4 mm to 18.5 mm. The horizontal displacement of control point 2 increased from 40.8 mm to 66.1 mm, and the vertical displacement increased from 13.5 mm to 33.2 mm. The horizontal displacement of control point 3 increased from 11.4 mm to 14.8 mm, and the vertical displacement of control point 3 increased from 18.1 mm to 26.1 mm. The horizontal displacement of control point 4 increased from 7.05 mm to 8.31 mm, and the vertical displacement increased from 18.4 mm to 22.8 mm. The increased value of the horizontal and vertical displacement of control point 2 was significantly greater than that of other control points. It is proved that the rainwater seepage in the bioretention facility concentrates on the slope foot through the seepage field and produces greater stress concentration at the slope foot, resulting in a greater displacement of the soil at the slope foot. In addition, the increase in displacement at point 3 is the same as that at point 4, which indicates that the influence of rainwater seepage from the bioretention facility on the horizontal and vertical displacement of the soil under the pavement and the roadway is the same.

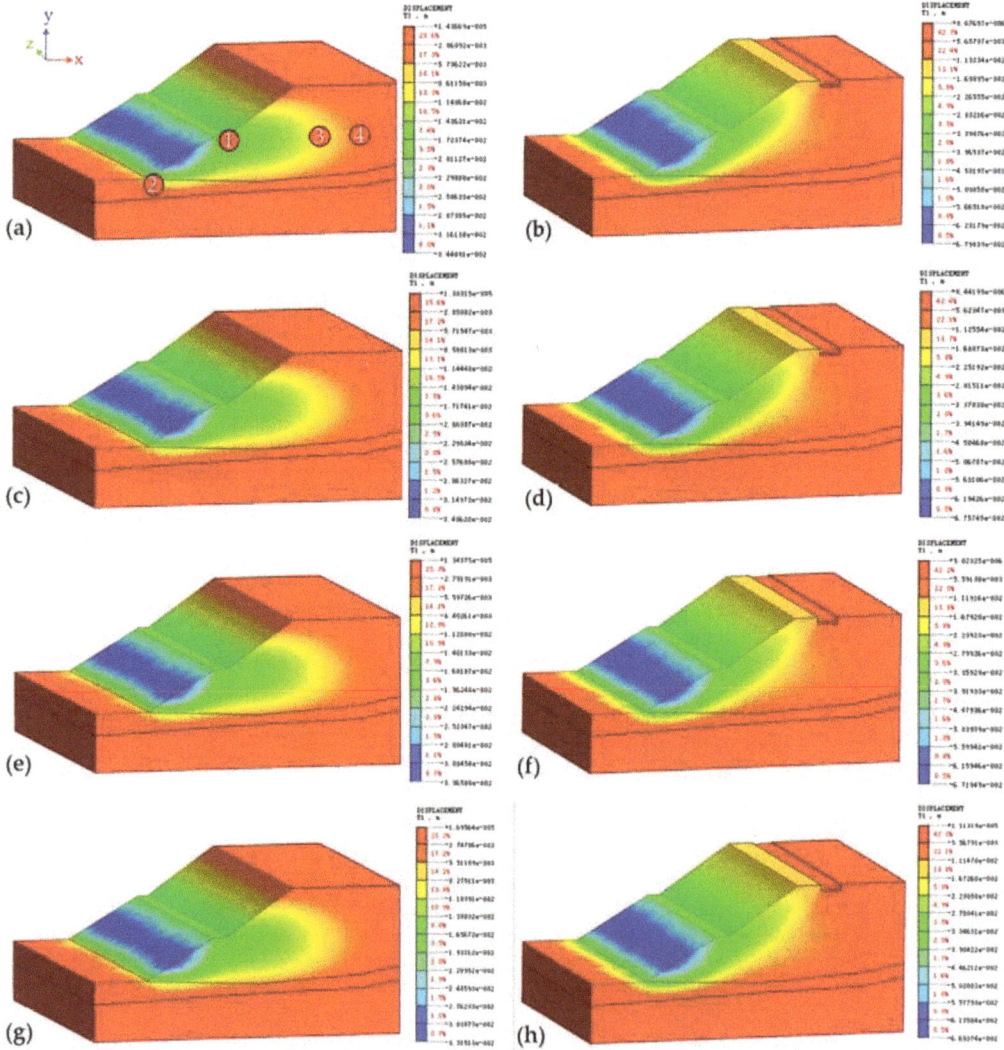

Figure 9. Comparison of horizontal displacement distribution in subgrade slopes without or with biological retention facilities within 48 h after rainfall: (**a**,**b**) 6 h; (**c**,**d**) 12 h; (**e**,**f**) 24 h; (**g**,**h**) 48 h.

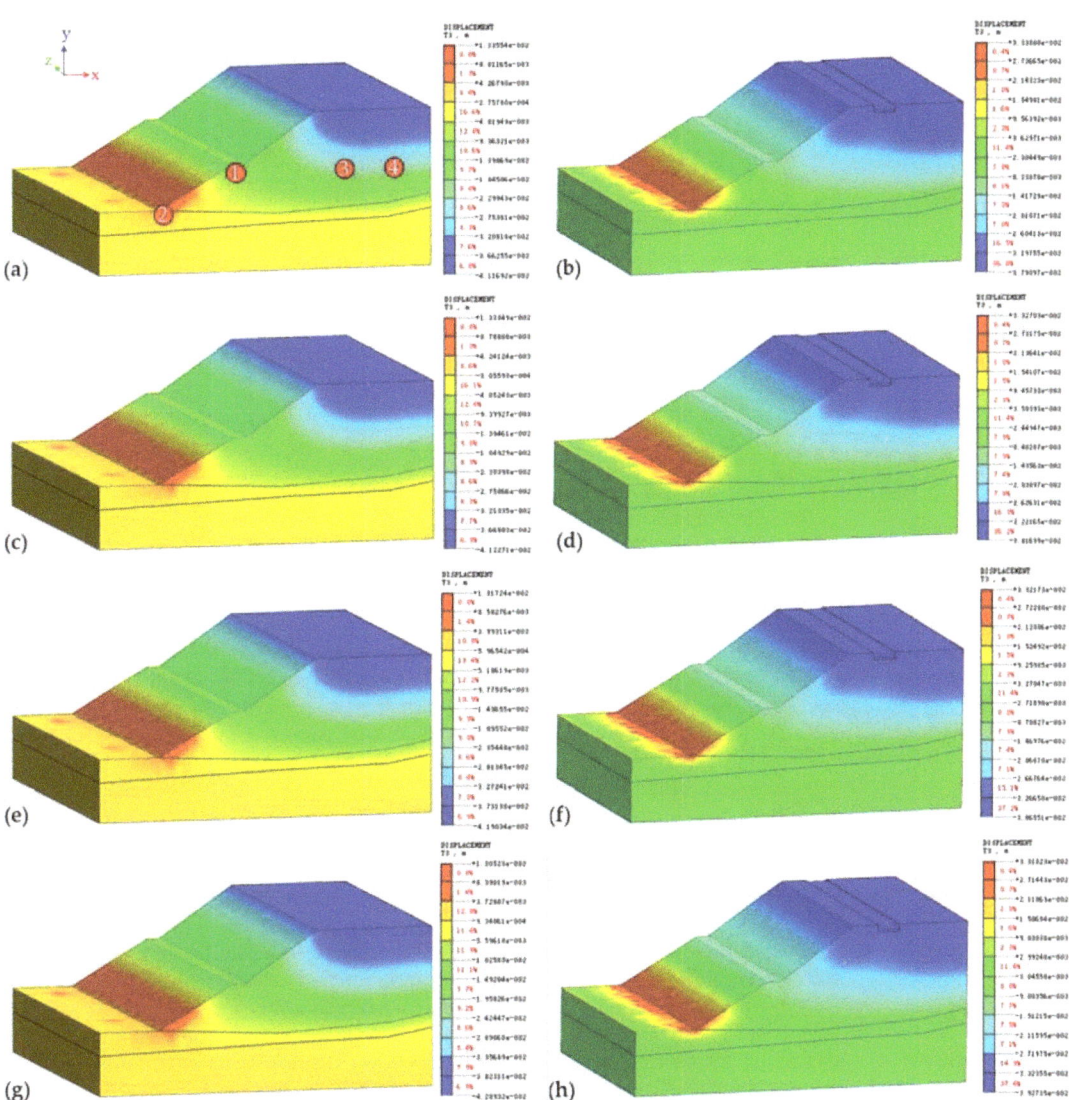

Figure 10. Comparison of vertical displacement distribution in subgrade slopes without or with biological retention facilities within 48 h after rainfall: (**a**,**b**) 6 h; (**c**,**d**) 12 h; (**e**,**f**) 24 h; (**g**,**h**) 48 h.

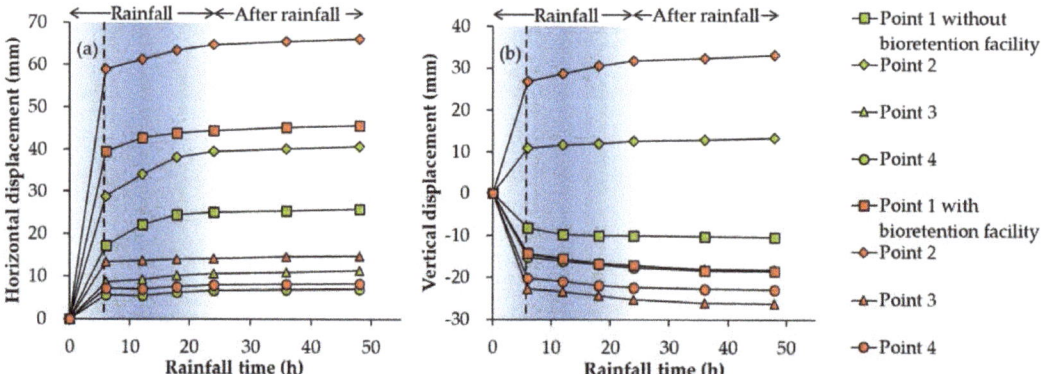

Figure 11. Horizontal (**a**) and vertical (**b**) displacement curves of different control points.

4. Discussion

Compared with the rainwater infiltration of traditional roads, obvious differences exist in that of sponge cities. Traditional rainwater runoff infiltration is dominated by areal infiltration. Rainwater runoff is scattered throughout the area, and the amount of infiltration rainwater accepted per unit area is small. However, rainwater runoff in the sponge city is dominated by concentrated infiltration in points and strips. Generally, the service area of the bioretention facility is 5–20 times that of itself, and the amount of infiltration water increases significantly. The confluence of traditional rainwater runoff gradually disappears after the rainfall ends, and the infiltration also ends quickly. However, the rainwater runoff centralized infiltration facility in the sponge city can store a large amount of runoff rainwater. After the rainfall is over, the infiltration continues for a certain time. During heavy rain or continuous rainfall, a large amount of rainwater intensively infiltrates the side slope of the roadbed, which changes the key factors such as the distribution of water pressure on the side slope of the roadbed, runoff velocity, and the hydraulic slope. Consequently, the displacement and stress of the slope change, leading to the change in the stability of the subgrade slope.

Combined with the project situation and the above analysis results, the rainwater infiltration from the bioretention facility increases the pore water pressure and flow velocity of soil in the subgrade slope, making the soil easier to reach saturation. In the saturated state, the substrate suction of the soil is significantly weakened, which leads to a rapid decrease in soil cohesion and internal friction angle and a large increase in horizontal and vertical displacement. Especially at the slope foot where the slope stress is concentrated, the uplift and destruction of the foot are more likely to occur, thus having a greater impact on the stability of the slope.

Different rainwater runoff regulation requirements, subgrade compaction degrees, slope heights, and slope ratios in sponge roads all affect the pore water pressure, seepage velocity, and displacement of the subgrade slope. When it is necessary to increase rainwater runoff infiltration of sponge road or change the size and type of filler of bioretention facility, the rainwater seepage in the bioretention facility leads to a larger impact range of the subgrade slope seepage field, higher pore water pressure and water flow velocity, faster saturation state of the roadbed soil, and higher slope displacement. When the compaction of the subgrade increases, the migration of rainwater seepage becomes more difficult, the expansion speed slows down [32], and the impact range of infiltration rainwater in bioretention facility, pore water pressure, flow velocity, and slope displacement is reduced, and vice versa. The higher slope height and steeper slope rate produce greater stress concentration, resulting in a greater displacement of the slope. The infiltration of rainwater brought by the bioretention facility aggravates this phenomenon, which makes the slope displacement larger, and has a more adverse effect on the stability of the slope.

According to the above analysis, with the spread of sponge cities nationwide, the impact of rainwater seepage from biological retention facilities varies in different regions, different geological environments, and different building types. For example, in sandy soil areas with a high soil hydraulic conductivity, rainwater can seep quickly, making the soil difficult to reach a saturated state, and the impact of rainwater seeping in biological retention facilities on subgrade slopes is relatively small [33]. In the loess area with strong water sensitivity, the infiltration of rainwater makes the loess soaked into collapsibility, the soil undergoes substantial settlement and deformation, and the shear strength of the soil decreases rapidly, causing serious instability and damage to the roadbed slope [34]. In addition, when the biological retention facilities are applied to the surrounding sites of the building, it also has an impact on the basic structure of the building. Infiltration of rainwater changes the effective stress of the foundation soil of the building, increases the additional settlement of the foundation, and causes the building to sink or tilt. Therefore, it is necessary to adjust the layout of biological retention facilities according to the requirements of building settlement [35].

5. Conclusions

The bioretention facility in the sponge city can effectively control the runoff of road rainwater and improve the utilization rate of water resources. At the same time, the rainwater seepage path, water pressure distribution, and water flow velocity of the subgrade slope are changed, and the slope displacement is affected. Some key findings are as follows:

(1) Compared with the condition without a bioretention facility, the rainwater seepage from the subgrade slope with a bioretention facility linearly increases the water pressure on the surface and foot of the subgrade slope. At the same time, rainwater seepage makes the soil on the slope, pavement, and roadway saturated, which has a significant impact on the distribution of pore water pressure on the subgrade slope;

(2) In terms of water flow velocity, the bioretention facility at the beginning of rainfall has little effect on the water flow velocity inside the soil of subgrade slope. In the middle and late stages of rainfall, the rainwater seepage through the seepage field is concentrated on the slope foot, which produces a greater flow velocity at the slope foot and is more likely to cause the soil to soften and uplift;

(3) The overall horizontal and vertical displacement of the subgrade slope gradually increases with the increase in rainfall time. The rainwater seepage in the bioretention facility concentrates on the slope foot through the seepage field, which produces greater stress concentration at the slope foot and results in a greater displacement of the soil at the slope foot.

The bioretention facility has an impact on the water pressure distribution, water flow velocity, and displacement of the subgrade slope. In particular, it causes a large uplift displacement at the foot of the slope, which adversely affects the stability of the slope. This study shows that the suitable width, depth, and structural layer of the bioretention facility should be selected to ensure the safety and stability of the subgrade slope in careful consideration of runoff control effect, subgrade compactness, slope height, and slope rate.

Supplementary Materials: The following are available online at https://www.mdpi.com/article/10.3390/w13233466/s1, Figure S1: Initial pore water pressure and water flow rate distribution diagram.

Author Contributions: Conceptualization, W.T. and H.C.; methodology, W.T. and H.C.; software, W.T.; validation, W.T., Z.S., and H.C.; formal analysis, W.T.; investigation, W.T.; resources, W.T. and H.C.; data curation, W.T. and H.M.; writing—original draft preparation, W.T.; writing—review and editing, W.T., H.M., and X.W.; visualization, W.T., X.W., and H.M.; supervision, H.M., Z.S., Q.H., and H.C.; project administration, H.C.; funding acquisition, Z.S. and H.C. All authors have read and agreed to the published version of the manuscript.

Funding: This research was funded by Chongqing Key R&D Program, Chongqing, PR China, grant number cstc2018jszx-zdyfxmX0010, and the support of Creative Research Groups in Colleges and

Universities of Chongqing (No. CXQT21001, Water Environment Protection and Management in Mountainous City).

Conflicts of Interest: The authors declare no conflict of interest.

References

1. Lee, J.Y.; Kim, H.; Kim, Y.; Han, M.Y. Characteristics of the event mean concentration (EMC) from rainfall runoff on an urban highway. *Environ. Pollut.* **2011**, *159*, 884–888. [CrossRef] [PubMed]
2. Helmreich, B.; Hilliges, R.; Schriewer, A.; Horn, H. Runoff pollutants of a highly trafficked urban road—Correlation analysis and seasonal influences. *Chemosphere* **2010**, *80*, 991–997. [CrossRef] [PubMed]
3. Paus, K.H.; Morgan, J.; Gulliver, J.S.; Hozalski, R.M. Effects of bioretention media compost volume fraction on toxic metals removal, hydraulic conductivity, and phosphorous release. *J. Environ. Eng.* **2014**, *140*, 04014033. [CrossRef]
4. Song, X.; Zhang, J.; Wang, G.; He, R.; Wang, X. Development and challenges of urban hydrology in a changing environment: II: Urban stormwater modeling and management. *Adv. Water Sci.* **2014**, *25*, 752–764.
5. Zhang, Q.; Xiangquan, L.; Wang, X.; Wan, W.; Ouyang, Z. Research advance in the characterization and source apportionment of pollutants in urban roadway runoff. *Ecol. Environ. Sci.* **2014**, *2*, 352–358.
6. USEPA. *National Water Quality Inventory: Report to Congress*; Office of Water Regulations and Standards: Washington, DC, United States, 1994; p. 344, ISBN 32437010554794.
7. Dikshit, A.; Loucks, D.P. Estimating Non-Point Pollutant Loadings—I: A Geographical-Information-Based Non-Point Source Simulation Model. *J. Environ. Syst.* **1996**, *24*, 395–408. [CrossRef]
8. Taylor, M.; Henkels, J. Stormwater Best Management Practices: Preparing for the Next Decade. *Stormwater* **2001**, *2*, 1–11.
9. Fletcher, T.D.; Shuster, W.; Hunt, W.F.; Ashley, R.; Butler, D.; Arthur, S.; Trowsdale, S.; Barraud, S.; Semadeni-Davies, A.; Bertrand-Krajewski, J.-L. SUDS, LID, BMPs, WSUD and more–The evolution and application of terminology surrounding urban drainage. *Urban Water J.* **2015**, *12*, 525–542. [CrossRef]
10. Mohurd, P. *Sponge City Construction Technical Guide*; China Construction Industry Press: Beijing, China, 2015; pp. 1–88.
11. Zhang, J.; Wang, Y.; Hu, Q.; He, R. Discussion and views on some issues of the sponge city construction in China. *Adv. Water Sci.* **2016**, *27*, 793–799.
12. Sun, X.; Davis, A.P. Heavy metal fates in laboratory bioretention systems. *Chemosphere* **2007**, *66*, 1601–1609. [CrossRef]
13. Mangangka, I.R.; Liu, A.; Egodawatta, P.; Goonetilleke, A. Performance characterisation of a stormwater treatment bioretention basin. *J. Environ. Manag.* **2015**, *150*, 173–178. [CrossRef] [PubMed]
14. Gülbaz, S.; Kazezyılmaz-Alhan, C.M. Experimental investigation on hydrologic performance of LID with rainfall-watershed-bioretention system. *J. Hydrol. Eng.* **2017**, *22*, D4016003. [CrossRef]
15. Davis, A.P.; Shokouhian, M.; Sharma, H.; Minami, C.; Winogradoff, D. Water quality improvement through bioretention: Lead, copper, and zinc removal. *Water Environ. Res.* **2003**, *75*, 73–82. [CrossRef]
16. Hunt, W.; Smith, J.; Jadlocki, S.; Hathaway, J.; Eubanks, P. Pollutant removal and peak flow mitigation by a bioretention cell in urban Charlotte, NC. *J. Environ. Eng.* **2008**, *134*, 403–408. [CrossRef]
17. Hatt, B.E.; Fletcher, T.D.; Deletic, A. Hydrologic and pollutant removal performance of stormwater biofiltration systems at the field scale. *J. Hydrol.* **2009**, *365*, 310–321. [CrossRef]
18. Stanković, S.M. Blue green component and integrated urban design. *Tehnika* **2016**, *71*, 365–371. [CrossRef]
19. Liang, H.-H.; Li, X.-L.; Zhang, X.; Ma, Y.; Ji, G.-Q.; Hu, Z.-P.; Lu, Y.-N. Optimization Analysis of Rainwater Infiltration in Bioretention Zone near Municipal Roads. *China Water Wastewater* **2020**, *36*, 107–112.
20. Meng, Y.; Wang, H.; Zhang, S.; Chen, J. Experiments on detention, retention and purifying effects of urban road runoff based on bioretention. *J. Beijing Norm. Univ. (Nat. Sci.)* **2013**, *49*, 286–291.
21. Pan, G.; Xia, J.; Zhang, X.; Wang, H.; Liu, E. Research on simulation test of hydrological effect of bioretention units. *Water Resour. Power* **2012**, *30*, 13–15.
22. Lin, H.-c.; Yu, Y.-z.; Li, G.; Peng, J. Influence of rainfall characteristics on soil slope failure. *Chin. J. Rock Mech. Eng.* **2009**, *28*, 198–204.
23. Pradel, D.; Raad, G. Effect of permeability on surficial stability of homogeneous slopes. *J. Geotech. Eng.* **1993**, *119*, 315–332. [CrossRef]
24. Ng, C.W.W.; Shi, Q. A numerical investigation of the stability of unsaturated soil slopes subjected to transient seepage. *Comput. Geotech.* **1998**, *22*, 1–28. [CrossRef]
25. Sun, Q.; Hu, X.; Wang, Y.; Li, M. Research on instability of slope composed of two strain-softening media. *Rock Soil Mech.* **2009**, *30*, 976–980.
26. Darcy, H. Determination of the laws of water flow through sand. *Public Fountains City Dijon Append. D Filtr. Victor Dalmont Paris* **1856**, *1*, 1–10.
27. Richards, L.A. Capillary conduction of liquids through porous mediums. *Physics* **1931**, *1*, 318–333. [CrossRef]
28. Guan, J.; Lyu, Y.; Zhao, G. Ecological Restoration and Slope Stability Analysis of Slag Discharge Site on the North Side of Yumen River. *Min. Res. Dev.* **2021**, *41*, 90–94.
29. Zhao, J.; Lin, J.; Gong, H.; Zheng, X. Effect of Biological Retention Ditch on Water Distribution in Subgrade of Sponge City. *Sci. Technol. Eng.* **2018**, *18*, 239–245.

30. Fredlund, D.G.; Krahn, J. Comparison of slope stability methods of analysis. *Can. Geotech. J.* **1977**, *14*, 429–439. [CrossRef]
31. Chugh, A.K. Variable interslice force inclination in slope stability analysis. *Soils Found.* **1986**, *26*, 115–121. [CrossRef]
32. Liu, Z. *The Evolution of Highway Subgrade Humidity Field in Western Arid and Semi Arid Regions*; China University of Mining and Technology: Xuzhou, China, 2012; pp. 1–146.
33. Zhang, J.; Zhang, S.; Shuai, P.; Ji, Y. Lateral Seepage Prevention Study of Sinking-Mode Greenbelt Based on Fluid-Solid Coupling. *China Munic. Eng.* **2017**, *1*, 85–88.
34. Wang, Q.; Li, X.; Zhang, X.; Wang, S. Risk Analysis of Leakage Location of Bioretention Zone in Municipal Roads. *Sci. Technol. Eng.* **2019**, *19*, 321–326.
35. Deng, Z.; Wen, X.; Hu, Z.; Chai, S.; Ma, Y.; Ji, G. Impact of Leakage of Sponge Facilities on Building at Loess Site in Xixian New Area of Shaanxi Province, China. *J. Earth Sci. Environ.* **2020**, *42*, 560–568.

Article

Field Study of the Road Stormwater Runoff Bioretention System with Combined Soil Filter Media and Soil Moisture Conservation Ropes in North China

Qian Li [1], Haifeng Jia [1,2,*], Hongkai Guo [3], Yunyun Zhao [4], Guohua Zhou [5], Fang Yee Lim [6], Huiling Guo [7], Teck Heng Neo [6], Say Leong Ong [6] and Jiangyong Hu [6]

[1] School of Environment, Tsinghua University, Beijing 100084, China; gullqa@126.com
[2] Jiangsu Collaborative Innovation Center of Technology and Material of Water Treatment, Suzhou University of Science and Technology, Suzhou 215009, China
[3] Glodon Co., Ltd., Beijing 100193, China; ghkjiaoaodehuo@163.com
[4] Academy of Agricultural Planning and Engineering, MARA, Beijing 100125, China; zhaoyunyun@aape.org.cn
[5] North China Municipal Engineering Design & Research Institute Co., Ltd., Tianjin 300381, China; 13821927908@163.com
[6] NUS Environmental Research Institute, National University of Singapore, Singapore 117576, Singapore; rlimtony@gmail.com (F.Y.L.); teckheng94@gmail.com (T.H.N.); say.leong_ong@nus.edu.sg (S.L.O.); hujiangyong@nus.edu.sg (J.H.)
[7] School of Life Sciences & Chemical Technology, Ngee Ann Polytechnic, 535 Clementi Road, Blk 83, Singapore 599489, Singapore; huiling@np.edu.sg
* Correspondence: jhf@tsinghua.edu.cn

Abstract: Growing concerns about urban runoff pollution and water scarcity caused by urbanization have prompted the application of bioretention facilities to manage urban stormwater. The purpose of this study was to evaluate the performance of proposed bioretention facilities regarding road runoff pollutant removal and the variation characteristics of the media physicochemical properties and microbial diversity in dry-cold regions. Two types of bioretention facilities were designed and then constructed in Tianjin Eco-city, China, on the basis of combined soil filter media screened by a laboratory-scale test with a modified bioretention facility (MBF) containing soil moisture conservation ropes. Redundancy analysis was performed to evaluate the relationships between the variation in media physicochemical properties and microbial communities. An increase in media moisture could promote an increase in the relative abundance of several dominant microbial communities. In the MBF, the relatively low nitrate-nitrogen (NO_3-N) (0.75 mg/L) and total nitrogen (TN) (4.71 mg/L) effluent concentrations, as well as better removal efficiencies for TN and NO_3-N in challenge tests, were mainly attributed to the greater relative abundance of *Proteobacteria* (25.2%) that are involved in the microbial nitrogen transformation process. The MBF also had greater media microbial richness (5253 operational taxonomic units) compared to the conventional bioretention facility and in situ saline soils. The results indicate that stormwater runoff treated by both bioretention facilities has potential use for daily greening and road spraying. The proposed design approach for bioretention facilities is applicable to LID practices and sustainable stormwater management in other urban regions.

Keywords: modified bioretention facility; road stormwater runoff; combined soil filter media; soil moisture conservation rope; field study; microbial diversity

Citation: Li, Q.; Jia, H.; Guo, H.; Zhao, Y.; Zhou, G.; Lim, F.Y.; Guo, H.; Neo, T.H.; Ong, S.L.; Hu, J. Field Study of the Road Stormwater Runoff Bioretention System with Combined Soil Filter Media and Soil Moisture Conservation Ropes in North China. *Water* 2022, 14, 415. https://doi.org/10.3390/w14030415

Academic Editors: Arash Massoudieh and Jose G. Vasconcelos

Received: 15 November 2021
Accepted: 26 January 2022
Published: 29 January 2022

Publisher's Note: MDPI stays neutral with regard to jurisdictional claims in published maps and institutional affiliations.

Copyright: © 2022 by the authors. Licensee MDPI, Basel, Switzerland. This article is an open access article distributed under the terms and conditions of the Creative Commons Attribution (CC BY) license (https://creativecommons.org/licenses/by/4.0/).

1. Introduction

With rapid urbanization, there has been a significant increase in impermeable surface areas such as urban roads and roofs, resulting in an increase in surface runoff and ultimately leading to environmental problems such as runoff pollution and water environment deterioration [1]. Table 1 shows the contamination of rainfall runoff from different types of urban land covers. It can be seen that the runoff indicators of total suspended solids (TSS), chemical oxygen demand (COD), total nitrogen (TN), and total phosphorus (TP) for roads are worse than those for other land cover types. This is mainly due to frequent traffic activities, automobile exhaust emissions, tire wear, and the corrosion of components, which lead to the accumulation of a large number of pollutants such as suspended particulate matter, nutrients, and organic matter on road surfaces. It can also be seen that urban road runoff is characterized by heavy and fluctuating pollution, especially for the indicators of TSS and COD [2,3]. Therefore, the control of road runoff pollution is of great significance to urban runoff pollution control and ecological environmental protection.

Table 1. Comparison of water quality of stormwater runoff from different types of land covers.

Land Cover Type	Region	TSS (mg/L)	COD (mg/L)	TN (mg/L)	TP (mg/L)	Reference
Road	China	53–1947	104–779	3.02–18.9	0.26–2.94	[4]
	US	9–466	19–2280	/	0.1–8.2	[5]
	Japan	60	49	/	/	[6]
	Italy	11–281	15–377	/	/	[7]
	Germany	18.3–3165	/	/	/	[8]
Grass land	China	28	21	4.20	0.24	[9]
Roof	China	43	52	4.30	0.11	[10]
	Italy	0–42	/	/	/	[7]

To address the problem of rainfall-runoff pollution, many countries have proposed concepts and measures such as low impact development (LID), best management practices (BMPs), and green infrastructure [11,12]. Sponge cities, a new paradigm of rainfall-runoff management with LID-BMP techniques as source control measures, have been promoted in China since 2013 [13]. At present, the measures used to deal with rainfall-runoff pollution from urban roads worldwide mainly include bioretention facilities, infiltration ponds, artificial wetlands, grassed swales, vegetated buffer strips, etc. In particular, bioretention facilities can effectively intercept, adsorb, and transform pollutants from rainfall-runoff through interactions between plants, soil filter media, and microorganisms, and they have been widely used in sponge city construction for rainfall-runoff control [14,15].

Due to the involvement of various pollutant removal processes, the water purification performance of bioretention facilities is influenced by many factors, such as the composition of filter media, plant configurations, anaerobic zone settings, hydraulic retention time, temperature, rainfall characteristics, etc. Developing strategies to improve their effectiveness in pollutant removal is still a research hotspot and a challenge [16,17]. Previous studies have attempted to adjust and optimize the structure of bioretention facilities, including by improving the composition and proportions of filter media [18–20], plant configurations [21,22], and increased submerged zones [23]. The addition of media with large surface areas and high adsorption capacities, such as zeolite, vermiculite, and biochar, to natural soil can improve the removal efficiency of nitrogen and phosphorus from runoff [24]. Given the economic cost of filter media and the utilization of solid waste resources, several studies have also been conducted to improve the removal efficiency by adding river sediments and water treatment plant sludge, among other activities [25]. In addition, the majority of studies have shown that plants play a role in promoting the removal of runoff pollutants [26,27], but some have found that plants do not significantly improve the removal efficiency, and there are even cases where pollutant removal efficiency deteriorates [28]. The modification of the bioretention facility structure also affects the growth and activity of microorganisms inside the facility. Microorganisms are involved in numerous reactions, such as ammonification, nitrification, denitrification, and phosphorus enrichment, during the pollutant removal process in bioretention facilities, facilitating denitrification and phosphorus removal from runoff. However, the above-mentioned studies were mainly conducted at the laboratory scale, and the outcomes were rarely transferred to on-site bioretention facilities for practical application and further evaluation. Previous field trials also lacked systematic design evaluation of bioretention facilities from physical, chemical, and microbiological perspectives [29]. These barriers will eventually limit the application of the bioretention technique to control stormwater runoff pollution.

In this work, a field study of bioretention facilities was conducted on the basis of preliminary laboratory tests, with an aim to evaluate the performance of the on-site bioretention facility regarding runoff pollutant removal and media variation characteristics from physical, chemical, and microbial perspectives as well as provide references for the optimization and promotion of stormwater runoff bioretention facilities. The composition and proportion of engineered soil filter media obtained from a laboratory-scale test were employed to design the filter media layer. Based on the design of the conventional bioretention facility (CBF), a modified bioretention facility (MBF) was developed by incorporating soil moisture conservation ropes into the filter media layer. Finally, by monitoring and evaluating the stormwater runoff purification efficiency of the bioretention facilities and the variation characteristics of the physicochemical properties and microbial diversity of the filter media, a preliminary analysis of the stormwater purification mechanism of on-site facilities was conducted.

2. Methods

2.1. Study Area

The study area is located on a municipal road in the Sino-Singapore Tianjin Eco-city, Tianjin, China, which has a continental semi-humid monsoon climate with rainy summers and cold and dry winters. The annual average temperature is 12.5 °C, with a maximum temperature of 39.9 °C and a minimum temperature of −18.3 °C. The annual average rainfall is 602.9 mm, mostly (60%) concentrated from July to August.

Figure 1B illustrates the on-site bioretention facilities and the boundary of the catchment area. The bioretention facilities mainly collect runoff from the driveway, sidewalk, and greenbelt areas, with a total catchment area of 1102 m^2 (Table 2). Bioretention facilities cover about 10% of the total catchment area. Two different types of bioretention facilities were designed and constructed independently on site. The north side is the CBF, and the south side is the MBF with soil moisture conservation ropes in the filter media layer. The two facilities have the same configuration except for the distinction of having soil moisture conservation ropes.

Table 2. Areas of different land covers in the study area.

Item	Area (m^2)
Bioretention facility	110
Driveway	456
Sidewalk	190
Greenbelts	346
Total	1102

2.2. Design and Construction of the On-Site Bioretention Facilities

2.2.1. Bioretention Structure and Stormwater Treatment Process

One of the special characteristics of the on-site bioretention facilities is that the soil filter media were screened by a preliminary laboratory-scale test. The optimized media composition was soil, river sand, peat soil, water treatment residual (WTR), vermiculite, and zeolite (Figure 1A) with a mass ratio of 30:24:6:20:10:10 [30]. River sand was used to enhance the infiltration properties of the filter media, and peat soil was used to optimize the media particle gradation for maintaining the water retention and infiltration properties of the filter media [20]. Iron and aluminum metal ions on the surface of WTR can react with dissolved phosphorus in runoff to form phosphate precipitation and result in phosphorus removal [6,31]. Further, WTR was obtained from the local water treatment plant, which reduced the cost of filter media and realized the utilization of solid waste resources. Vermiculite and zeolite are filter media with large surface areas and enhanced adsorption capacity, which can improve the removal efficiency of nitrogen and phosphorus [24]. Table 3 shows the physical properties of the filter media applied in this study.

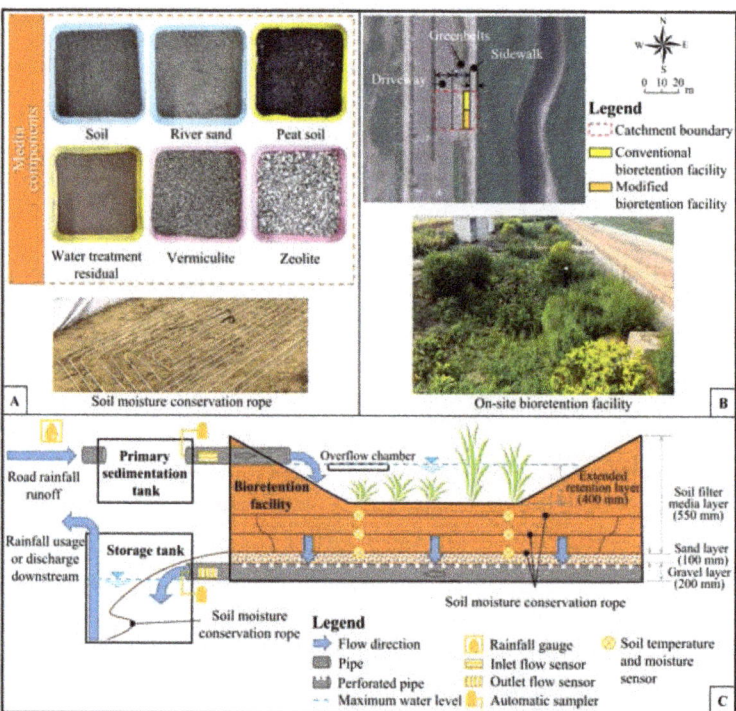

Figure 1. Composition of the combined soil filter media and soil moisture conservation ropes in the field (**A**), constructed bioretention facilities and catchment areas (**B**), and schematic diagram of bioretention facilities and stormwater runoff treatment process (**C**).

Table 3. Physical properties of the bioretention soil filter media.

Media Type	Particle Size (mm)	Dry Density (kg/m^3)	Porosity (%)
Natural soil	<0.5	1.15	52.2
River sand	0.06–0.5	1.20	39.5
Peat soil	1–5	0.61	54.6
Vermiculite	0.5–1	0.33	65.5
Zeolite	5–10	0.93	51.8
Water treatment residual	<0.01	0.65	25.5

Another special characteristic of the on-site bioretention facilities is that the soil moisture conservation rope, the performance of which was proved by a laboratory-scale pilot test, was applied to the systematic design of the on-site bioretention system (Figure 1A). The soil moisture conservation rope draws on the concept of moisture conservation in agriculture to maintain the moisture content of the filter media layer [32]. The rope was arranged in a circular pattern in the horizontal direction with a horizontal spacing of 10 cm and a spacing of 20 cm between different horizontal layers. The rope eventually entered the storage tank, and the stored rainwater could flow back through the soil moisture conservation rope into the bioretention facility during the dry period.

The design height of the bioretention facility is 400 mm for the extended retention layer, 550 mm for the soil filter media layer, 100 mm for the sand layer (0.8–1.0 mm particle size), and 200 mm for the gravel layer (5–10 mm particle size) (Figure 1C). The height of the soil filter media layer is greater than the height of 0.4 m used in other studies [33], thus facilitating the growth of plants and the residence time of runoff in the facility and, to a certain extent, facilitating the removal of runoff pollutants.

The rainwater collected in the primary sedimentation tank was divided and evenly distributed to the two bioretention facilities with equal volumes to ensure the same inflow and pollutant loads in both bioretention facilities. The downstream section of the pipe connecting the primary sedimentation tank to the bioretention facility was placed on the side slope of the bioretention facility. The pipe was slotted on the side facing the facility to guarantee uniform inflow distribution along the inlet pipe. The rainwater in the storage tank, which meets relevant standards for water reuse [34,35], can be used for greening and/or road spraying. Otherwise, it is discharged into downstream municipal rainwater networks. The excess rainwater will be discharged directly into municipal rainwater networks through the overflow chamber (Figure 1C). Based on the proposed design, on-site bioretention facilities were constructed from May 2020 to September 2020. The main construction process included the construction of the primary sedimentation tank and storage tanks, site excavation and slope trimming, arrangement of inlet and outlet pipes and plants, etc. The constructed bioretention facilities are shown in Figure 1B.

2.2.2. Monitoring Methods and Equipment

A series of devices were installed around the bioretention facility, including an automatic sampler, flow sensor, rain gauge, and soil temperature and moisture sensor, enabling automatic on-site sampling of inflow and outflow, along with real-time online monitoring of rainfall and the temperature and moisture conditions of the soil filter media. The monitoring period covered a total of nine months from September 2020 to May 2021, spanning from fall to summer in the study area.

The automatic sampler (Beijing Grasp Technology Development Co., Beijing, China) was used to collect 1 L of rainwater samples from the inlet and outlet of the bioretention facility at 5 min intervals. The purpose of collecting rainfall runoff samples every 5 min was to reveal how runoff pollutant concentrations vary over the rainfall period and the severity of runoff pollution compared to water quality standards. Such monitoring will obtain the event mean concentration (EMC), which is often used to analyze the pollutant treatment efficiency of bioretention facilities [36]. It is also expected to provide a reference for the analysis of rainfall-runoff pollution characteristics and the operation and maintenance of facilities. A 5 min interval was determined according to rainfall characteristics and was consistent with those of flow sensors and rainfall monitoring, which contributed to the accuracy of results. Water quality testing indicators for rainwater samples include TSS, COD, TN, ammonia-nitrogen (NH_3-N), nitrate-nitrogen (NO_3-N), and TP. The standard methods and main instruments used for water quality testing are shown in the Supplementary Materials (Table S1). A flow sensor (Wuhan Newfiber Optics Electron Co., Ltd., Wuhan, China) was used to record the flow rates of inflow and outflow at 5 min intervals. A rainfall gauge (Wuhan Newfiber Optics Electron Co., Ltd., Wuhan, China) was used to monitor the rainfall condition on-site at 5 min intervals. A soil temperature and humidity sensor (Shandong Renke Control Technology Co., Ltd., Jinan, China) was used to monitor the change patterns of temperature and moisture in the filter media layer. The measured moisture refers to the volumetric moisture content (%) of the soil filter media, which was obtained by measuring the soil dielectric constant and analyzing the relationship between the soil dielectric constant and the volumetric moisture content [37]. Monitoring sites for soil temperature and moisture are shown in Figure 2, with the monitoring sensors numbered from 1 to 12.

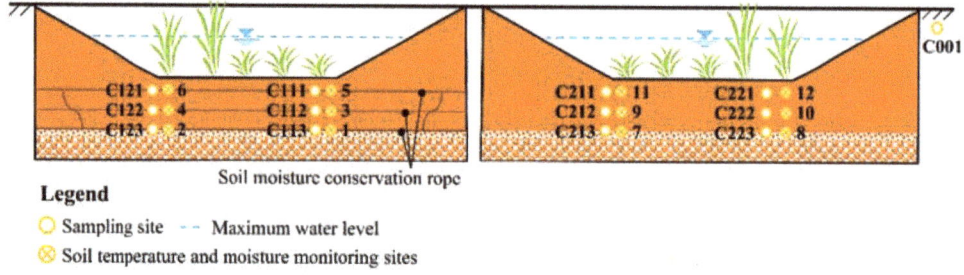

Figure 2. Schematic location and serial number of sampling and monitoring sites.

In addition, the monitoring of the chemical properties and microbial communities of the filter media layer was carried out by manual sampling. The sampling sites (numbered from C111 to C223) were next to the location of the temperature- and moisture-monitoring sites (Figure 2), except that the control group (C001) was sampled from the original on-site saline soil next to the bioretention facility. The chemical properties of the samples were tested using 6 indicators: pH, cation-exchange capacity (CEC), organic matter (OM), soil total nitrogen (STN), soil total phosphorus (STP), and total salt content (TSC). The standard methods and main instruments used for testing media chemical properties are shown in the Supplementary Materials (Table S2). The microbial diversity of the filter media was examined using the samples collected. Sampling was performed in November 2020, March 2021, and June 2021, and 19 samples, including 18 bioretention facility filter media samples and 1 control on-site saline soil sample, were obtained. High-throughput sequencing was performed on these samples. The data were analyzed on the free online Majorbio Cloud Platform (www.majorbio.com accessed on 25 August 2021). The accession number of these DNA sequences is PRJNA773235. In order to avoid differences in the microbial biomass of each sample, the sample sequences were extracted equally by the minimum number of sequences, and the standardized data were obtained for subsequent statistical analysis. Dilution curves based on the Sobs index on the operational taxonomic unit (OTU) level were used to evaluate the reliability of the amount of sequencing data depending on whether the curves had reached a plateau. The Sobs index reflects the observed richness of the microbial community [38] and was calculated by counting the number of observed OTUs in each sample sequence. The OTU is a common basic unit used in numerical taxonomy to count the number of microorganisms. The effects of environmental factors on the abundance and composition of the microbial community were quantified based on the redundancy analysis (RDA) method. RDA is considered a principal component analysis with instrumental environmental factors, which can reflect relationships between the distribution of microorganisms and relevant environmental factors [39,40].

2.2.3. Challenge Tests

A challenge test, which involves the dosing of synthetic stormwater runoff into the bioretention facilities in controlled conditions, was conducted in September and October 2021. Challenge tests are the most widely accepted method to confirm the removal contribution of a treatment system [29]. The target influent pollutant concentration was based on 8 storm events in Tianjin between March and May 2019 (Table 4). The event volume was used to determine the amount of synthetic stormwater that would be dosed into the bioretention facilities. In this study, the amount of water added per challenge test was around 3 m³ for each bioretention facility (approximately 0.4 times the pore volume). Based on this volume, the bioretention facilities took roughly 3 h until the last drop of water was observed to exit the bioretention facilities. During the challenge test, water from an adjacent canal was pumped into a mixing tank, and prepared chemicals were mixed using an electric mixer. The mixed-dosed water was then released to the bioretention facilities. For sampling, three inflow samples were taken from the outlet hose of the mixing tank. For the outflow samples, time-based sampling every 5 min was conducted using the autosamplers on-site. Once the water samples were collected, they were stored in an icebox before delivery to the laboratory for water quality testing.

Table 4. Target influent pollutant concentration in challenge test.

Water Quality Parameter	Targeted Influent Pollutant Concentration
TSS	150 mg/L
TP	1 mg/L
TN	6 mg/L
NH_3-N	2 mg/L
COD_{Cr}	100 mg/L

3. Results and Discussion

3.1. Stormwater Purification Performance

The performance of the on-site bioretention facility in removing runoff pollutants during the rainfall periods was evaluated. During a rainfall event in September 2020 (Figure 3a), the cumulative rainfall at the site was 30.5 mm, with a maximum rain intensity of 28.5 mm/30 min and an antecedent dry period of 21 days. The water quality of the collected samples of road rainfall runoff was evaluated on the basis of the Environmental Quality Standards for Surface Water [2], the Reuse of Urban Recycling Water-Water Quality Standard for Urban Miscellaneous Use [35], and Engineering Technical Code for Rain Utilization in Building and Sub-district [34] (Figure 3b–g). The results indicated that the concentrations of COD, TN, NH_3-N, and NO_3-N showed decreasing trends to varying extents, while TP and TSS showed fluctuating upward trends. This is attributed to the rainfall patterns and associated influent flow processes. The standard for surface water quality evaluation imposes higher requirements on the quality of collected water [2]. Based on the influent flow hydrograph (Figure 3a), the EMC of influent COD_{Cr} was 36 mg/L, which belongs to Class V ($\leq$40 mg/L) of surface water quality [2]. The EMC of TP was 0.28 mg/L, which belongs to Class IV ($\leq$0.3 mg/L). The EMCs of TN and NH_3-N were 5.59 and 2.90 mg/L, respectively, both exceeding Class V of surface water quality ($\leq$2 mg/L), although NH_3-N met the standard for non-potable water reuse [35]. The EMC of NO_3-N was 1.41 mg/L. The EMC of SS was 498 mg/L, which far exceeds the standard for rainwater reuse [34].

The road stormwater runoff was treated through the combination of physical, chemical, and microbial processes in the bioretention facilities, and the effluent was eventually collected in the storage tank. In the storage tank of the CBF, 6 mg/L TSS, 22 mg/L COD_{Cr}, 0.11 mg/L TP, 5.25 mg/L TN, 0.13 mg/L NH_3-N, and 3.86 mg/L NO_3-N were detected. The water quality of the collected water generally showed satisfactory improvement compared to that of the influent, where TSS was reduced by around 98.8%. Indicators of COD_{Cr}, TP, TN, and NH_3-N were also reduced, with COD_{Cr} reaching Class IV ($\leq$30 mg/L), TP reaching Class III ($\leq$0.2 mg/L), and NH_3-N reaching Class I ($\leq$0.15 mg/L). Due to the lack of a denitrification environment in the filter media layer, TN was less likely to be removed and still exceeded Class V [2]. This also resulted in an increase in NO_3-N in the collected water. Previous studies have also shown that there is uncertainty in TN and NO_3-N removal, which could be improved by designing an anaerobic zone at the bottom of the bioretention facility and/or adding extra carbon sources [41–43]. According to these indicators, the effluent meets the standards for the usage of rainwater and non-potable water (TSS $\leq$ 10 mg/L, $COD_{Cr} \leq$ 30 mg/L, NH_3-N $\leq$ 15 mg/L) [34,35] and hence has the potential for water resources reuse, such as daily greening, road spraying, and ornamental water supply.

In the storage tank in the MBF, 8 mg/L TSS, 42 mg/L COD_{Cr}, 0.07 mg/L TP, 4.71 mg/L TN, 3.09 mg/L NH_3-N, and 0.75 mg/L NO_3-N were detected. The results indicated that TSS was significantly reduced by 98.4%, and TP, TN, and NO_3-N were also reduced. In contrast, the COD_{Cr} and NH_3-N of the effluent were higher than those of the influent, which both exceed Class V (≤40 mg/L). Compared with the CBF, the higher COD_{Cr} and NH_3-N in the MBF were mainly attributed to the high soil moisture content, which reduced the degradation rate of organic matter and impeded nitrification reactions. This eventually led to a reduction in the removal rate of NH_3-N, and the results were consistent with the previous laboratory-scale pilot test [32]. These indicators show that the effluent from the MBF meets the standards for the usage of rainwater and non-potable water (TSS ≤ 10 mg/L, NH_3-N ≤ 15 mg/L) [34,35] and hence has the potential for water resources reuse.

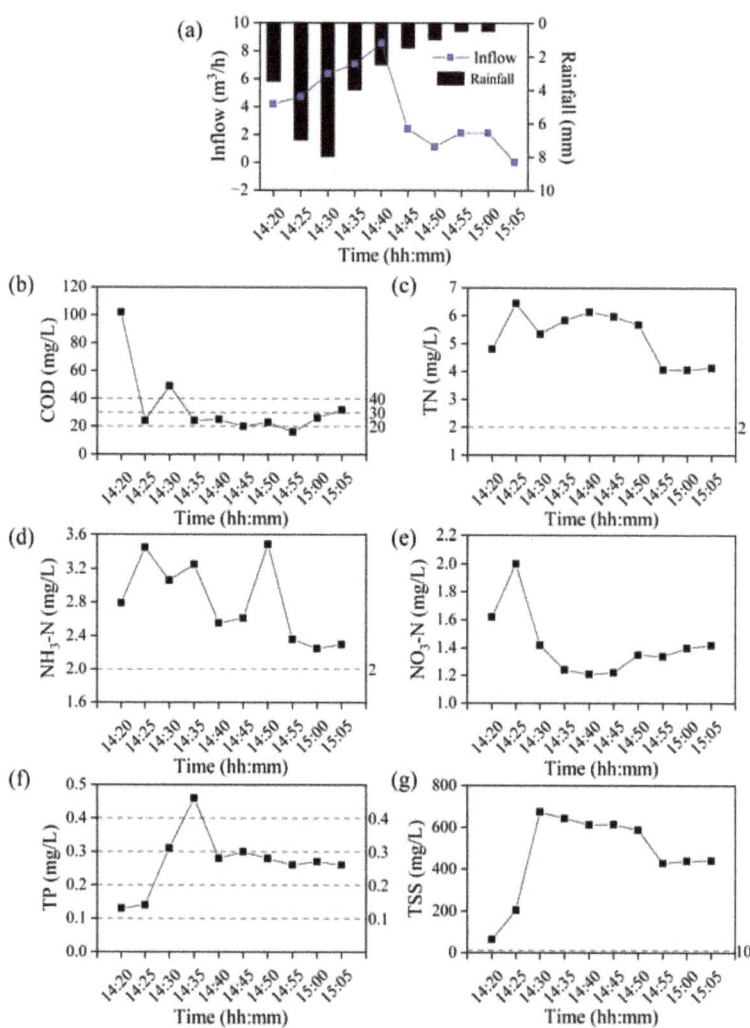

Figure 3. Rainfall and influent flow hydrographs (a) and change patterns of inflow water quality (b–g) (dashed lines are classification thresholds for relevant water quality standards).

3.2. Challenge Test Results of Bioretention Facilities

The water quality results of the challenge test are compiled and tabulated in Table 5. In the challenge test, TSS, TP, and COD_{Cr} had an effective average removal efficiency of 62.4%, 66.3%, and 78.2%, respectively, for the MBF and 64.4%, 70.3%, and 78.9%, respectively, for the CBF. The removal performance of these three water quality parameters was similar between the two bioretention facilities, and this was due to the similar composition of the filter media. Previous literature studies also concluded that amendments in the filter media such as WTR and zeolite can help to remove TP and COD_{Cr} from the runoff [29,44]. For TN and nitrogen species, the MBF had an average removal efficiency of 59.4%, 70.6%, and −68.6% for TN, NH_3-N, and NO_3-N, respectively, while the CBF had an average removal efficiency of 55.1%, 75.3%, and −140.2%, respectively. Nitrate was not removed effectively, and thus, the majority of the TN removal may be due to the decrease in ammonia nitrogen by the nitrification process or plant uptake and particulate organic nitrogen by sedimentation or filtration processes. As reported by Lopez-Ponnada et al. [16], the removal of NO_3-N can be highly variable, ranging from −630% to 46%.

Table 5. Water quality results of challenge test.

Water Quality Parameter	Influent (mg/L)	Effluent EMC (mg/L)		Removal Efficiency (%)	
		MBF	CBF	MBF	CBF
TSS	595	224	212	62.4%	64.4%
TN	8.70	3.53	3.91	59.4%	55.1%
NH_3-N	2.31	0.68	0.57	70.6%	75.3%
NO_3-N	1.02	1.72	2.45	−68.6%	−140.2%
TP	1.01	0.34	0.30	66.3%	70.3%
COD_{Cr}	422	92	89	78.2%	78.9%

3.3. Change Patterns of Media Moisture and Temperature

Figure 4 illustrates the spatiotemporal changes in the media temperature and moisture of the bioretention facilities. The numbers M1-2 and T1-2 represent the average values of moisture contents and temperature for soil temperature and moisture sensors No. 1 and No. 2 (Figure 2), respectively, and the same scheme applies to the other numbers. For the MBF, the soil moisture content of the upper layer was significantly influenced by external climatic conditions during the monitoring period. With the decrease in external temperature from October to January, the soil temperature gradually decreased to below 0 °C, and the moisture content decreased from 23% to around 9.5% (Figure 4a). The relatively significant moisture decline was mainly related to significant changes in the soil dielectric constant accompanied by changes in soil structure (e.g., water freezing). Afterward, the soil temperature gradually increased, and the moisture content recovered to around 18% by the end of May. In contrast, the moisture contents in the middle and bottom layers (M3-4 and M1-2) were maintained at about 24.5% and 28%, respectively, during the monitoring period. The media moisture content of the upper layer of the CBF was less affected by external climatic conditions (Figure 4b). It was close to that of the MBF at around 24% when the soil temperature was above 0 °C and dropped to a minimum of around 16.5% when the soil temperature reached below 0 °C. The moisture content of the middle layer was close to that of the MBF, while the bottom layer was about 2–3% lower than the MBF.

In terms of media temperature, the bottom and middle temperatures were at similar levels for both types of bioretention facilities in the autumn, but by winter, the CBF was about 0.5–1.0 °C cooler than the MBF. Further, the MBF was able to recover to similar levels more rapidly than the CBF during the spring and summer periods (Figure 4c). The results suggest that the MBF is more conducive to the maintenance and restoration of the media temperature under cold climate conditions.

The variation in media moisture content with precipitation was analyzed for the rainfall event during September 2020, as shown in Figure 4d. The results indicated that during the dry period, the moisture content of the upper layer of the MBF was similar to that of the CBF at about 22%, while the moisture content of the middle and bottom layers of the MBF was 2–3% higher than that of the CBF on average. During the rainy period, the upper layer water content of the MBF was significantly higher, rising to 40–46%, while it only rose to about 29% for the CBF. The results indicated that the media moisture content of the MBF could be maintained at a high level under dry climate conditions.

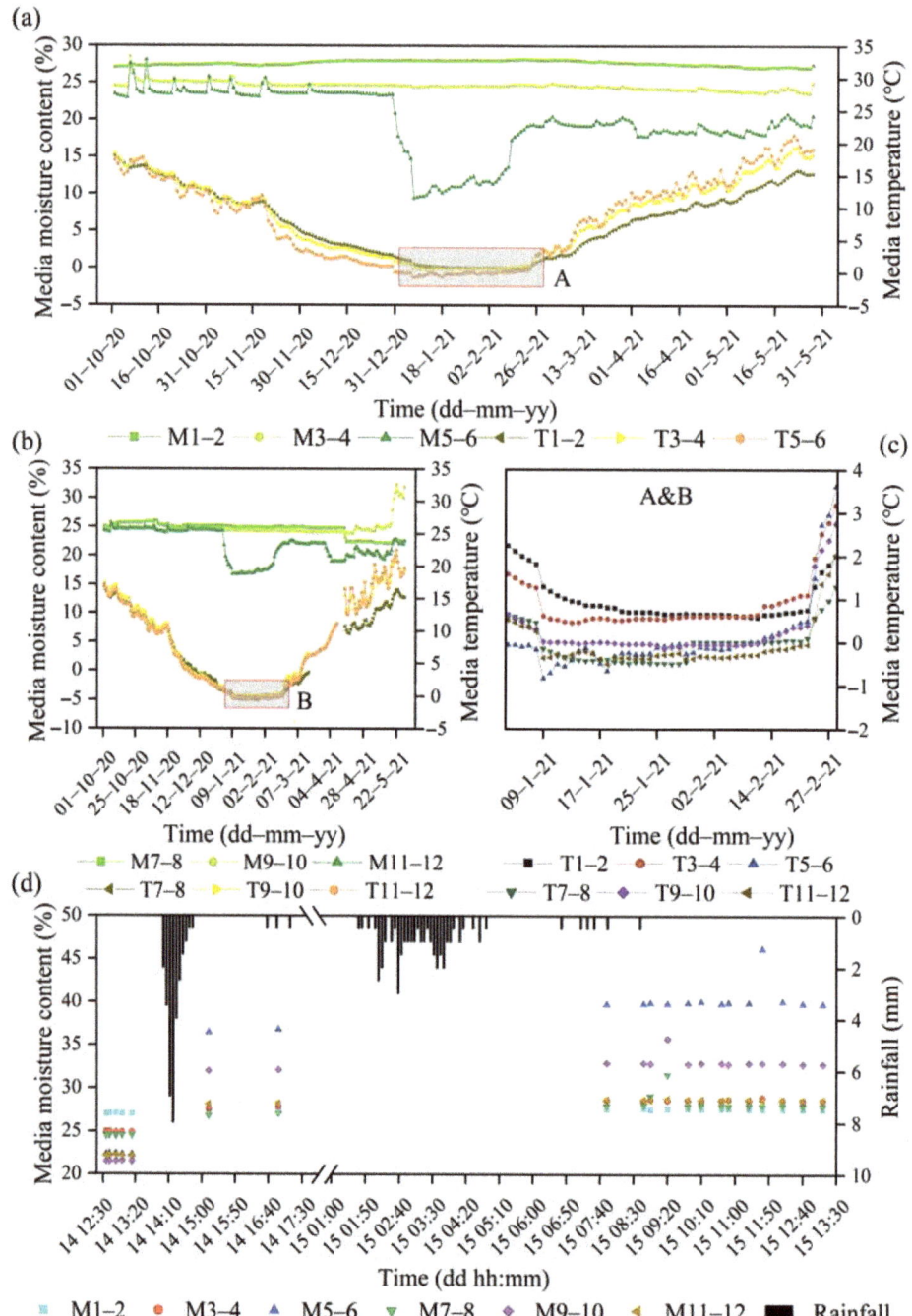

Figure 4. Temporal change patterns of media moisture and temperature of the modified bioretention facility (**a**) and the conventional bioretention facility (**b**), comparison of temperature in winter between facilities (**c**), and variation in media moisture content with precipitation of the facilities (**d**) (the letter M represents media moisture and the letter T represents media temperature).

3.4. Change Patterns of the Media Chemical Properties

Figure 5 illustrates the change patterns of media chemical properties since the early stage (T0) of the bioretention facilities. T1, T2, and T3 represent the sampling times in November 2020, March 2021, and June 2021, respectively. It is assumed that the chemical properties of the control group remained constant over the study period. At the initial stage, the media pH value was 7.83 and the TSC was 4.26 g/kg, indicative of an alkaline media with intense salinization in the study area (Figure 5a,b). Such a medium will affect the nitrification reaction of microorganisms and reduce the effectiveness of phosphorus content for plant growth. By the end of the study period, the media pH and TSC were reduced by 2.7% and 63.3%, respectively, to a neutral alkaline media with moderate salinization. This indicated that the internal microenvironment of the filter media changed in a direction favorable for plant growth and microbial activities. The increase in media CEC indicated that bioretention facilities were capable of creating a relatively stable medium environment for plant growth and soil microbial activities. This can also be concluded from the changes in OM, which was much higher than that of the on-site saline soil. There was a slight increase in STN from the initial stage, while the STP gradually decreased to a level close to that of the control group, indicating that there is less risk of nitrogen and phosphorus leaching with stormwater runoff in the soil filter media layer.

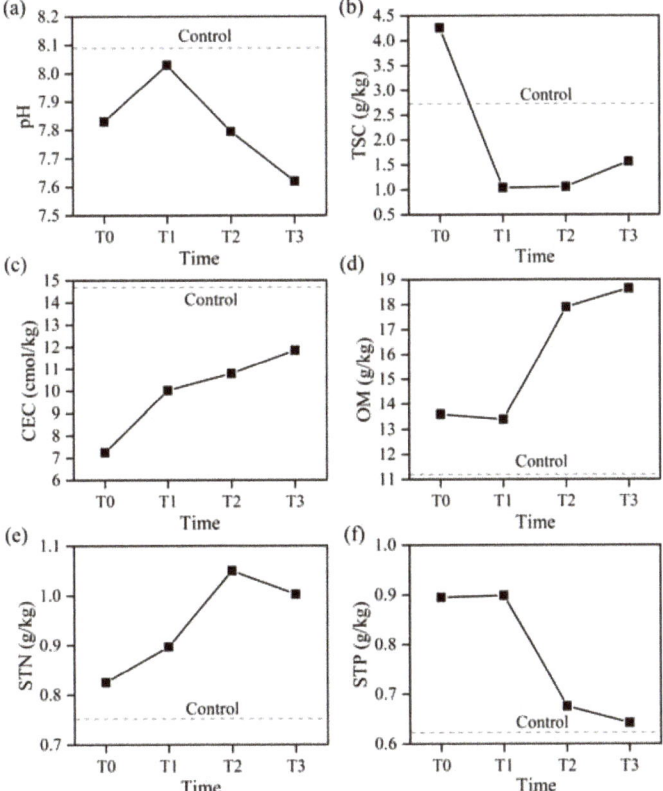

Figure 5. Change patterns of media chemical properties: (**a**) pH, (**b**) TSC (total salt content), (**c**) CEC (cation-exchange capacity), (**d**) OM (organic matter), (**e**) STN (soil total nitrogen), and (**f**) STP (soil total phosphorus) (the dotted line represents the value of the control group and the letter T represents the sampling time).

3.5. Characteristics of the Media Microbial Diversity

3.5.1. Distribution of Microbial Communities

The number of valid sequences of each sample was 24,183, and there were 51 phyla, 135 orders, 278 orders, 529 families, 1106 genera, 2286 species, and 6796 OTUs. Figure 6 shows the dilution curves of each sample based on the Sobs index at the OTU level. The results indicated that most of the microorganisms in the soil filter media had been analyzed, and the results could realistically reflect the microbial community composition of the study area.

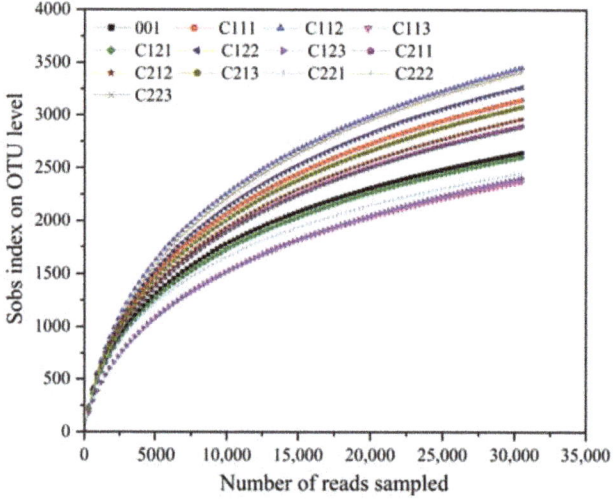

Figure 6. Dilution curves of each sample based on Sobs index at the OTU level.

A Venn diagram was used to visually capture the number of shared and unique OTUs in the soil filter media (Figure 7a). The results indicated that the MBF and the CBF had 5253 and 5184 OTUs, respectively, with 4722 OTUs appearing in both facilities. The soil filter media of both facilities had many more species than the on-site saline soil (control). There were 447, 353, and 63 OTUs unique to the MBF, CBF, and control, respectively, accounting for 8.5%, 6.8%, and 2.4% of their total OTUs, respectively. These data indicated that the bioretention facility filter media significantly enhanced the species diversity of soil microorganisms. At the same time, the MBF had more OTUs and more unique OTUs than the CBF, indicating that the soil moisture conservation ropes could increase the microbial diversity in the filter media layer of the bioretention facility.

The microbial diversity of different vertical layers of the bioretention facility was analyzed (Figure 7b–d). The results indicated that all vertical layers of the bioretention facility had more OTUs than the control saline soil. The number of OTUs was higher in the upper layer (UMBF) and middle layer (MMBF) of the MBF than in the upper (UCBF) and middle (MCBF) layers of the CBF, but it was the opposite in the bottom layer. The bottom layer (BCBF) of the CBF had the highest number of OTUs and more unique OTUs (21.3%) than the other layers, while the bottom layer (BMBF) of the MBF had the lowest number of OTUs and fewer unique OTUs (12.7%). The results suggest that there are differences in the composition and distribution characteristics of OTUs in the vertical direction of bioretention facilities. This is mainly due to the difference in physical properties such as media moisture and temperature in the vertical direction.

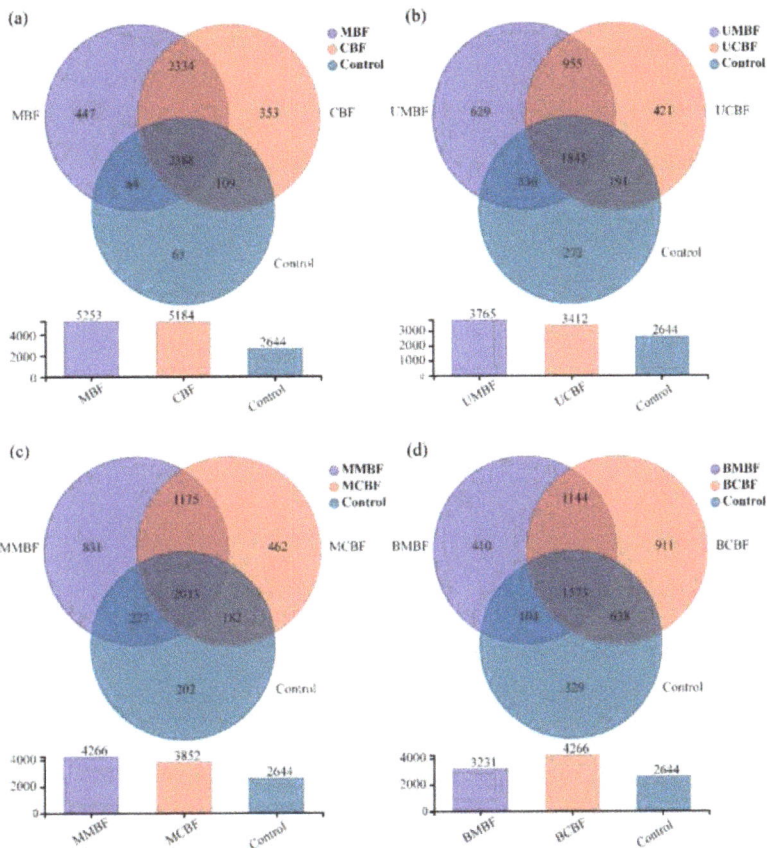

Figure 7. Venn diagram of OTUs for the modified bioretention facility (MBF), the conventional bioretention facility (CBF) and control group (**a**), and Venn diagram of OTUs for the upper layer (UMBF, UCBF), middle layer (MMBF, MCBF), and bottom layer (BMBF, BCBF) of the MBF and the CBF, respectively (**b–d**).

3.5.2. Composition of Microbial Communities

There were 10 major phyla in the study area (Figure 8), namely, *Proteobacteria*, *Chloroflexi*, *Actinobacteria*, *Acidobacteria*, *Bacteroidetes*, *Gemmatimonadota*, *Firmicutes*, *Myxococcota*, *Patescibacteria*, and *Desulfobacterota*. *Proteobacteria* was the dominant phylum in the MBF, accounting for 25.2% of the microbial community. It has been demonstrated that *Proteobacteria* make a major contribution to nitrogen removal, and microorganisms in this phylum are often involved in nitrogen fixation, nitrification, and denitrification processes [45]. In contrast, the most dominant phylum in the CBF was *Chloroflexi*, with a relative abundance of 21.7% in the microbial community, slightly higher than the second most abundant *Proteobacteria* (21.2%), and *Actinobacteria* (20.7%) ranked third. *Chloroflexi* are often involved in the process of microbial denitrification for nitrogen removal [46] and also have a relatively high capacity for biological phosphorus removal [47]. A large number of *Actinobacteria* genera are present in simultaneous denitrification and phosphorus removal systems [48]. The most dominant phylum in the control saline soils was *Chloroflexi*, which accounted for 25.4% of the microbial community.

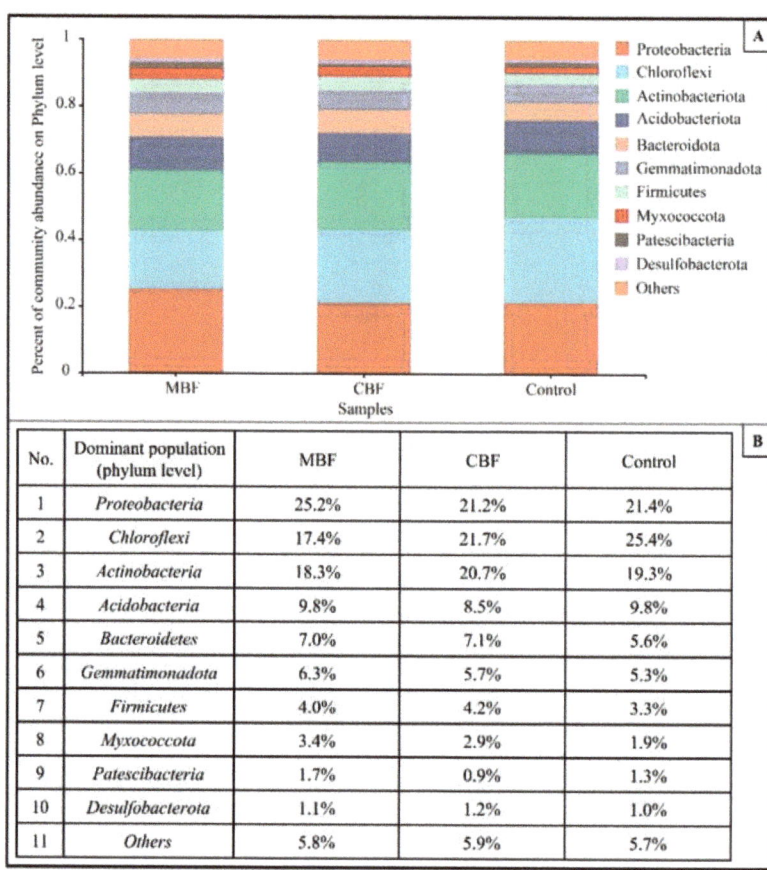

Figure 8. Histogram of the composition and relative abundance of dominant microbial communities in the bioretention facilities and control group at the phylum level (**A**) and the summary statement (**B**).

Acidobacteria were relatively abundant in the MBF, CBF, and control group (Figure 8), with relative abundances ranging from 8.5% to 9.8%, and are often involved in the denitrification reactions [49]. In addition, the relative abundance of *Bacteroidota* was also relatively high, ranging from 5.6% to 7.1%. Microorganisms belonging to this phylum mostly obtain energy through the degradation of COD and are involved in the process of nitrogen and phosphorus removal [48]. *Gemmatimonadota* was also one of the dominant phyla, with a relative abundance ranging from 5.3% to 6.3%, followed by *Firmicutes* (3.3–4.2%), *Myxococcota* (1.9–3.4%), *Patescibacteria* (0.9–1.7%), and *Desulfobacterota* (0.9–1.2%). Microorganisms belonging to the phylum *Firmicutes* can participate in denitrification processes under anaerobic conditions [50]. The results indicated that there were differences in the composition of the microbial community between the MBF and CBF, mainly for *Proteobacteria* and *Chloroflexi*. *Proteobacteria* was the most metabolically active microbial population in bioretention facilities, which is consistent with the findings of other related studies [38,51].

3.6. Redundancy Analysis of Environmental Factors and Microbial Communities

The relationship between environmental factors and the relative abundance of microbial communities was quantified, as shown in Figure 9. Microbial communities are illustrated by blue arrows, and red arrows represent environmental factors. The length of the environmental factor arrows reflects how much of the variation in the relative abundance of microbial communities can be explained by variation in environmental factors. The angle between the arrows represents positive and negative correlations, where an acute angle represents a positive correlation, an obtuse angle represents a negative correlation, and a right angle represents no correlation. The results indicated that the first axis explained 33.6% of the relationship, and the second axis explained 19.2%, amounting to around 53%, which can reliably reflect the relationship between environmental factors and microbial communities. The extent to which environmental factors explained the relative abundance of microbial communities in descending order was media temperature > STP > OM > pH > STN > moisture > CEC.

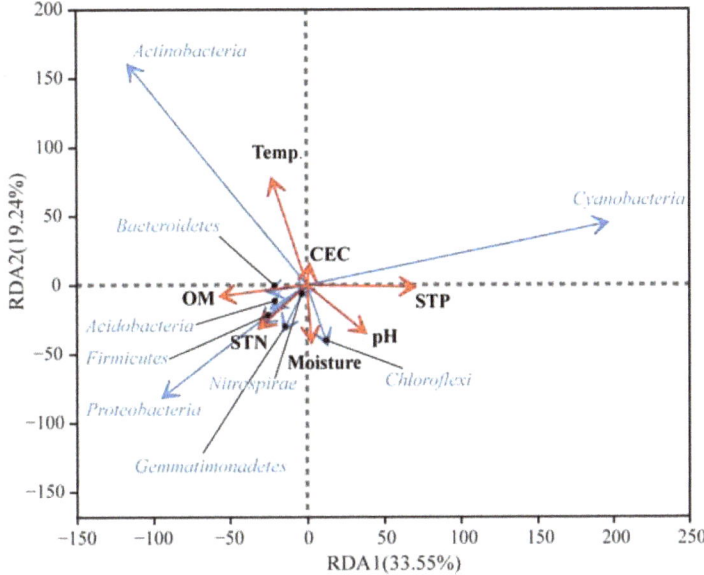

Figure 9. Redundancy analysis of environmental factors and microbial community at the phylum level.

Media temperature was negatively correlated with *Chloroflexi*; i.e., the relative abundance of *Chloroflexi* increased with decreasing media temperature. The temperature was also negatively correlated with *Gemmatimonadota, Proteobacteria, Firmicutes,* and *Nitrospirae*, whereas positive correlations were found with *Actinobacteria*. As seen in Section 3.3, the difference in temperature between the MBF and CBF was relatively small in comparison with the difference between the vertical layers of the bioretention facilities. Hence, the variation in temperature mainly contributed to the differences in the relative abundance of microorganisms between layers. STP was positively correlated with *Cyanobacteria*, while it was negatively correlated with *Bacteroidetes, Proteobacteria, Acidobacteria,* and *Firmicutes*. OM was positively correlated with all microbial communities except *Chloroflexi* and *Cyanobacteria*. pH was negatively correlated with *Actinobacteria* and positively correlated with *Chloroflexi* and *Gemmatimonadota*. STN was positively correlated with *Proteobacteria, Acidobacteria, Bacteroidetes,* and *Gemmatimonadota*; i.e., the relative abundance of these microbial communities increased with STN. Media moisture was positively correlated with the relative abundance of *Chloroflexi, Gemmatimonadota, Proteobacteria, Nitrospirae, Firmicutes,* and *Acidobacteria*, which agrees with the findings of previous studies [39]. *Nitrospira* in the phylum *Nitrospirae* can facilitate the nitrification reaction [52]. The minimal extent of explanation of the CEC may be related to its decreased variation in the study area.

3.7. Preliminary Analysis of Stormwater Purification Mechanisms

The media configuration and internal microenvironment of bioretention facilities are key factors affecting the effectiveness of runoff pollution control [53]. The difference between the two types of bioretention facilities was the arrangement of the soil filter media layer with soil moisture conservation ropes in the MBF. The RDA indicated that the relative abundance of *Chloroflexi*, *Proteobacteria*, *Nitrospirae*, *Firmicutes*, and *Acidobacteria*, which participate in the denitrification process, could be increased with the media moisture. This suggested that the MBF facilitated the increase in the relative abundance of nitrogen removal microbial communities, promoting the removal of nitrogen from stormwater runoff. Moreover, from the Venn diagram, it can be seen that the MBF had a greater abundance of OTUs. The aggregated results indicated that the MBF had a larger number of microbial communities involved in nitrogen removal. Comparing the collected water in the storage tanks of the bioretention facilities and the results from the challenge tests, the MBF had better removal efficiency for TN and NO_3-N than the CBF. The poor efficiency of NH_3-N removal of the MBF was correlated with the larger moisture content of the filter media.

In addition, the RDA also reflected the relationship between microbial communities. For example, *Proteobacteria* and *Firmicutes* were positively correlated since the arrow angle between them is close to 0°, which is in line with the literature [54]. That is, these bacterial communities can coexist in bioretention facilities and hence promote the process of microbial nitrogen fixation, nitrification, and denitrification. In contrast, the RDA indicated that the increase in media moisture could reduce the relative abundance of *Actinobacteria*, a microbial community widely observed in simultaneous nitrogen and phosphorus removal systems. That is, to some extent, the increase in the media moisture resulted in higher COD_{Cr} and TP in the effluent of the MBF in comparison with that of the CBF.

4. Conclusions

The proposed bioretention facilities in this study contribute to the control of road stormwater runoff pollution and have the potential to realize the utilization of rainwater resources in the process of developing an eco-city. A field study on the basis of preliminary laboratory tests was implemented in Tianjin Eco-city, and the performance of the bioretention facilities with respect to stormwater runoff purification and variation characteristics of the media physicochemical properties and microbial diversity were evaluated. The significant wet-dry variation and temperature changes in north China affected the internal microenvironment of bioretention facilities. With the implementation of soil moisture conservation ropes in the MBF, the media moisture could remain at a higher level (2–3% higher than the bottom layer of the CBF), and the media temperature could be maintained (0.5–1.0 °C higher than the CBF in winter) and recovered immediately during dry-cold periods. Such a soil microenvironment with relatively high moisture content and small temperature variations created by the MBF is conducive to the viability of the soil microbial communities. This resulted in greater abundance (5253 OTUs) of microbial communities and more unique OTUs (447) in the MBF, as well as a greater relative abundance of *Proteobacteria* (25.2%) that are involved in microbial nitrogen transformation processes. This also indicated that the MBF had a larger number of *Proteobacteria* and other coexisting nitrogen removal microorganisms and were thus more efficient in nitrogen removal.

The combined soil filter media of on-site bioretention facilities are characterized by the localization of media and the recycling of solid wastes. In addition, the design of soil moisture conservation ropes helps to reduce the maintenance costs of bioretention facilities in dry-cold regions. These characteristics are conducive to providing favorable economic benefits and, at the same time, facilitate the application of facilities in other urban regions. Future research will concentrate on improving the synergistic treatment of runoff pollutants, e.g., by planting robust plants and extending the hydraulic retention time of runoff. Long-term monitoring is also needed to better evaluate and optimize the effectiveness of bioretention facilities for sponge city construction.

Supplementary Materials: Supplementary materials to this article can be found in the file "Supplementary Material". The following are available online at https://www.mdpi.com/article/10.3390/w14030415/s1. Table S1: The standard methods and main instruments used for water quality testing; Table S2: The standard methods and main instruments used for media chemical property testing.

Author Contributions: Conceptualization, Q.L., H.J., and J.H.; formal analysis, Q.L., H.G. (Hongkai Guo), F.Y.L., and T.H.N.; investigation, G.Z., H.G. (Huiling Guo), and S.L.O.; methodology, Q.L., H.G. (Hongkai Guo), and Y.Z.; project administration, H.J. and J.H.; supervision, H.J. and J.H.; writing—original draft, Q.L., Y.Z., and J.H.; writing—review and editing, H.J. and J.H. All authors have read and agreed to the published version of the manuscript.

Funding: This work was supported by the National Nature Science Foundation of China (Grant No. 52070112 and No. 41890823). This work was funded by the Ministry of National Development, Singapore.

Institutional Review Board Statement: Not applicable.

Informed Consent Statement: Not applicable.

Data Availability Statement: The data presented in this study is available on request from the corresponding author.

Conflicts of Interest: The authors declare no conflict of interest.

References

1. Jacobson, C. Identification and quantification of the hydrological impacts of imperviousness in urban catchments: A review. *J. Environ. Manag.* **2011**, *92*, 1438–1448. [CrossRef] [PubMed]
2. SEPA. *Environmental Quality Standards for Surface Water (GB3838-2002)*; State Environmental Protection Administration of the People's Republic of China: Beijing, China, 2002. (In Chinese)
3. SEPA. *Discharge Standard of Pollutants for Municipal Wastewater Treatment Plant (GB18918-2002)*; State Environmental Protection Administration of the People's Republic of China: Beijing, China, 2002. (In Chinese)
4. Zhang, N.; Zhao, L.; Li, T.; Jin, Z. Characteristics of pollution and monitoring of water quality in Tianjin. *Ecol. Environ. Sci.* **2009**, *19*, 2127–2131. (In Chinese)
5. Han, Y.; Lau, S.L.; Kayhanian, M.; Stenstrom, M.K. Characteristics of highway stormwater runoff. *Water Environ. Res.* **2006**, *78*, 2377–2388. [CrossRef] [PubMed]
6. Tanaka, Y.; Matsuda, T.; Shimizu, Y.; Matsui, S.; Lee, B.C. A new installation for treatment of road runoff: Up-flow filtration by porous polypropylene media. *Water Sci. Technol.* **2005**, *52*, 225–232. [CrossRef]
7. Gnecco, I.; Berretta, C.; Lanza, L.G.; La Barbera, P. Storm water pollution in the urban environment of Genoa, Italy. *Atmos. Res.* **2005**, *77*, 60–73. [CrossRef]
8. Helmreich, B.; Hilliges, R.; Schriewer, A.; Horn, H. Runoff pollutants of a highly trafficked urban road–correlation analysis and seasonal influences. *Chemosphere* **2010**, *80*, 991–997. [CrossRef]
9. Xie, J.; Hu, Z.; Xu, T.; Han, H.; Yin, D. Water quality characteristics of rainfall runoff in Hefei City. *China Environ. Sci.* **2012**, *32*, 1018–1025. (In Chinese)
10. Wang, S.; He, Q.; Ai, H.; Wang, Z.; Zhang, Q. Pollutant concentrations and pollution loads in stormwater runoff from different land uses in Chongqing. *J. Environ. Sci.* **2013**, *25*, 502–510. [CrossRef]
11. Davis, A. Green Engineering Principles Promote Low-impact Development. *Environ. Sci. Technol.* **2005**, *39*, 338A–344A. [CrossRef]
12. Jia, H.; Yao, H.; Tang, Y.; Yu, S.L.; Field, R. LID-BMPs planning for urban runoff control and the case study in China. *J. Environ. Manag.* **2015**, *149*, 65–76. [CrossRef]
13. Jia, H.; Wang, Z.; Zhen, X.; Clar, M.; Yu, S.L. China's sponge city construction: A discussion on technical approaches. *Front. Environ. Sci. Eng.* **2017**, *11*, 18. [CrossRef]
14. Li, Q.; Wang, F.; Yu, Y.; Huang, Z.; Li, M.; Guan, Y. Comprehensive performance evaluation of LID practices for the sponge city construction: A case study in Guangxi, China. *J. Environ. Manag.* **2019**, *231*, 10–20. [CrossRef] [PubMed]
15. Jia, H.; Yao, H.; Tang, Y.; Yu, S.L.; Zhen, J.X.; Lu, Y. Development of a multi-criteria index ranking system for urban runoff best management practices (BMPs) selection. *Environ. Monit. Assess* **2013**, *185*, 7915–7933. [CrossRef] [PubMed]
16. Lopez-Ponnada, E.V.; Lynn, T.J.; Ergas, S.J.; Mihelcic, J.R. Long-term field performance of a conventional and modified bioretention system for removing dissolved nitrogen species in stormwater runoff. *Water Res.* **2020**, *170*, 115336. [CrossRef] [PubMed]
17. Li, L.; Davis, A.P. Urban stormwater runoff nitrogen composition and fate in bioretention systems. *Environ. Sci. Technol.* **2014**, *48*, 3403–3410. [CrossRef] [PubMed]
18. Palmer, E.T.; Poor, C.J.; Hinman, C.; Stark, J.D. Nitrate and phosphate removal through enhanced bioretention media: Mesocosm study. *Water Environ. Res.* **2013**, *85*, 823–832. [CrossRef] [PubMed]
19. Xiong, J.; Ren, S.; He, Y.; Wang, X.C.; Bai, X.; Wang, J.; Dzakpasu, M. Bioretention cell incorporating Fe-biochar and saturated zones for enhanced stormwater runoff treatment. *Chemosphere* **2019**, *237*, 124424. [CrossRef]

20. Singh, R.; Zhao, F.; Ji, Q.; Saravanan, J.; Fu, D. Design and Performance Characterization of Roadside Bioretention Systems. *Sustainability* **2019**, *11*, 2040. [CrossRef]
21. Muerdter, C.; Özkök, E.; Li, L.; Davis, A.P. Vegetation and Media Characteristics of an Effective Bioretention Cell. *J. Sustain. Water Built Environ.* **2016**, *2*, 04015008. [CrossRef]
22. Bratieres, K.; Fletcher, T.D.; Deletic, A.; Zinger, Y. Nutrient and sediment removal by stormwater biofilters: A large-scale design optimisation study. *Water Res.* **2008**, *42*, 3930–3940. [CrossRef]
23. Wu, J.; Cao, X.; Zhao, J.; Dai, Y.; Cui, N.; Li, Z.; Cheng, S. Performance of biofilter with a saturated zone for urban stormwater runoff pollution control: Influence of vegetation type and saturation time. *Ecol. Eng.* **2017**, *105*, 355–361. [CrossRef]
24. Ashoori, N.; Teixido, M.; Spahr, S.; LeFevre, G.H.; Sedlak, D.L.; Luthy, R.G. Evaluation of pilot-scale biochar-amended woodchip bioreactors to remove nitrate, metals, and trace organic contaminants from urban stormwater runoff. *Water Res.* **2019**, *154*, 1–11. [CrossRef] [PubMed]
25. O'Neill, S.W.; Davis, A.P. Water Treatment Residual as a Bioretention Amendment for Phosphorus. II: Long-Term Column Studies. *J. Environ. Eng.* **2012**, *138*, 328–336. [CrossRef]
26. Glaister, B.J.; Fletcher, T.D.; Cook, P.L.; Hatt, B.E. Co-optimisation of phosphorus and nitrogen removal in stormwater biofil-ters: The role of filter media, vegetation and saturated zone. *Water Sci. Technol.* **2014**, *69*, 1961–1969. [CrossRef]
27. Turk, R.P.; Kraus, H.T.; Hunt, W.F.; Carmen, N.B.; Bilderback, T.E. Nutrient Sequestration by Vegetation in Bioretention Cells Receiving High Nutrient Loads. *J. Environ. Eng.* **2017**, *143*, 06016009. [CrossRef]
28. Liu, J.; Sample, D.J.; Owen, J.S.; Li, J.; Evanylo, G. Assessment of selected bioretention blends for nutrient retention using mesocosm experiments. *J. Environ. Qual.* **2014**, *43*, 1754–1763. [CrossRef]
29. Lim, F.Y.; Neo, T.H.; Guo, H.; Goh, S.Z.; Ong, S.L.; Hu, J.; Lee, B.C.Y.; Ong, G.S.; Liou, C.X. Pilot and Field Studies of Modular Bioretention Tree System with Talipariti tiliaceum and Engineered Soil Filter Media in the Tropics. *Water* **2021**, *13*, 817. [CrossRef]
30. Zhao, Y.; Li, Q.; Chen, Z.; Zhou, G.; Jia, H. Optimization of bioretention facility media for municipal road runoff pollution control based on multi-objective evaluation. *Water Resour. Prot.* **2021**, *37*, 96–101. (In Chinese)
31. Xu, D.; Lee, L.Y.; Lim, F.Y.; Lyu, Z.; Zhu, H.; Ong, S.L.; Hu, J. Water treatment residual: A critical review of its applications on pollutant removal from stormwater runoff and future perspectives. *J. Environ. Manag.* **2020**, *259*, 109649. [CrossRef]
32. Guo, H.; Zhou, G.; Zhao, Y.; Zhan, J.; Li, Q.; Jia, H. Experimental study on soil moisture conservation and nitrogen removal of the modified bioretention facility. *Water Wastewater Eng.* **2021**, *47*, 66–71. (In Chinese)
33. Goh, H.W.; Lem, K.S.; Azizan, N.A.; Chang, C.K.; Talei, A.; Leow, C.S.; Zakaria, N.A. A review of bioretention components and nutrient removal under different climates-future directions for tropics. *Environ. Sci. Pollut. Res.* **2019**, *26*, 14904–14919. [CrossRef] [PubMed]
34. MOHURD. *Engineering Technical Code for Rain Utilization in Building and Sub-District (GB50400-2016)*; Ministry of Housing and Urban-Rural Development of the People's Republic of China: Beijing, China, 2016. (In Chinese)
35. SAMR. *Reuse of Urban Recycling Water-Water Quality Standard for Urban Miscellaneous Use (GB/T 18920-2020)*; State Administration for Market Regulation of the People's Republic of China: Beijing, China, 2020. (In Chinese)
36. Sungji, K.; Jiwon, L.; Kyungik, G. Inflow and outflow event mean concentration analysis of contaminants in bioretention facilities for non-point pollution management. *Ecol. Eng.* **2020**, *147*, 105757. [CrossRef]
37. Cao, Q.; Song, X.; Wu, H.; Gao, L.; Liu, F.; Yang, S.; Zhang, G. Mapping the response of volumetric soil water content to an intense rainfall event at the field scale using GPR. *J. Hydrol.* **2020**, *583*, 124605. [CrossRef]
38. Zuo, X.; Guo, Z.; Wu, X.; Yu, J. Diversity and metabolism effects of microorganisms in bioretention systems with sand, soil and fly ash. *Sci. Total Environ.* **2019**, *676*, 447–454. [CrossRef] [PubMed]
39. Yu, W.; Lawrence, N.C.; Sooksa-nguan, T.; Smith, S.D.; Tenesaca, C.; Howe, A.C.; Hall, S.J. Microbial linkages to soil biogeochemical processes in a poorly drained agricultural ecosystem. *Soil Biol. Biochem.* **2021**, *156*, 108228. [CrossRef]
40. Li, Y.; Yang, Y.; Zhang, J.; Zhang, Z.; Li, J. Experimental Study on the Effect of the Physicochemical Properties of Contaminated Fillers in Bioretention System on Microbial Community Structure. *Water Air Soil Pollut.* **2021**, *232*, 1–16. [CrossRef]
41. Collins, K.A.; Lawrence, T.J.; Stander, E.K.; Jontos, R.J.; Kaushal, S.S.; Newcomer, T.A.; Grimm, N.B.; Cole Ekberg, M.L. Opportunities and challenges for managing nitrogen in urban stormwater: A review and synthesis. *Ecol. Eng.* **2010**, *36*, 1507–1519. [CrossRef]
42. Davis, A.P.; Hunt, W.F.; Traver, R.G.; Clar, M. Bioretention Technology: Overview of Current Practice and Future Needs. *J. Environ. Eng.* **2009**, *135*, 109–117. [CrossRef]
43. Hunt, W.F.; Smith, J.T.; Jadlocki, S.J.; Hathaway, J.M.; Eubanks, P.R. Pollutant removal and peak flow mitigation by a bioretention cell in urban Charlotte, NC. *J. Environ. Eng. -Asce* **2008**, *134*, 403–408. [CrossRef]
44. Qiu, F.; Zhao, S.; Zhao, D.; Wang, J.; Fu, K. Enhanced nutrient removal in bioretention systems modified with water treatment residuals and internal water storage zone. *Environ. Sci. Water Res. Technol.* **2019**, *5*, 993–1003. [CrossRef]
45. Wagner, M.; Loy, A. Bacterial community composition and function in sewage treatment systems. *Curr. Opin. Biotechnol.* **2002**, *13*, 218–227. [CrossRef]
46. Xing, W.; Li, D.; Li, J.; Hu, Q.; Deng, S. Nitrate removal and microbial analysis by combined micro-electrolysis and autotrophic denitrification. *Bioresour. Technol.* **2016**, *211*, 240–247. [CrossRef] [PubMed]

47. Kragelund, C.; Levantesi, C.; Borger, A.; Thelen, K.; Eikelboom, D.; Tandoi, V.; Kong, Y.; van der Waarde, J.; Krooneman, J.; Rossetti, S.; et al. Identity, abundance and ecophysiology of filamentous Chloroflexi species present in activated sludge treatment plants. *FEMS Microbiol. Ecol.* **2007**, *59*, 671–682. [CrossRef] [PubMed]
48. Waller, L.J.; Evanylo, G.K.; Krometis, L.H.; Strickland, M.S.; Wynn-Thompson, T.; Badgley, B.D. Engineered and Environmental Controls of Microbial Denitrification in Established Bioretention Cells. *Environ. Sci. Technol.* **2018**, *52*, 5358–5366. [CrossRef] [PubMed]
49. Li, M.; Wei, D.; Zhang, Z.; Fan, D.; Du, B.; Zeng, H.; Li, D.; Zhang, J. Enhancing 2,6-dichlorophenol degradation and nitrate removal in the nano-zero-valent iron (nZVI) solid-phase denitrification system. *Chemosphere* **2021**, *287*, 132249. [CrossRef] [PubMed]
50. Wang, H.; He, Q.; Chen, D.; Wei, L.; Zou, Z.; Zhou, J.; Yang, K.; Zhang, H. Microbial community in a hydrogenotrophic denitrification reactor based on pyrosequencing. *Appl. Microbiol. Biotechnol.* **2015**, *99*, 10829–10837. [CrossRef] [PubMed]
51. Hong, J.; Geronimo, F.K.; Choi, H.; Kim, L.H. Impacts of nonpoint source pollutants on microbial community in rain gardens. *Chemosphere* **2018**, *209*, 20–27. [CrossRef]
52. Daims, H.; Lebedeva, E.V.; Pjevac, P.; Han, P.; Herbold, C.; Albertsen, M.; Jehmlich, N.; Palatinszky, M.; Vierheilig, J.; Bulaev, A.; et al. Complete nitrification by Nitrospira bacteria. *Nature* **2015**, *528*, 504–509. [CrossRef]
53. You, Z.; Zhang, L.; Pan, S.Y.; Chiang, P.C.; Pei, S.; Zhang, S. Performance evaluation of modified bioretention systems with alkaline solid wastes for enhanced nutrient removal from stormwater runoff. *Water Res.* **2019**, *161*, 61–73. [CrossRef]
54. Zhao, X.; Huang, J.; Lu, J.; Sun, Y. Study on the influence of soil microbial community on the long-term heavy metal pollution of different land use types and depth layers in mine. *Ecotoxicol. Environ. Saf.* **2019**, *170*, 218–226. [CrossRef]

Article

Pollutant Removal Efficiency of a Bioretention Cell with Enhanced Dephosphorization

Chia-Chun Ho * and Yi-Xuan Lin

Department of Civil and Construction Engineering, National Taiwan University of Science and Technology, Taipei 106, Taiwan; smile917006@gmail.com
* Correspondence: cchocv@mail.ntust.edu.tw; Tel.: +886-2-2730-1073

Abstract: Low impact development can contribute to Sustainable Development Goals (SDGs) 2, 6, 7, 11, and 13, and bioretention cells are commonly used to reduce nonpoint source pollution. However, although bioretention is effective in reducing ammonia nitrogen and chemical oxygen demand (COD) pollution, it performs poorly in phosphorus removal. In this study, a new type of enhanced dephosphorization bioretention cell (EBC) was developed; it removes nitrogen and COD efficiently but also provides excellent phosphorus removal performance. An EBC (length: 45 m; width: 15 m) and a traditional bioretention cell (TBC) of the same size were constructed in Anhui, China, to treat rural nonpoint source pollution with high phosphorus concentration levels. After almost 2 years of on-site operation, the ammonium nitrogen removal performance of the TBC was 81%, whereas that of the EBC was 78%. The COD removal rates of the TBC and EBC were 51% and 65%, and they removed 51% and 92% of the total phosphorus, respectively. These results indicate that the TBC and EBC have similar performance in the removal of ammonium nitrogen and COD, but the EBC significantly outperforms the TBC in terms of total phosphorus removed.

Keywords: low impact development; Sustainable Development Goals; non-point source pollution; enhanced dephosphorization bioretention

Citation: Ho, C.-C.; Lin, Y.-X. Pollutant Removal Efficiency of a Bioretention Cell with Enhanced Dephosphorization. *Water* **2022**, *14*, 396. https://doi.org/10.3390/w14030396

Academic Editor: Kyoung Jae Lim

Received: 31 December 2021
Accepted: 25 January 2022
Published: 28 January 2022

Publisher's Note: MDPI stays neutral with regard to jurisdictional claims in published maps and institutional affiliations.

Copyright: © 2022 by the authors. Licensee MDPI, Basel, Switzerland. This article is an open access article distributed under the terms and conditions of the Creative Commons Attribution (CC BY) license (https://creativecommons.org/licenses/by/4.0/).

1. Introduction

Rapid urbanization is increasingly affected by extreme weather events. The frequent occurrence of short-duration intense rainfall creates pollution sources such as surface runoff, which lead to urban nonpoint source pollution [1]. China is affected not only by urban nonpoint source pollution but also by the environmental pollution caused by runoff rainwater in rural areas. Rural runoff rainwater contains nonpoint source pollution from human activities, livestock, and agriculture. Therefore, the pollution concentration in rural runoff is higher than that of urban runoff, particularly in terms of nutrients [2].

Low impact development (LID) practices can purify and reintegrate contaminated runoff into the hydrological cycle by increasing infiltration, reducing runoff velocity, and reducing pollutant load [3]. Macedo et al. [4] proposed that a new-generation LID model with the capacity to contribute to Sustainable Development Goals (SDGs) 2 (zero hunger), 6 (clean water and sanitation), 7 (affordable and clean energy), 11 (sustainable cities and communities), and 13 (climate action) [5]. The implementation of LID practices that align with SDGs is guided by the phrase "Think globally, act locally," which is increasingly used in the lexicon of sustainable development [6]. Bioretention cells are commonly used to reduce nonpoint source pollution because of their runoff and pollution control capabilities and their role as a landscape feature.

Bioretention was developed in Prince George County, Maryland, USA, in the 1990s [7]. Bioretention is an LID approach that can be used to address nonpoint source pollution [8]. The composition of a typical bioretention system includes plantings, a mulch layer, planting soil, filter fabric, a sand layer, a gravel bed, an outflow pipe, and geotextiles (Figure 1).

Bioretention mainly uses processes such as filtration, adsorption, plant uptake, and biological transformation to reduce the pollutants generated by runoff [9]. Most studies that aimed to improve the pollution removal ability of bioretention have focused on changing the composition of the planting soil (e.g., changing the ratio of perlite to vermiculite) [10]. Le Coustumer et al. [11] reported that adding compost to planting soil can increase the water permeability and pollutant removal rate of bioretention; the improvement in performance achieved through this method is greater than that achieved by adding 10% vermiculite and 10% perlite to planting soil. In general, the effect of bioretention on phosphorus reduction is relatively unstable, and it can easily cause clogging in the long term [12].

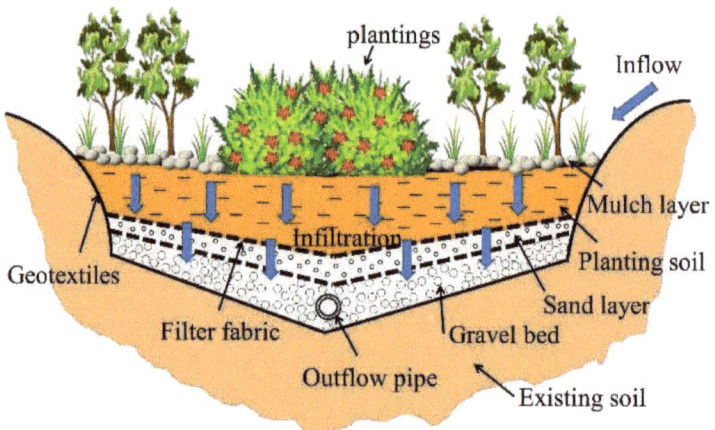

Figure 1. Typical bioretention system.

Fan et al. [13] proposed that if a proper submerged space is established inside a bioretention cell, the denitrification effect can be significantly increased. Li et al. [14] developed two groups of lab-scale bioretention cells. The first group of bioretention cells was filled with coarse gravel, medium gravel, fine gravel, and planting soil in order from bottom to top according to the model used for traditional bioretention cells (TBCs). For the second group of bioretention cells, the order of the fillers was reversed, such that the coarse gravel became the top layer (Figure 2). The test results indicated that the TBC was slightly more effective than the inverted cell in removing ammonium nitrogen (NH_4^+), with the removal percentages of the traditional cell being between 96.6% and 99.7% and those of the inverted cell being between 80.5% and 97.4%. However, the inverted bioretention cell had a significantly higher nitrate (NO_3^--N) removal efficiency than the traditional cell. This is because the inverted bioretention cell can form an anoxic zone at its bottom and accelerate the denitrification reaction. Moreover, a series of bioretention box experiments involving various runoff inflow characteristics were performed by Davis et al. [15]. Their experimental results indicated that increasing or decreasing the hydrogen ion concentration (pH) from a neutral level leads to the release of phosphorus in the upper soil, resulting in the poor phosphorus removal performance of bioretention cells.

Numerous studies have demonstrated that bioretention cells provide favorable nitrogen removal performance but poor phosphorus removal performance. For algal growth, nitrogen is generally a limiting nutrient in coastal and oceanic waters, whereas phosphorus tends to be a limiting nutrient in freshwater systems [16]. If runoff rainwater with a high phosphorus concentration flows directly into rivers or lakes, it may cause water eutrophication. To date, no study has demonstrated that TBCs have sufficient phosphorus removal efficiency. Therefore, the present study modified the materials inside a bioretention cell and developed an enhanced dephosphorization bioretention cell (EBC).

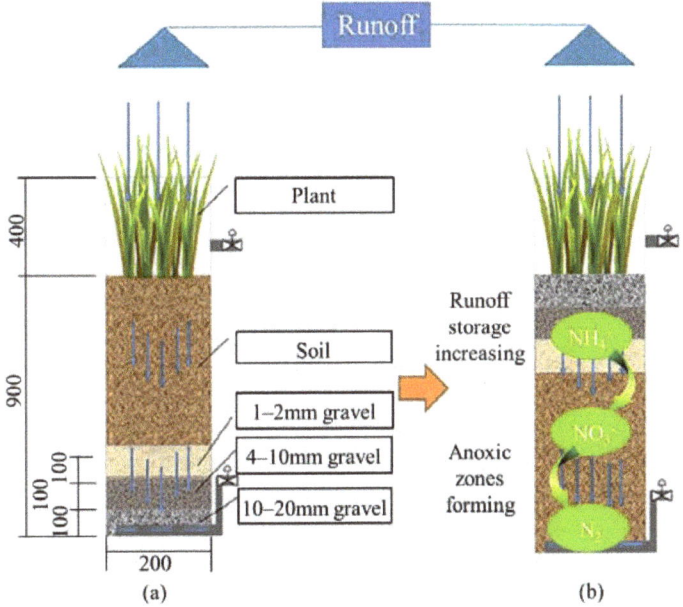

Figure 2. Profile of the retention cell (**a**) traditional, (**b**) inverted.

2. Materials and Methods

2.1. Enhanced Dephosphorization Bioretention Cell

The lowermost layer of the EBC is a mixed filter material layer instead of a gravel bed, as is the case in TBCs. The mixed filter material layer primarily comprises soil mixture layers (SMLs) and permeable layers (PLs). Figure 3 illustrates the enhanced dephosphorization bioretention system. The SMLs are 40 cm wide, 60 cm long, and 10 cm high, and they are stacked in layers. The SMLs are separated from each other by a 10 cm gap (at the top, bottom, left, and right sides of each SML) that is filled by the PL. The SMLs comprise approximately 70% to 80% of on-site soil mixed with approximately 20% to 30% of additional materials (e.g., active charcoal powder, organic matter, and iron).

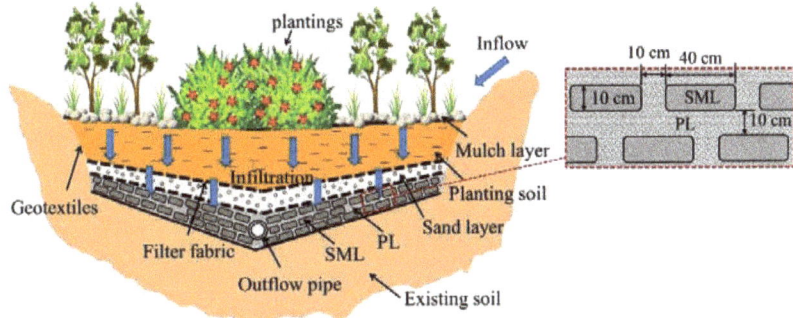

Figure 3. Enhanced dephosphorization bioretention cell (EBC). SML, soil mixture layer; PL, permeable layer.

Among the various materials that form SMLs, the soil serves as a habitat for microorganisms, and the active charcoal powder adsorbs high amounts of organic matter in wastewater, thus enhancing the efficiency of organic matter decomposition. The organic matter (e.g., sawdust, straw, corn cobs, and kenaf) serves as nutrients for microorganisms, and the iron materials effectively adsorb phosphates. The materials are mixed and packed into fiber bags, which are then stacked to form the SMLs, with each layer being separated by a PL. A PL comprises aggregates of gravel, pumice, or zeolite measuring approximately 1–5 mm in diameter. Aggregates should be of consistent size to reduce the risk of clogging and to facilitate the dispersion of water in the system. Moreover, the surface of the aggregates that constitute a PL also serves as a habitat for nitrobacteria and adsorbs the organic matter in wastewater. Therefore, both layers actively remove pollutants from wastewater [17].

2.2. Study Site

Chaohu Lake, situated in the central region of Anhui Province, is the fifth largest shallow freshwater lake in China with a surface area of 775 km^2 and watershed area of 12,938 km^2. Chaohu Lake plays a key role in the social, economic, and ecological functions of the local basin. However, the excessive discharge of industrial and municipal wastewater caused by rapid industrialization and urbanization has led to the heavy pollution of Chaohu Lake [18–21]. To improve the water quality of Chaohu Lake, the nutrients flowing into the lake must first be reduced.

The study site is located to the northeast of Chaohu Lake (Figure 4). GPS coordinates of the study site are 31°40.01′ N, 117°40.73′ E. The sub-watershed area of the study site is approximately 520,000 m^2. It includes 440,000 m^2 of farmland and 80,000 m^2 of rural land. When it rains, the runoff flushes agricultural and rural nonpoint source pollution into the Jiyu River (which flows into Chaohu Lake), resulting in severe eutrophication in Chaohu Lake.

Figure 4. Location of study site in Chaohu Lake.

Agricultural nonpoint source pollution and urban domestic sewage are the main sources of nitrogen and phosphorus in Chaohu Lake [22,23]. In the experience of the authors, bioretention cells are often used to purify nonpoint source pollution. Therefore, in the present study, a TBC and an EBC were constructed on site, and their effectiveness in removing nutrients was compared. The two bioretention cells had the same dimensions (45 m length, 15 m width, and 1.2 m depth). Detailed cross-sectional views of the two cells are illustrated in Figure 5.

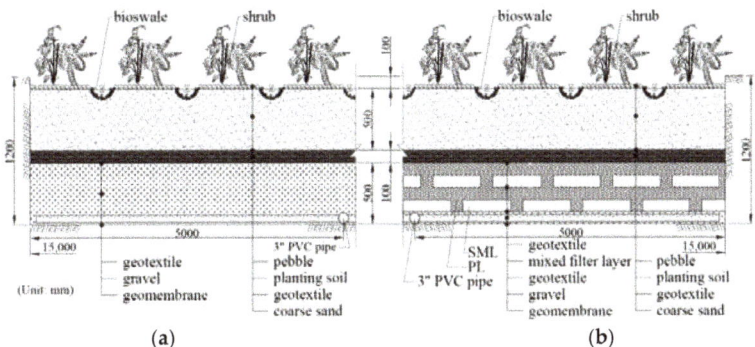

Figure 5. Cross-sectional view of bioretention cells: (a) traditional bioretention cell (TBC) and (b) EBC.

2.3. Construction Steps

A 675 m² TBC and a 675 m² EBC were constructed to treat the nonpoint source pollution and agricultural irrigation tailwater generated in an area measuring 520,000 m². The treatment rate of this case was 1:770, and the construction steps were as follows.

1. Excavation: Two adjacent pits with the same depth of 1.2 m were excavated at the site. Because the soil provided sufficient strength, the excavation could be performed vertically and no earth-retaining facility was required (see Figure 6a).
2. Installation of geomembrane: A 2-mm-thick geomembrane was spread around the pit of each cell to prevent groundwater from entering the cells (see Figure 6b).
3. Filling of cells with gravel: The TBC and EBC were, respectively, filled with 500-mm- and 100-mm-thick gravel layers (see Figure 6c).
4. Installation of outflow pipes: A polyvinyl chloride (PVC) pipe with a 3-in diameter was placed in the gravel layer and connected by a porous PVC pipe with a 2-in diameter (see Figure 6d).
5. Filling of SMLs and PLs: The inside of the EBC was filled with two layers of SMLs and two layers of PLs; the thickness of each layer was 100 mm (see Figure 6e).
6. Installation of geotextiles and filling of coarse sand: After geotextiles were laid on top of the gravel or PL, the cells were filled with a 100-mm-thick coarse sand layer (see Figure 6f).
7. Installation of geotextiles: Geotextiles were laid over the coarse sand layer to separate the planting soil and coarse sand layer (see Figure 6g).
8. Filling of planting soil: The two cells were filled with a 500-mm-thick layer of planting soil (see Figure 6h).
9. Planting: *Photinia serratifolia*, *Ficus microcarpa* cv. Golden Leaves, and *Ehretia microphylla* Lamk were planted in the two bioretention cells (see Figure 6i).
10. Installation of a weir: To allow the water in the channel to flow into the retention tank, a weir was installed to raise the water level (see Figure 6j).
11. Installation of inlet pipes: 4-in PVC pipes were used to divert water from the weir into the bioretention cells (see Figure 6k).
12. Operation: Contaminated water was drained into the bioswales of the bioretention cells to allow for infiltration and purification. The purified water was then discharged into the river using gravity (see Figure 6l).

(a)

(b)

(c)

(d)

(e)

(f)

Figure 6. *Cont.*

Figure 6. Construction steps: (**a**) excavation; (**b**) installation of geomembrane; (**c**) filling of gravel; (**d**) installation of outflow pipes; (**e**) filling of SMLs and PLs; (**f**) filling of coarse sand; (**g**) installation of geotextiles; (**h**) filling of planting soil; (**i**) planting; (**j**) installation of a weir; (**k**) installation of inlet pipes; (**l**) operation.

3. Results and Discussion

The two bioretention cells were completed in March 2019. To elucidate their effectiveness in removing pollutants, a 2-year sampling survey was conducted. Sampling was performed once a month, and a total of 24 samples were collected for analysis. Suspended solids (SS), chemical oxygen demand (COD), ammonium nitrogen (NH_4^+-N), total nitrogen (TN), total phosphorus (TP), and phosphate (PO_4) were tested in the present study.

3.1. Suspended Solid

Table 1 and Figure 7 shows the test results for SS that were obtained through a 2-year experiment that was conducted between May 2019 and April 2021. The results indicate that the concentration of SS was higher during the rainy season from June to October. When contaminated water flows into the bioretention cells, SS decrease significantly. Because of the filtering function of the material inside the bioretention cells, they have a significant effect on SS removal. The average removal percentages of the TBC and EBC were 85% and 83%, respectively. The TBC and EBC have similar average removal percentages, indicating their favorable performance for SS reduction. However, the bioswales and pebbles on the surface of the bioretention cells required regular cleaning to avoid clogging due to the excessive accumulation of mud. Moreover, Student's t-test was used to was used to determine whether the performance of TBC and EBC were different. For SS removal, the T-value of TBC and EBC is 47.5%. If the significance level (α = 5%) was used, the test result showed that there was no difference in SS removal performance between TBC and EBC.

Table 1. The concentration and removal percentage of SS during the 2-year experiment.

Date	Concentration (mg/L)			Removal Percentage (%)	
	C_i [1]	$C_{o,TBC}$ [2]	$C_{o,EBC}$ [3]	TBC	EBC
May 2019	51	7	10	86	80
June 2019	34	4	5	88	85
July 2019	61	5	7	92	89
August 2019	44	4	7	91	84
September 2019	32	4	4	88	88
October 2019	31	4	8	87	74
November 2019	21	0	4	100	81
December 2019	22	4	0	82	100
January 2020	19	2	4	89	79
February 2020	7	5	4	29	43
March 2020	18	0	2	100	89
April 2020	18	0	2	100	89
May 2020	42	5	2	88	95
June 2020	77	10	8	87	90
July 2020	83	8	4	90	95
August 2020	69	11	8	84	88
September 2020	66	8	12	88	82
October 2020	54	10	6	81	89
November 2020	48	4	8	92	83
December 2020	32	8	4	75	88
January 2021	29	5	4	83	86
February 2021	21	8	2	62	90
March 2021	37	4	8	89	78
April 2021	19	4	4	79	79
Average	39	5	5	85	83
T-test value				47.5	

[1] Inflow concentration. [2] Outflow concentration of TBC. [3] Outflow concentration of EBC.

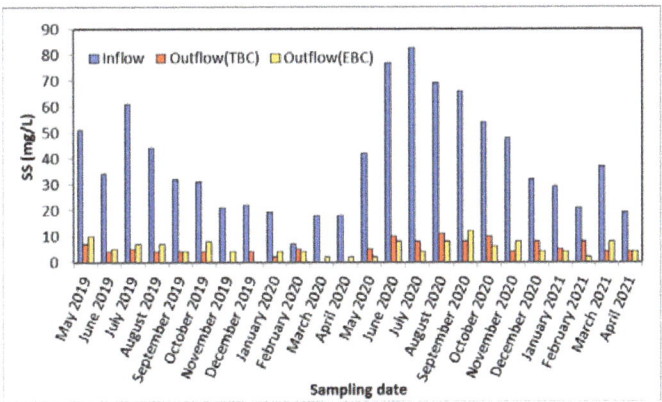

Figure 7. Suspended solid (SS) levels during the 2-year experiment.

3.2. Chemical Oxygen Demand

The test results for COD are presented in Table 2 and Figure 8. Both the TBC and EBC performed poorly for COD removal in the early stage of the facility's operation. This is because, at that stage, the microbial system within the bioretention cells was not fully established, and the ability of the microbes to biodegrade COD was not fully developed. However, after approximately half a year, stable pollution removal results were observed. The TBC's average pollution removal efficiency for COD was approximately 51%, and that of the EBC was 65%. The EBC outperformed the TBC, and the outflow concentration of the EBC was relatively stable. The first reason for this performance disparity is that the PLs in the EBC comprise zeolite, which can perform the function of adsorption. The second reason is that the SMLs in the EBC contains natural organic materials (e.g., rice stalks and rice husks), and the decomposed plant fiber can provide a carbon source and cultivate cellulose-degrading bacteria that biodegrades COD. However, both the TBC and EBC exhibited favorable COD biodegradation performance because the planting soil was sufficiently conducive to the cultivation of a microbial system. The *T*-value of COD removal between TBC and EBC was 0.2%. It showed that the performance of TBC and EBC in COD removal was different.

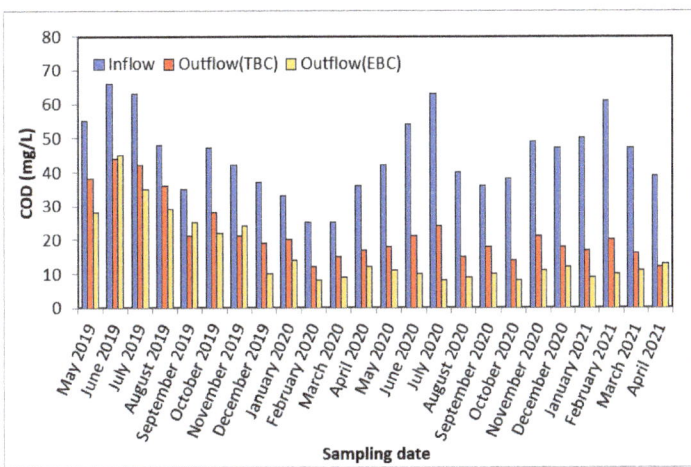

Figure 8. Chemical oxygen demand (COD) during the 2-year experiment.

Table 2. The concentration and removal percentage of COD during the 2-year experiment.

Date	Concentration (mg/L)			Removal Percentage (%)	
	C_i [1]	$C_{o,TBC}$ [2]	$C_{o,EBC}$ [3]	TBC	EBC
May 2019	55	38	28	31	49
June 2019	66	44	45	33	32
July 2019	63	42	35	33	44
August 2019	48	36	29	25	40
September 2019	35	21	25	40	29
October 2019	47	28	22	40	53
November 2019	42	21	24	50	43
December 2019	37	19	10	49	73
January 2020	33	20	14	39	58
February 2020	25	12	8	52	68
March 2020	25	15	9	40	64
April 2020	36	17	12	53	67
May 2020	42	18	11	57	74
June 2020	54	21	10	61	81
July 2020	63	24	8	62	87
August 2020	40	15	9	63	78
September 2020	36	18	10	50	72
October 2020	38	14	8	63	79
November 2020	49	21	11	57	78
December 2020	47	18	12	62	74
January 2021	50	17	9	66	82
February 2021	61	20	10	67	84
March 2021	47	16	11	66	77
April 2021	39	12	13	69	67
Average	45	22	16	51	65
T-test value				0.2	

[1] Inflow concentration. [2] Outflow concentration of TBC. [3] Outflow concentration of EBC.

3.3. Ammonium Nitrogen

The bioretention cells efficiently removed NH_4^+-N from inflow water. Table 3 and Figure 9 presents the test results for NH_4^+-N. The concentration of NH_4^+-N in inflow water was noticeably higher during the farming season from April to October. However, the concentration of NH_4^+-N in inflow water was lower from August to October 2019 because the area was affected by a drought and reduced rainfall during that time. For the removal of NH_4^+-N, the TBC and EBC exhibited similar performance. The average removal percentages of the TBC and EBC were 81% and 78%, respectively. Thus, both of these systems have favorable performance for NH_4^+-N removal. The T-value of NH_4^+-N removal between TBC and EBC was 26.0%. The test result showed that there was no difference in NH_4^+-N removal performance between TBC and EBC.

Table 3. The concentration and removal percentage of NH_4^+-N during the 2-year experiment.

Date	Concentration (mg/L)			Removal Percentage (%)	
	C_i [1]	$C_{o,TBC}$ [2]	$C_{o,EBC}$ [3]	TBC	EBC
May 2019	1.61	0.17	0.15	90	90
June 2019	0.87	0.22	0.12	75	86
July 2019	1.43	0.48	0.42	66	70
August 2019	0.34	0.06	0.11	84	72
September 2019	0.12	0.04	0.06	69	52
October 2019	0.44	0.10	0.09	78	79
November 2019	0.71	0.09	0.07	88	91
December 2019	0.10	0.05	0.03	48	69

Table 3. Cont.

Date	Concentration (mg/L)			Removal Percentage (%)	
	C_i [1]	$C_{o,TBC}$ [2]	$C_{o,EBC}$ [3]	TBC	EBC
January 2020	0.12	0.03	0.03	77	75
February 2020	0.06	0.02	0.05	68	17
March 2020	0.27	0.03	0.08	90	71
April 2020	0.39	0.07	0.10	81	74
May 2020	0.71	0.14	0.17	81	76
June 2020	0.93	0.18	0.19	81	79
July 2020	1.63	0.13	0.12	92	92
August 2020	1.40	0.19	0.11	87	92
September 2020	2.49	0.54	0.35	78	86
October 2020	1.83	0.27	0.18	85	90
November 2020	0.62	0.11	0.16	82	75
December 2020	0.91	0.08	0.12	91	87
January 2021	1.08	0.17	0.14	84	87
February 2021	0.76	0.10	0.10	87	87
March 2021	0.84	0.09	0.09	89	89
April 2021	1.39	0.10	0.10	93	93
Average	0.88	0.14	0.13	81	78
T-test value				26	

[1] Inflow concentration. [2] Outflow concentration of TBC. [3] Outflow concentration of EBC.

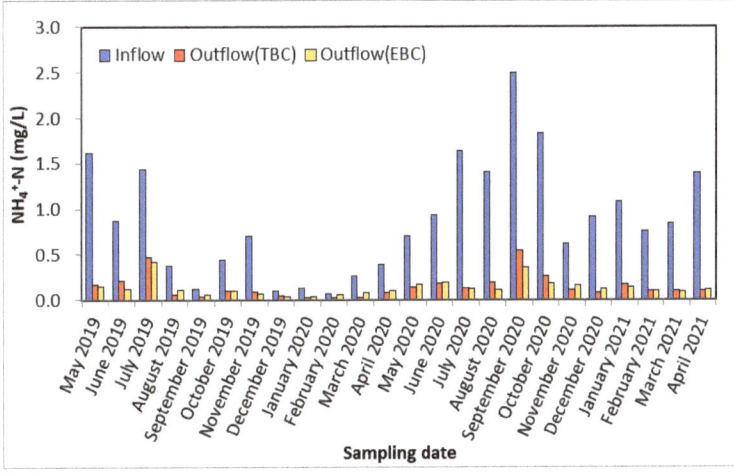

Figure 9. Ammonium nitrogen (NH_4^+-N) levels during the 2-year experiment.

3.4. Total Nitrogen

As indicated by Table 4 and Figure 10, under an average TN inflow of 2.23 mg/L, the average TN removal efficiency was 50% and 67% for the TBC and EBC, respectively. During the farming season, the concentration of TN was higher relative to other seasons. Nitrification coupled with denitrification is the main removal process for TN. When wastewater flows into the bioretention cells, the microorganisms in their cell systems begin to produce nitrification and denitrification reactions to reduce TN. For the TBC, a slight denitrification reaction only occurs in the planting soil layer. However, for the EBC, the denitrification reaction not only occurs in the planting soil layer but also in the mixed filter material layer. Luanmanee et al. [24] reported that the low C/N ratio of pretreated wastewater tends to result in low TN removal due to a lack of carbon available for denitrification. Consequently, in the present study, the TBC was less effective at removing TN. The addition of active charcoal powder and organic matter to the SML potentially provided a sufficient carbon source for microorganisms. Under aerobic conditions, NH_4^+-N is converted into NO_3^--N through biological nitrification, and the resulting NO_3^--N then infiltrates into the SML, where nitrogen gas is formed through biological denitrification under anaerobic conditions and contributes to efficient TN removal (Figure 11). Therefore, the TN removal performance of the EBC is superior to that of the TBC. The T-value of TN removal showed that the performance of TBC and EBC in TN removal was different.

Table 4. The concentration and removal percentage of TN during the 2-year experiment.

Date	Concentration (mg/L)			Removal Percentage (%)	
	C_i [1]	$C_{o,TBC}$ [2]	$C_{o,EBC}$ [3]	TBC	EBC
May 2019	3.99	1.98	1.24	50	69
June 2019	2.80	1.94	1.50	31	46
July 2019	4.35	3.84	2.90	12	33
August 2019	1.90	1.66	1.19	13	37
September 2019	1.54	0.41	0.68	73	56
October 2019	2.01	0.79	0.48	61	76
November 2019	1.88	1.37	1.01	27	46
December 2019	1.97	0.97	0.26	51	87
January 2020	1.68	0.54	0.24	68	86
February 2020	0.97	0.42	0.38	57	61
March 2020	1.88	0.53	0.36	72	81
April 2020	1.24	0.84	0.46	32	63
May 2020	3.18	1.27	0.89	60	72
June 2020	2.87	1.08	0.50	62	83
July 2020	2.68	0.81	0.46	70	83
August 2020	3.74	1.67	0.51	55	86
September 2020	3.04	0.94	0.42	69	86
October 2020	2.43	1.07	0.48	56	80
November 2020	1.66	0.96	0.52	42	69
December 2020	1.76	0.89	0.34	49	81
January 2021	0.98	0.43	0.25	56	74
February 2021	1.16	1.02	0.87	12	25
March 2021	1.84	0.83	0.61	55	67
April 2021	2.03	0.83	0.59	59	71
Average	2.23	1.13	0.71	50	67
T-test value				0.1	

[1] Inflow concentration. [2] Outflow concentration of TBC. [3] Outflow concentration of EBC.

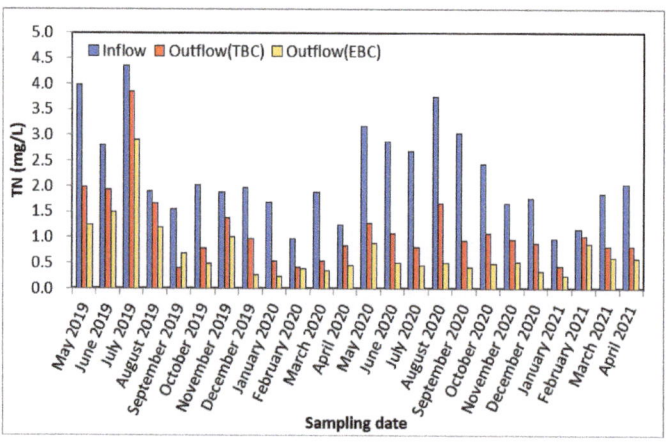

Figure 10. Total nitrogen (TN) levels during the 2-year experiment.

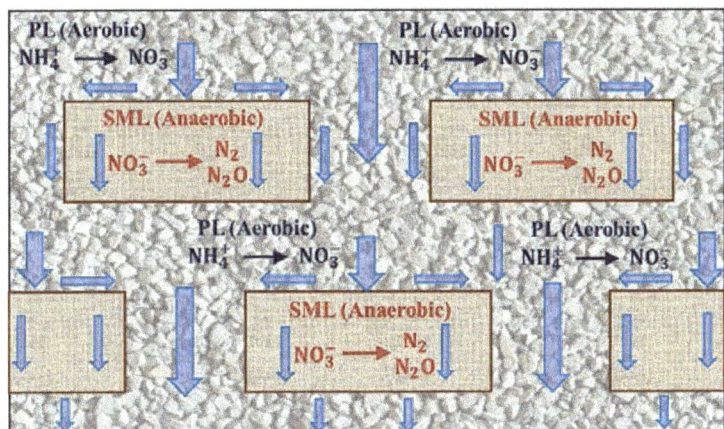

Figure 11. Nitrification and denitrification process in the mixed filter material layer.

To determine the nitrification and denitrification reactions in the EBC system, samples from PL and SML materials were collected (through drilling) for microbial analysis. Because the 16S rRNA gene, which is present in all bacteria, is highly conserved and can be easily amplified using universal primers, environmental microbial analyses are often performed using 16S rRNA amplicon sequencing [25]. Moreover, sequences are clustered into bins called operational taxonomic units (OTUs) on the basis of similarity. In the absence of traditional systems of biological classification, such as those available for macroscopic organisms, OTUs serve as pragmatic proxies for species (microbial or metazoan) at different taxonomic levels. For several years, OTUs have been the most commonly used units of diversity, especially for the analysis of small subunit 16S (for prokaryotes) or 18S rRNA (for eukaryotes) [26] marker gene sequence datasets. Figure 12 presents the microbial analysis results for PLs and SMLs.

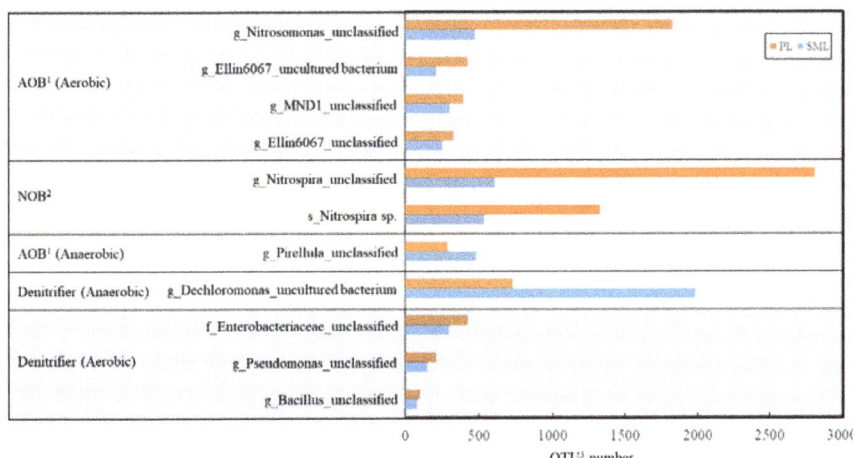

Figure 12. Microbial analysis results for PLs and SMLs. [1] AOB: ammonia oxidizing bacteria. [2] NOB: nitrite oxidizing bacteria. [3] OTU: Operational taxonomic unit.

The experimental results indicate that the main dominant bacteria of PLs are aerobic nitrifying bacteria. Ammonia oxidizing bacteria are responsible for oxidizing NH_4^+-N to NO_2^--N, and nitrite oxidizing bacteria (NOB) are responsible for oxidizing NO_2^--N to NO_3^--N. In PLs, the nitration reaction can be completed through the two aforementioned reactions. Therefore, both the TBC and EBC provide a favorable nitration reaction for NH_4^+-N. However, the denitrification reaction is mainly influenced by anaerobic denitrifying bacteria. The test results indicate that the anaerobic bacteria in SMLs consisted of 2818 OTUs, which is 4.6 times the amount detected in the PLs (612 OTUs). When the agricultural runoff water with NH_4^+-N flows into the mixed filter material layer, the aerobic bacteria of the PLs first nitrify the NH_4^+-N into NO_2^--N and NO_3^--N. Subsequently, the water flows into the SMLs (anaerobic zone), where it undergoes denitrification, which reduces NO_3^--N and NO_2^--N to N_2. Therefore, the TN removal performance of the EBC is superior to that of the TBC.

3.5. Total Phosphorus and Phosphate

The phosphorus in the runoff from agricultural land is a key component of nonpoint source pollution, and it can accelerate the eutrophication of lakes and streams [27]. Therefore, the effective reduction of the phosphorus in agricultural runoff is a crucial task. Table 5 and Figure 13 chart the TP concentrations of the inflow into the cells, the outflow from the TBC, and the outflow from the EBC. The nonpoint source pollution during the fertilization period contained a relatively high TP concentration relative to the subsequent period. The average concentration of TP in inflowing water was 0.761 mg/L. After TBC purification, the average concentration of TP in the outflow was 0.362 mg/L, indicating a removal percentage of 51%. However, the average concentration of TP after EBC purification was 0.059 mg/L, indicating a removal percentage of 92%. These results indicate that the TP removal effect of the EBC is superior to that of the TBC. The T-value of TP removal between TBC and EBC was <0.1%. It showed that the performance of TBC and EBC in TP removal was significantly different.

Table 5. The concentration and removal percentage of TP during the 2-year experiment.

Date	Concentration (mg/L)			Removal Percentage (%)	
	C_i [1]	$C_{o,TBC}$ [2]	$C_{o,EBC}$ [3]	TBC	EBC
May 2019	0.56	0.27	0.05	52	91
June 2019	0.81	0.53	0.05	35	94
July 2019	0.67	0.32	0.07	52	90
August 2019	0.91	0.26	0.07	71	92
September 2019	0.80	0.37	0.05	54	94
October 2019	0.94	0.27	0.07	71	93
November 2019	0.72	0.42	0.09	42	88
December 2019	0.68	0.33	0.05	51	93
January 2020	0.51	0.29	0.04	43	92
February 2020	0.84	0.40	0.02	52	98
March 2020	0.89	0.32	0.02	64	98
April 2020	0.70	0.41	0.04	41	98
May 2020	0.99	0.48	0.05	52	95
June 2020	1.03	0.52	0.07	50	93
July 2020	0.63	0.31	0.06	51	90
August 2020	0.74	0.37	0.10	50	86
September 2020	0.94	0.42	0.05	55	95
October 2020	0.83	0.31	0.08	63	90
November 2020	0.56	0.33	0.05	41	91
December 2020	0.67	0.29	0.05	57	93
January 2021	0.58	0.34	0.05	41	91
February 2021	0.60	0.41	0.07	32	88
March 2021	0.74	0.33	0.10	55	86
April 2021	0.93	0.38	0.09	59	90
Average	0.76	0.36	0.06	51	92
T-test value				<0.1	

[1] Inflow concentration. [2] Outflow concentration of TBC. [3] Outflow concentration of EBC.

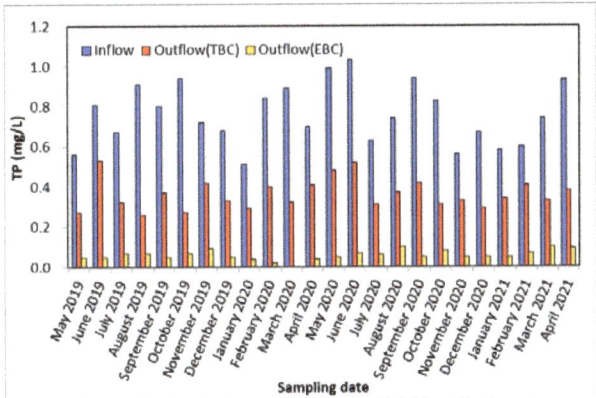

Figure 13. Total phosphorus (TP) levels during the 2-year experiment.

Table 6 and Figure 14 presents the performance of the TBC and EBC for the removal of phosphate (PO_4^{3-}) in water. The average concentration of phosphate in inflow was 0.625 mg/L. However, the phosphate removal percentage of the TBC (49%) was significantly higher than that of the EBC (90%), indicating that the EBC was more effective than the TBC in removing phosphate. In the TBC, the phosphate removal percentage differed significantly over time. In the EBC, the phosphate removal percentage did not exhibit significant changes over time. This indicates that the EBC is more stable than the TBC in removing phosphate. The T-value of PO_4^{3-} removal also showed that the performance of TBC and EBC in PO_4^{3-} removal was significantly different.

Table 6. The concentration and removal percentage of PO_4^{3-} during the 2-year experiment.

Date	Concentration (mg/L)			Removal Percentage (%)	
	C_i [1]	$C_{o,TBC}$ [2]	$C_{o,EBC}$ [3]	TBC	EBC
May 2019	0.67	0.38	0.05	43	93
June 2019	0.67	0.34	0.05	49	93
July 2019	0.72	0.41	0.07	43	90
August 2019	0.72	0.43	0.07	39	90
September 2019	0.75	0.45	0.05	41	93
October 2019	0.71	0.40	0.07	43	90
November 2019	0.62	0.31	0.05	50	92
December 2019	0.55	0.25	0.04	55	93
January 2020	0.40	0.21	0.06	48	85
February 2020	0.74	0.42	0.03	43	96
March 2020	0.67	0.28	0.02	58	97
April 2020	0.59	0.31	0.04	47	93
May 2020	0.72	0.37	0.06	49	92
June 2020	0.75	0.42	0.07	44	91
July 2020	0.54	0.27	0.05	50	91
August 2020	0.59	0.31	0.08	47	86
September 2020	0.74	0.31	0.05	58	93
October 2020	0.71	0.26	0.08	63	89
November 2020	0.41	0.23	0.05	44	88
December 2020	0.51	0.22	0.06	57	88
January 2021	0.41	0.31	0.05	24	87
February 2021	0.44	0.24	0.07	45	84
March 2021	0.59	0.22	0.08	63	86
April 2021	0.79	0.28	0.09	65	89
Average	0.63	0.32	0.06	49	90
T-test value				<0.1	

[1] Inflow concentration. [2] Outflow concentration of TBC. [3] Outflow concentration of EBC.

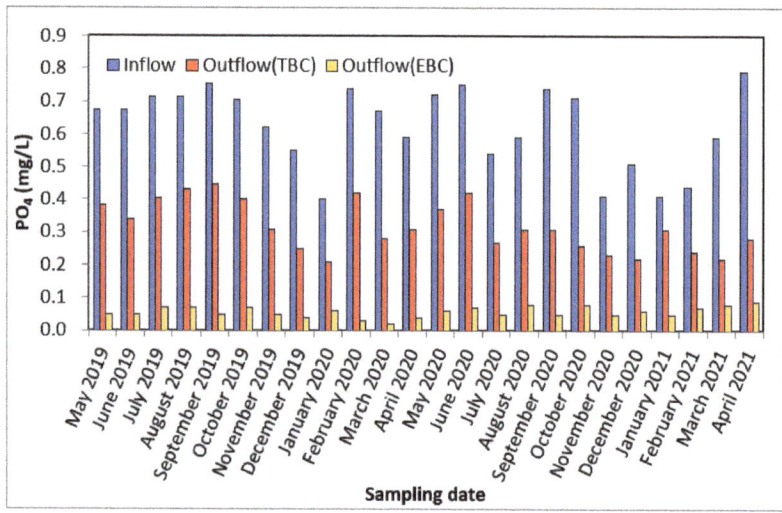

Figure 14. Phosphate (PO_4^{3-}) levels during the 2-year experiment.

Normally, the phosphorus in agricultural runoff contains particulate and dissolved forms. The particulate phosphorus in water can be removed through filtering. Therefore, both the TBC and EBC can remove particulate phosphorus through the filtration of water in the upper planting soil. However, dissolved phosphorus is generally difficult to remove through filtration. In the EBC, a specific ratio of iron particles is mixed into its SMLs. Phosphorus can be adsorbed by the Al and Fe hydroxides in the soil. As discussed in an earlier section of this paper, the inside of SMLs presents an anaerobic state (Figure 15). In such an environment, the iron added to SMLs transforms into ferrous iron (Fe^{2+}), which is subsequently translocated to PLs and oxidized to ferric ion (Fe^{3+}); Fe^{3+} aids the coprecipitation of PO_4^{3-} from percolating wastewater [28]. Under anaerobic conditions, the reaction mechanism through which PO_4^{3-} produces $Fe_3(PO_4)_2$ is expressed by Equation (1). Equations (2) and (3) express the aforementioned reaction mechanism of PO_4^{3-} under aerobic conditions. The final reaction product of iron ions and phosphate precipitates at the bottom of a bioretention cell. Sato et al. [29] monitored a case in Japan using SMB to treat domestic sewage and used mass balance to calculate TP removal efficiency. It shows that 110 kg of metal iron fix 52 kg of P, and its service life is more than 10 years. The SML of this study used the same percentage of metal iron content as in the previous study. As the TP concentration of agriculture non-point source pollution is much lower than that of domestic sewage, the service life of this study site can be expected to be longer than 10 years.

$$3Fe^{2+} + 2PO_4^{3-} \rightarrow Fe_3(PO_4)_2 (\downarrow) \qquad (1)$$

$$Fe^{3+} + PO_4^{3-} \rightarrow FePO_4 (\downarrow) \qquad (2)$$

$$Fe(OH)_3 + H_3PO_4 \rightarrow FePO_4 (\downarrow) + 3H_2O \qquad (3)$$

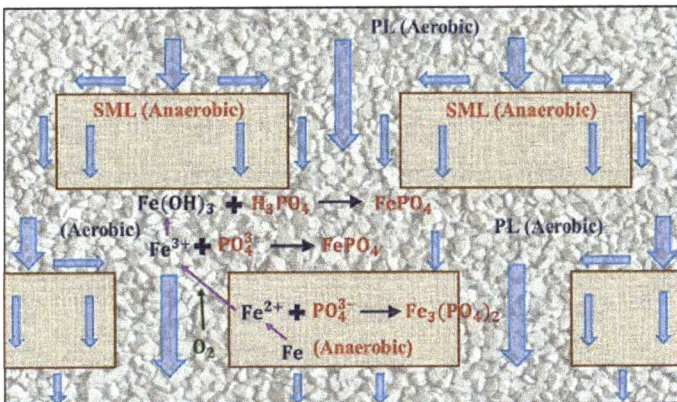

Figure 15. Iron ions and phosphate reaction process in the mixed filter material layer.

4. Conclusions

The results of the present study indicate that both the TBC and EBC can purify agricultural nonpoint source pollution. The removal percentages for SS, COD, and NH_4^+-N were similar in the TBC and EBC. However, for TN, TP, and PO_4^{3-} removal, the EBC significantly outperforms the TBC. Regarding the composition of the two bioretention cells, the only difference is the bottom material. The TBC uses a gravel bed, whereas the EBC uses a mixed filter material layer. The two layers that form the mixed filter material layer comprise an aerobic layer and an anaerobic layer. The aerobic layers (PLs) consist of zeolite, and they alternate with the anaerobic layers (SMLs). The zeolite in the aerobic layers increases the adsorption of NH_4^+-N and reacts to nitrate through the reaction of nitrifying bacteria. Moreover, zeolite can enhance the oxidation and precipitation of mobile ferrous iron to high-surface-area ferric oxide, which improves phosphorus sorption. In the anaerobic layer formed by SMLs, nitrate is converted into nitrous oxide and nitrogen (denitrification), and ferrous iron is oxidized to mobile ferric iron, which exits the anaerobic layer and reacts to $FePO_4$ with the PO_4^{3-} in the aerobic layer. As a result, relative to the TBC, the EBC provides superior removal performance for TN, TP, and PO_4^{3-}. The use of the EBC to purify agricultural nonpoint source pollution results in greater pollution removal efficiency, and consequently, contributes to SDGs such as environmental sustainability, clean water, and food security.

Author Contributions: C.-C.H.: Conceptualization, writing manuscript draft and supervision; Y.-X.L.: Investigation, formal analysis and data curation. All authors have read and agreed to the published version of the manuscript.

Funding: This research received no external funding.

Informed Consent Statement: Informed consent was obtained from all subjects involved in the study.

Data Availability Statement: Data used in this study are duly available from the first authors on reasonable request.

Acknowledgments: The authors are grateful to the Editor and anonymous reviewers for their constructive comments and suggestions.

Conflicts of Interest: The authors declare no conflict of interest.

References

1. Liu, Y.; Engel, B.A.; Flanagan, D.C.; Gitau, M.W.; McMillan, S.K.; Chaubey, I.A. Review on effectiveness of best management practices in improving hydrology and water quality: Needs and opportunities. *Sci. Total Environ.* **2017**, *601*, 580–593. [CrossRef] [PubMed]
2. Nordman, E.E.; Isely, E.; Isely, P.; Denning, R. Benefit-cost analysis of stormwater green infrastructure practices for Grand Rapids, Michigan, USA. *J. Clean. Prod.* **2018**, *200*, 501–510. [CrossRef]
3. Fletcher, T.; Andrieu, H.; Hamel, P. Understanding, management and modelling of urban hydrology and its consequences for receiving waters: A state of the art. *Adv. Water Resour.* **2013**, *51*, 261–279. [CrossRef]
4. Macedo, M.B.; Oliveira, T.R.P.; Oliveira, T.H.; Gomes Junior, M.N.; Brasil, J.A.T.; Lago, C.A.F.; Mendionda, E.M. Evaluating low impact development practices potentials for increasing flood resilience and stormwater reuse through lab-controlled bioretention systems. *Water Sci. Technol.* **2021**, *84*, 1103–1124. [CrossRef] [PubMed]
5. Macedo, M.B.; Gomes Junior, M.N.; Oliveira, T.R.P.; Giacomoni, M.H.; Imani, M.; Zhang, K.; Lago, C.A.F.; Mendiondo, E.M. Low impact development practices in the context of united nations sustainable development goals: A new concept, lessons learned and challenges. Critical Reviews in Environ. *Sci. Technol.* **2021**. [CrossRef]
6. Charlesworth, S.M. A review of the adaptation and mitigation of global climate change using sustainable drainage in cities. *J. Water Clim. Chang.* **2010**, *1*, 165–180. [CrossRef]
7. Prince George's County Department of Environmental Resources (PGDER). *Design Manual for Use of Bioretention in Stormwater Management*; Division of Environmental Management, Water-Shed Protection Branch: Landover, MD, USA, 2007.
8. Goh, H.W.; Lem, K.S.; Azizan, N.A.; Chang, C.K.; Talei, A.; Leow, C.S.; Zakaria, N.A.A. Review of bioretention components and nutrient removal under different climates—Future directions for tropics. *Environ. Sci. Pollut. Res.* **2019**, *26*, 14904–14919. [CrossRef]
9. Davis, A.P.; Shokouhian, M.; Sharma, H.; Minami, C. Laboratory study of biological retention for urban stormwater management. *Water Environ. Res.* **2001**, *73*, 5–14. [CrossRef]
10. Davis, A.P.; Hunt, W.F.; Traver, R.G.; Clar, M. Bioretention technology: Overview of current practice and future needs. *J. Environ. Eng.* **2009**, *135*, 109–117. [CrossRef]
11. Le Coustumer, S.; Fletcher, T.D.; Deletic, A.; Barraud, S.; Poelsma, P. The influence of design parameters on clogging of stormwater biofilters: A large-scale column study. *Water Res.* **2012**, *46*, 6743–6752. [CrossRef]
12. Hatt, B.E.; Fletcher, T.D.; Deletic, A. Hydraulic and pollutant removal performance of fine media stormwater filtration systems. *Enviro. Sci. Technol.* **2008**, *42*, 2535–2541. [CrossRef] [PubMed]
13. Fan, G.; Ning, R.; Huang, K.; Wang, S.; You, Y.; Du, B.; Yan, Z. Hydrologic characteristics and nitrogen removal performance by different formulated soil medium of bioretention system. *J. Clean. Prod.* **2021**, *290*, 125873. [CrossRef]
14. Li, Y.; Zhang, Y.; Yu, H.; Han, Y.; Zuo, J. Enhancing nitrate removal from urban stormwater in an inverted bioretention system. *Ecol. Eng.* **2021**, *170*, 106315. [CrossRef]
15. Davis, A.P.; Shokouhian, M.; Sharma, H.; Minami, C. Water quality improvement through bioretention media: Nitrogenand phosphorus removal. *Water Environ. Res.* **2006**, *78*, 284–293. [CrossRef] [PubMed]
16. Howarth, R.W.; Marino, R. Nitrogen as the limiting nutrient for eutrophication in coastal marine ecosystems: Evolving views over three decades. *Limnol. Oceanogr.* **2006**, *51*, 364–376. [CrossRef]
17. Ho, C.C.; Wang, P.H. Efficiency of a Multi-Soil-Layering System on Wastewater Treatment Using Environment-Friendly Filter Materials. *Int. J. Environ. Res. Public Health* **2015**, *12*, 3362–3380. [CrossRef]
18. Tong, Y.; Lin, G.; Ke, X. Comparison of microbial community between two shallow freshwater lakes in middle Yangtze basin, East China. *Chemosphere* **2005**, *60*, 85–92. [CrossRef]
19. Yang, L.; Lei, K.; Meng, W. Temporal and spatial changes in nutrients and chlorophyll-α in a shallow lake, Lake Chaohuhu, China: An 11-year investigation. *J. Environ. Sci.* **2013**, *25*, 1117–1123. [CrossRef]
20. Kong, X.; Jørgensen, S.; He, W. Predicting the restoration effects by a structural dynamic approach in Lake Chaohuhu, China. *Ecol. Model.* **2013**, *266*, 73–85. [CrossRef]
21. Yang, X.; Cui, H.B.; Liu, X.S.; Wu, Q.A.; Zhang, H. Water pollution characteristics and analysis of Chaohuhu Lake basin by using different assessment methods. *Environ. Sci. Pollut. Res.* **2020**, *27*, 18168–18181. [CrossRef]
22. Liu, E.; Shen, J.; Birch, G. Human-induced change in sedimentary trace metals and phosphorus in Chaohu Lake, China, over the past half-millennium. *J. Paleolimnol.* **2012**, *47*, 677–691. [CrossRef]
23. Gao, H.; Lv, C.; Song, Y. Chemometrics data of water quality and environmental heterogeneity analysis in Pu River, China. *Environ. Earth Sci.* **2015**, *73*, 5119–5129. [CrossRef]
24. Luanmanee, S.; Attanandana, T.; Masunaga, T. The efficiency of a multi-soil-layering system on domestic wastewater treatment during the ninth and tenth years of operation. *Ecol. Eng.* **2001**, *18*, 185–199. [CrossRef]
25. Nguyen, N.P.; Warnow, T.; Pop, M.; White, R. A perspective on 16S rRNA operational taxonomic unit clustering using sequence similarity. *NPJ Biofilms Microbiomes* **2016**, *2*, 16004. [CrossRef] [PubMed]
26. Sommer, S.A.; Woudenberg, L.V.; Lenz, P.H.; Cepeda, G.; Goetze, E. Vertical gradients in species richness and community composition across the twilight zone in the North Pacific Subtropical Gyre. *Mol. Ecol.* **2017**, *26*, 6136–6156. [CrossRef] [PubMed]

27. Daniel, T.C.; Sharpley, A.N.; Lemunyon, J.L. Agricultural Phosphorus and Eutrophication: A Symposium Overview. *J. Environ. Qual.* **1998**, *27*, 251–257. [CrossRef]
28. Wakatsuki, T.; Esumi, H.; Omura, S. High Performance and N&P Removal On-Site Municipal Wastewater Treatment System by Multi-Soil-Layering Method. *Water Sci. Technol.* **1993**, *27*, 31–40.
29. Sato, K.; Iha, Y.; Luanmanee, S.; Masunaga, T.; Wakatsuki, T. Long term on-site experiments and mass balances in waste water treatment by multi-soil-layering system. In Proceedings of the 17th World Congress of Soil Science, Bangkok, Thailand, 14–21 August 2002; Volume 1261, pp. 1–10.

Article

Pilot and Field Studies of Modular Bioretention Tree System with *Talipariti tiliaceum* and Engineered Soil Filter Media in the Tropics

Fang Yee Lim [1], Teck Heng Neo [1], Huiling Guo [2], Sin Zhi Goh [1], Say Leong Ong [1], Jiangyong Hu [1,*], Brandon Chuan Yee Lee [1], Geok Suat Ong [3] and Cui Xian Liou [3]

[1] Department of Civil and Environmental Engineering, National University of Singapore, Block E1A, #07-03, 1 Engineering Drive 2, Singapore 117576, Singapore; rlimtony@gmail.com (F.Y.L.); erinth@nus.edu.sg (T.H.N.); gohsinzhi@gmail.com (S.Z.G.); say.leong_ong@nus.edu.sg (S.L.O.); ceeblcy@nus.edu.sg (B.C.Y.L.)

[2] School of Life Sciences & Chemical Technology, Ngee Ann Polytechnic, 535 Clementi Road, Blk 83, Singapore 599489, Singapore; guo_huiling@np.edu.sg

[3] PUB, Singapore's National Water Agency, 40 Scotts Road, #09-01 Environment Building, Singapore 228231, Singapore; ONG_Geok_Suat@pub.gov.sg (G.S.O.); LIOU_Cui_Xian@pub.gov.sg (C.X.L.)

* Correspondence: ceehujy@nus.edu.sg; Tel.: +65-6516-4540

Citation: Lim, F.Y.; Neo, T.H.; Guo, H.; Goh, S.Z.; Ong, S.L.; Hu, J.; Lee, B.C.Y.; Ong, G.S.; Liou, C.X. Pilot and Field Studies of Modular Bioretention Tree System with *Talipariti tiliaceum* and Engineered Soil Filter Media in the Tropics. *Water* **2021**, *13*, 1817. https://doi.org/10.3390/w13131817

Academic Editor: Jose G. Vasconcelos

Received: 10 May 2021
Accepted: 28 June 2021
Published: 30 June 2021

Publisher's Note: MDPI stays neutral with regard to jurisdictional claims in published maps and institutional affiliations.

Copyright: © 2021 by the authors. Licensee MDPI, Basel, Switzerland. This article is an open access article distributed under the terms and conditions of the Creative Commons Attribution (CC BY) license (https://creativecommons.org/licenses/by/4.0/).

Abstract: Stormwater runoff management is challenging in a highly urbanised tropical environment due to the unique space constraints and tropical climate conditions. A modular bioretention tree (MBT) with a small footprint and a reduced on-site installation time was explored for application in a tropical environment. Tree species used in the pilot studies were *Talipariti tiliaceum* (TT1) and *Sterculia macrophylla* (TT2). Both of the MBTs could effectively remove total suspended solids (TSS), total phosphorus (TP), zinc, copper, cadmium, and lead with removal efficiencies of greater than 90%. Total nitrogen (TN) removal was noted to be significantly higher in the wet period compared to the dry period ($p < 0.05$). Variation in TN removal between TT1 and TT2 were attributed to the nitrogen uptake and the root formation of the trees species. A field study MBT using *Talipariti tiliaceum* had a very clean effluent quality, with average TSS, TP, and TN effluent EMC of 4.8 mg/L, 0.04 mg/L, and 0.27 mg/L, respectively. Key environmental factors were also investigated to study their impact on the performance of BMT. It was found that the initial pollutant concentration, the dissolved fraction of influent pollutants, and soil moisture affect the performance of the MBT. Based on the results from this study, the MBT demonstrates good capability in the improvement of stormwater runoff quality.

Keywords: urban runoff remediation; *Talipariti tiliaceum*; modular bioretention tree; field study; tree-pit

1. Introduction

Rapid urbanization has resulted in an increase of impervious surface area with a drastic loss of green spaces;, ensuing in a reduction in surface infiltration and evapotranspiration, causing surface runoff to increase in volume [1]. Due to these surface modifications, the urban environment often faces flash floods and problems of nutrient contamination in the surface water [2,3]. Hence, there is a pressing need for a solution to mitigate flooding events and manage nutrient pollution. Urban tropical countries like Singapore pose particular challenges due to their high amount of rainfall, land shortage, and rapid urbanisation. Singapore has a mean annual rainfall of more than 2000 mm [4], with an increase in the occurrence of reported flash floods contemporarily (post-2000), compared to preceding (1984–1999) periods [5]. Land scarcity is also a perennial challenge for highly urbanised countries [6].

A bioretention system (BRS) is a viable treatment option that is able to reduce stormwater runoff volume by reducing infiltration runoff and is aesthetically pleasing [7]. BRS makes use of an engineered environment that integrates with the natural biota to maximize infiltration and support vegetative growth [1]. BRS has shown great potential for the removal of a large variety of pollutants such as suspended solids, heavy metals, and nutrients through the trapping of particle-bound metals, filtration, and sorption methods [8–11]. Factors that affect BRS performance include environmental factors (e.g., the intensity of rain events and evapotranspiration), site selections, and design parameters such as submerged zone depth, filter material, and plant selection [1,12]. A typical filter media mixture includes materials such as sand, silt, clay, and waste materials [1]. The composition of the filter media influences both the hydraulic characteristics (e.g., infiltration rate, detention volume) and the biogeochemical processes that take place in the BRS for surface runoff treatment [12], whereas the vegetation absorbs nutrients and heavy metals from the infiltrated stormwater runoff [13].

Interactions between plants and soil were reported to influence the soil structure, hydrologic processes, and nutrient cycling of the entire ecosystem [14–17]. The plants used in the BRS had no clear distinction between woody and leafy species [17]. Tirpak et al. [18] studied the performance of a bioretention mesocosm planted with trees native to the USA (*Acer rubrum*, *Pinus taeda* and *Quercus palustris*) and found that nutrient uptake via the tree roots is minimal compared to the soil/microbial process. No significant differences were observed for ammonium, nitrite, nitrate, and phosphate in the effluent between all of studied the tree species. The authors explained that the removal of ammonium and phosphate were attributed to the aerobic nature of the bioretention media and the chemical sorption mechanism, respectively. Frosi et al. [19] monitored street tree-pits in Canada and concluded that the systems were able to remove a mass flux of contaminants (such as Na, Cu, and Zn) effectively. The authors also noted that tree pits with high soil organic matter (SOM) could decrease the mass flux of Na and Cu. For instance, with depth, the mass flux of Na and Cu decreased by 66% and 73% in tree pits with less SOM and by 87% and 86% in tree pits with more SOM. The authors recommended the increase of surface permeability and SOM in street tree pits for the improvement of runoff quality and quantity. Elliot et al. [20] measured forty tree pits representing the variety of physical conditions commonly seen in New York City and found that higher infiltration rates in tree pits were associated with larger pit areas, built-up surface elevations, and the combined presence of ground cover planting and mulch. The Ecosol^{TM} tree pit is a commercial MBT that can provide the tertiary treatment of stormwater flows in one compact device [21]. It can achieve suspended solids removal (95%), particulate heavy metals removal (90%), and total petroleum hydrocarbon removal (99%). Stockholm Treepits uses structural soils that can provide a solid base for surfacing while allowing large voids to remain for water movement [22]. A Tree pit (or tree wall) designed by StormTree could provide healthy and thriving trees and stormwater management. It has a unique open design that allows unrestricted tree root growth [23]. However, the reported studies were performed in the temperate regions with limited research on BRS in the tropics. Climate conditions are a key factor that affects the performance of low impact development practices [24]. Climate could affect a wide range of parameters such as plant metabolism, the adsorption capacity of media, microbial composition, and metabolism rate. Tropical climates have a higher rainfall, and higher rainfall frequency, and higher average temperature than temperate regions. In addition, Blecken et al. [25] have demonstrated that extended dry periods worsen the performance of biofilters. Rahman et al. [26] reported that a shorter antecedent dry period (ADP) led to nitrate and nitrite removals, and a longer ADP resulted in nitrate and nitrite export. Denitrification rates were also reported to be lower at low temperatures [27,28]. Clearly, these variations could impact the design and performance of bioretention tree systems in tropical areas and has a need to be further studied.

Furthermore, the design of a conventional BRS also requires longer construction time and larger land area for construction, which is unsuitable for highly urbanized environment such as Singapore. A BRS is also generally constructed on-site, and limited studies have explored integrating existing urban landscapes with BRS. For instance, Brown et al. [29]

combined a BRS with pervious concrete with an internal water storage (IWS) layer and found improvement in hydrologic performance. A volume reduction approaching 100% was noted for sand cells, with reductions of 87% (1.03 m IWS depth) and 75% (0.73 m IWS depth) for the sandy clay loam in the underlying soil. The authors concluded that (a) the hydraulic conductivity of the underlying soil, (b) the IWS zone depth, and (c) the surface infiltration rate were the primary factors that controlled the outflow volume. Kazemi et al. [30] coupled BRS with permeable pavement, which showed beneficial downstream irrigation impacts. In this study, the sodium adsorption index (SAR) for high salinity runoff (1500 mg/L) was reduced from 196.45 meq/L to 6.68 meq/L when discharged from the bioretention basin. The authors concluded that retention time in both the permeable pavement and the bioretention basin storage zone are important, and they recommended that local design guidelines should include data on plants, with focus on their salinity threshold and salt stress.

Hence, this study proposes the use of a unique modular bioretention tree system (MBT) in an urban tropical context. MBT is a treatment device that consists of a tree planted in soil media in a compact module that can be applied to manage non-point source pollution from stormwater runoff. The MBT unique deep soil media helps to sustain a healthy tree and to provide extended stormwater runoff treatment. An MBT has a small footprint and amalgamates with existing urban infrastructure for the treatment of stormwater. To the best of our knowledge, this is the first study that systematically investigated bioretention trees with engineered soil in tropical conditions. In-depth and comprehensive measurements of the performance of an MBT under various 'wet' and 'dry' periods would provide valuable information for water professionals in conceptualizing stormwater runoff management strategies in the tropics. This study also monitored the MBT on a pilot- and field-scale to obtain key insights for the pollutant removal capabilities of the MBT in various scaled operations.

2. Materials and Methods

2.1. Methodology

In this work, there were 3 phases to study the performance of the Modular Bioretention Tree (MBT) system. In the first phase, the interactive effects of the engineered soil and trees were investigated for their system performance with synthetic storm events. The second phase involved the testing of the pilot-scale biofilters with real stormwater runoff. Finally, the last phase was the real application of a scaled-up MBT system with actual urban stormwater runoff. The selected tree used in Phase 3 of the study was screened during the first two phases for its performance.

2.2. Design of Modular Bioretention Tree (MBT) System

For Phases 1 and 2, the pilot MBT biofilters consisted of an impermeable concrete tank with a cross-sectional area of 0.36 m^2, with and had an extended detention depth of 0.2 m (Figure 1a). The MBT biofilter consisted of a filter media layer, a transition layer, and a drainage layer. The top filter media was composed of engineered soil, and the preparation of it followed the patent "An Engineered Soil Composition and a Method of Preparing the Same" patent [31]. Engineered soil consists of a proprietary blend of coconut fibres, water treatment residue (WTR), soil, and sand. Coconut fibre acts as a source of organic matter and encourages denitrification to occur within the bioretention system. Studies by Barrett and Burke [32] showed that coconut fibre increases TN removal by increasing the immobilization of nitrogen within the fibres. WTR consists of aluminium oxides, which provide adsorption sites for pollutants removal [33]. Dissolved phosphorus can be adsorbed readily and reversibly to the surface of aluminium oxides during rain events. This slow and usually irreversible adsorption occurs as the phosphorus that is adsorbed on the surface diffuses deeper into the matrix of the aluminium oxides and becomes deposited [34,35].

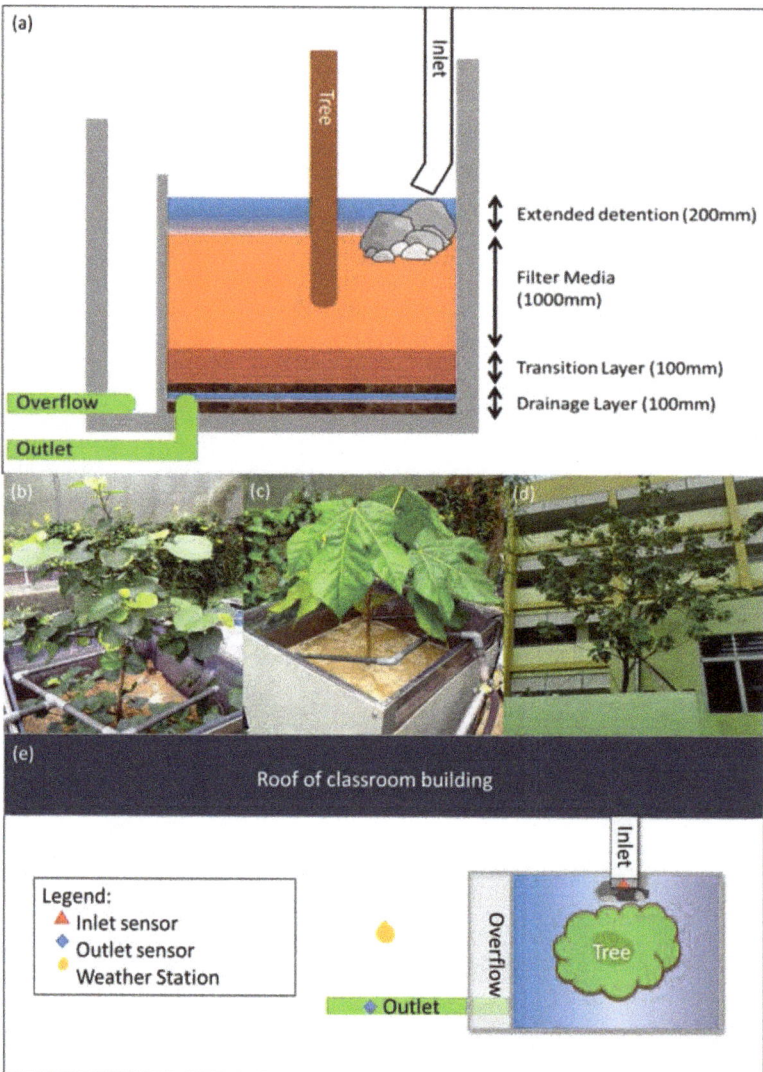

Figure 1. (**a**) Cross-sectional area for MBT system, (**b**) *Talipariti tiliaceum* used in the pilot study (**c**) *Sterculia macrophylla* used in the pilot study, (**d**) MBT system in the field study, (**e**) overview of MBT on site.

The unique design of the MBT system's unique design aims to increase the storage capacity of the BRS per surface area by having a deeper filter media layer compared to conventional systems, allowing it to treat more water in the system. Generally, the media layer was reported to be around 0.4 m [9,36]. A deeper filter media layer of 1.0 m also allows sufficient space and soil depth for the root growth of the tree species planted in the MBT system. The drainage layer consists of 0.1 m of gravel, and the transition layer consists of 0.1 m of coarse sand. The drainage layer facilitates the collection of the effluent from the system, while the transition layer minimizes the probability of the filter medium being washed out of the system during each dosing event.

For the pilot-study, water was directed using a pump through the inlet of the system and dispersed using a small layer of rocks on the top of the filter media. The tree species used in the pilot studies were *Talipariti tiliaceum* (TT1), a coastal species (Figure 1b)), and *Sterculia macrophylla*, a forest tree species (TT2) (Figure 1c). TT1 and TT2 were previously screened in a prior study for the best performance and retention rate in a tropical context [37]. A detailed breakdown of the screened tree species is shown in the Appendix A.

In this study, 3-sided transparent tanks were constructed so that root growth could be monitored, based on a proposed method by Judd et al. [38]. New root shoots and their corresponding depth were observed and recorded. The spatial distribution of root depth within the bioretention system was also recorded in terms of the percentage of new roots formed. The spatial distribution of roots has been shown to affect the removal efficiency of nitrogen species, with deeper roots favoring higher removal efficiency [39,40]. The potted systems used in BRS limit the growth of trees [41,42], hence the decision of keeping the system compact and modular, especially for space-constrained locations. Traditional BRS requires regular maintenance and favors fast-growing plants, as they are assumed to take up nutrients quickly and efficiently to support their fast growth [37,43]. Due to the space constraints in MBT, it is important to balance the growth rate to keep the system modular while having an adequate growth rate for high removal efficiency.

For the field study, the study area was located in an educational institute in the west of Singapore. The MBT system was situated between an open field, school field, and classroom blocks. *Talipariti tiliaceum* has been selected for the field study due to its better TN removal performance. Roof catchment from the surrounding classroom buildings (100% imperviousness, 90 m^2) was channeled into the MBT system. The MBT system had a surface area of 3.7 m^2 and a filter media depth of 1.0 m. The transition and drainage layers were 0.1 m and 0.3 m, respectively. The overflow manhole allowed for a maxi-mum detention depth of 0.2 m (Figure 1d). This unique engineering design allowed for a small compact system, which limited the spread of roots within the system, allowing for a deeper root system. The MBT system was also constructed offsite and was brought to the study area for installation, which reduced the time needed for on-site construction. As such, the MBT could be positioned and retrofitted easily in an urban setting dominated by impervious infrastructures.

A series of sensors were placed around the MBT system to monitor and assess the system's performance (Figure 1e). Automatic water samplers (900 MAX, Sigma, CO, USA) were used to collect 1L of stormwater runoff samples with 6-min intervals for both the inlet and the outlet of the MBT. Area-velocity sensors (AV sensor, Sigma, CO, USA) were installed at the inlet and outlet (subsoil drainage pipe) to monitor the flow velocity and volume. Weather conditions at the site were also monitored using a rain gauge (Sigma, Colorado, USA). Rainfall information from the rain gauge was logged at 1-min intervals. The soil moisture and water potential of the soil were also measured with a soil measure sensor (EC5, Decagon, Pullman, WA, USA) and tensiometer (Tensiometer with pressure transducer, SMS, CA, USA). The monitoring study was conducted for 3 months after the commission of the system in order to take the initial stabilisation period of the system into account, wherein nutrient content was leached from the filter media. The entire monitoring spanned over 15 months and covered both the dry and wet seasons in Singapore, which are caused by the northeast and southwest monsoon.

2.3. Preparation of Synthetic Stormwater Runoff and Actual Stormwater Runoff

In Phase 1 of the study, synthetic stormwater was made based on the average range of urban stormwater pollutant concentrations found in the Singaporean environment [44,45]. This allowed for minimal fluctuation of the inflow concentration while ensuring a realistic composition for the study [45]. Whereas for Phase 2, the stormwater runoff was collected from a canal located in the north of Singapore and dosed into the pilot system. For both phases, the flow rate and volume of the dosed water were based on the average monthly Singapore rainfall data from the years 2003 to 2007. As study [46] has shown that different

ADPs could affect the water quality performance of BRS, the pilot biofilter studies were conducted in 2 different dosing regimens (e.g., 2 times per week for the dry dosing event and 3 times per week for the wet dosing event). The dosing of synthetic water lasted for approximately 8 weeks and 10 weeks for Phase 1 and Phase 2, respectively. Doses of synthetic stormwater runoff were introduced a total of 20 times (10 dry dosing events and 10 wet dosing events) in Phase 1 of the study. Each pilot MBT biofilter received 24 dosing events (12 dry dosing events and 12 wet dosing events) in Phase 2 of the study. The average pollutant concentration and dosing regimen is shown in Table 1.

Table 1. Average pollutant concentration and dosing regimens for Phase 1 and Phase 2.

Pollutant	Phase 1	Phase 2
TSS (mg/L)	100 ± 0.54	121.8 ± 43.8
TP (mg/L)	1.80 ± 0.32	2.20 ± 0.57
TN (mg/L)	2.5 ± 0.12	3.45 ± 1.21
NO_3^- (mg/L)	0.4 ± 0.30	0.50 ± 0.25
NH_3 (mg/L)	0.6 ± 0.20	0.33 ± 0.21
Zinc, Zn (µg/L)	1127 ± 4.20	1034.35 ± 367.99
Cadmium, Cd (µg/L)	4.57 ± 1.30	3.74 ± 2.53
Copper, Cu (µg/L)	1127 ± 2.30	232.28 ± 145.02
Lead, Pb (µg/L)	90.25 ± 3.30	100.26 ± 65.05
Flow rate (L/hr)	77 (Dry), 95 (Wet)	
Volume of synthetic runoff dosed (L)	95 (Dry), 131 (Wet)	
Frequency of dosing	Twice weekly (Dry), Thrice weekly (Wet)	

The chemicals dosed in Phase 1 and the detailed pollutant concentration used in Phase 2 are summarised in Appendix B. The peak flow rate was computed based on the formula shown in Appendix C, which summarizes the frequency of dosing, flow rate, and volume of synthetic runoff, which were dosed to reflect the dry and wet periods of Singapore.

2.4. Analytical Procedures

For each dosing event in Phase 1 and Phase 2, 20 L inflow samples were taken to form a composite sample. The water quality analytical tests were done following the *Standard Methods for the Examination of Water and Wastewater* [47]. Both influent and effluent water samples were tested for key water quality parameters such as total phosphorus (TP) (DR 6000, Hach, CO, USA), total nitrogen (TN) (TOC-L, Shimadzu, Kyoto, Japan), and total suspended solids (TSS). For heavy metal analysis, samples were filtered through 0.45 um Millipore PTFE filter paper, and 2% ultrapure nitric acid was added to the samples. Inductively coupled plasma mass spectrometry (ICP-MS) (7700, Agilent Technologies, CA, USA) was used to analyse the concentration of copper (Cu), lead (Pb), cadmium (Cd), and zinc (Zn).

After a storm event during Phase 3, water samples were collected and transported to the laboratory inside a cooler box with ice packs. Only TSS, TP, and TN were tested for the water samples collected from the field study. Heavy metals in the field study were found to be in negligible concentrations (although not reported in this study), as the study area did not have a source of heavy metal pollutants. Heavy metals pollutants are generally sourced from industrial emissions or activities and are commonly found in road runoff, which are mainly contributed to by vehicle emissions or the wear and tear of tyres or brake linings [48]. In terms of plant growth characteristics, Fv:Fm ratio, chlorophyll meter, the height of tree growth, and leaf growth were used as indicators of the health of the tree in the MBT. The Fv:Fm ratio was measured using a chlorophyll fluorometer (PAM-210, Walz, Effeltrich, Germany) to estimate the photosynthetic performance of the leaves [9,49,50]. A

Fv:Fm ratio of above 0.75 indicates healthy leaves. The chlorophyll leaf colour changes were measured using the SPAD-502 Plus chlorophyll meter (Konica-Minolta, Tokyo, Japan), which can show the relationship between leaf chlorophyll-a and nitrogen levels [51,52]. SPAD provides a rapid method to determine the chlorophyll content of the plants while the Fv:Fm ratio is able to indicate plant health via chlorophyll fluorescence [53]. The hydraulic conductivity of the soil (K) was measured based on the method adapted from the Facility for Advancing Water Biofiltration (FAWB) [54]. Event mean concentration (EMC) analysis was performed and used to benchmark performance in this study.

3. Results

3.1. Pilot Biofilter Study (Phase 1)

The detailed breakdown of the overall performance efficiency of the system for Phases 1 and 2 is shown in Appendix D. Both systems demonstrated good removal efficiencies of TSS, TP, and TN, as shown in Figure 2. On an overall note, there was not much difference between the performance of TT1 and TT2 in terms of removing the TSS, TP, and TN from synthetic stormwater. The average difference of EMC removal for TSS, TP, and TN (including both the wet and dry periods) between the two biofilters was 2.2%, 0.1%, and 1.8%, respectively. Detailed statistical difference analysis is shown in Appendix E.

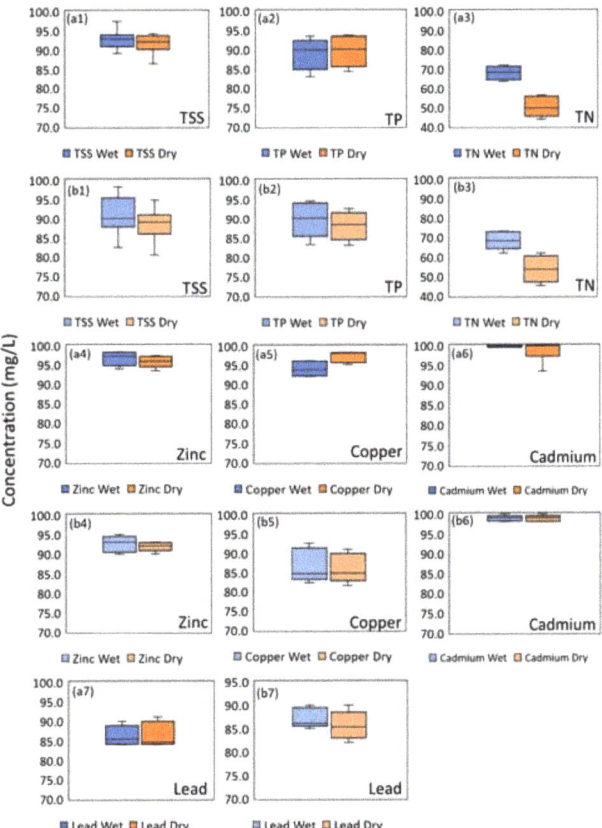

Figure 2. Phase 1 EMC removal % for (**a1**) TT1 TSS, (**a2**) TT1 TP, (**a3**) TT1 TN, (**a4**) for TT1 Zinc, (**a5**) TT1 Copper, (**a6**) TT1 Cadmium, (**a7**) TT1 Lead, (**b1**) EMC removal % for TT2 TSS, (**b2**) TT2 TP, (**b3**) TT2 TN, (**b4**) TT2 Zinc, (**b5**) TT2 Copper, (**b6**) TT2 Cadmium, and (**b7**) TT2 Lead.

Both biofilters were highly effective in removing TSS, and their performance was consistent over time, as reported was also found in other reported studies [55,56]. The EMC removal of all of the sampling runs was mostly greater than 90%, with a coefficient of variation (CV) below 5%. No significant difference ($p > 0.05$) was found between the effluent of both biofilters. The removal of TSS however is attributed more to simple filtration from the media itself within the BRS itself. The plant uptake by plants was reported to not be significant [55,57]. This was similarly reported by Bratieres et al. [55], who reported that the TSS removal for bioretention systems with plants was similar to the soil only control [41]. The EMC removal for TP was noted with a mean value of 89.4% and 89.3% for TT1 and TT2, respectively (CV less than 5%). The exceptional performance for TP removal was attributed to the application of WTR in the filter media. A study found that systems using Al-WTR showed a PO_4^{3-} removal of approximately 99% [58].

For TN, EMC removal was relatively worse, with a mean value of 61.0% and 59.2% for TT1 and TT2, respectively. TN constitutes highly mobile and soluble species like nitrites and nitrates, which are reported to be less efficiently removed as compared to TSS [28]. Dissolved nitrogen species are reported to have sorption to filter media [59–61]. Tirpak et al. [18] concluded that TN removal in a large mesocosm was largely attributed to soil adsorption or microbial processes (nitrification and denitrification). In addition, previous studies also noted that plant uptake is an important mechanism for TN removal [62,63]. The slight difference in the TN removal between the two pilot biofilter studies could be due to the different root structures or traits of the plants that were used [64,65]. The role of plant uptake on TN removal will be further discussed in Section 3.2.

The EMC removal percentages of TSS and TP were likely to be independent of the wet/dry periods. The difference in the mean EMC removal percentages for TSS and TP were 1.0% and 0.7% (both $p > 0.05$) for TT1 between the different periods, while the difference in the mean EMC removal percentages for TSS and TP were 2.5% and 2.0% (both $p > 0.05$) for TT2 between the different periods. This suggested that the removal mechanisms for TP and TSS were not affected by the frequency of rainfall events or the volume and intensity of the rainfall. The adsorption of phosphorous was also likely to be unaffected by the weather conditions. The removals of TN have the most distinctive difference in terms of removal efficiency between the wet and dry periods. A significant difference (14.9% for TT1, 17.7% for TT2, both $p < 0.05$) was found between the EMC removal percentage of TN in the wet and dry periods. When wet period testing was conducted, the larger volume of influent resulted in a higher volume of retained water in the system. The retention of water in the system created pockets of anoxic submerged zones, which promoted denitrification to occur [66]. NO_3^- undergoes denitrification and is hence removed from the system through its conversion to nitrogen gas.

In terms of heavy metal removal, the findings are agreeable with various laboratory studies in the literature [25,67]. The heavy metals were removed effectively by both pilot biofilters. In this case, the average EMC removal of heavy metals for both biofilters was more than 85%. A relatively high variable of cadmium (Cd) removals was obtained (65.5–100.0% for TT1 and 67.2–100.0% for TT2) and may be attributed to the low effluent concentration, which was always below the detection limit (3 and 5 events are detectable, respectively for TT1 and TT2). Overall, there were no obvious differences between TT1 and TT2 heavy metals removal. Heavy metals in particulate form are mostly intercepted by the filter media surface layer, while the dissolved form of heavy metals is mostly removed via the sorption process by the filter media layer [68]. Heavy metals uptake by plants (mainly via the roots) is relatively lower [58,69], which may explain the similarities in heavy metals removal between the two pilot biofilters.

For this study, there was no significant difference ($p > 0.05$) in EMC removal efficiency of heavy metals between the wet and dry periods. A statistical test for Cd was not conducted due to limited datasets (most dosing events have effluent concentrations below the detection limit). Hatt et al. [70] also found no difference in heavy metals during the wetting and drying regime for their column tests. However, this phenomenon was different from the intermittent wet and dry column study by Blecken et al. [25], which reported that heavy metal removal efficiency was significantly lower (relative to wet period performance) after a prolonged drying period. Possible reasons for this include the mobilisation of fine sediments, preferential flow paths, and the reduced metal uptake by plants during the dry period. The difference between this study and the study by Blecken et al. [25] could be due to the span of the dry period for their experiments, which ranged from 1 to 7 dry weeks.

3.2. Pilot Biofilter Study (Phase 2)

Similar to the trend found in the previous section, the MBT system had a comparable EMC removal of pollutants for real stormwater runoff. Figure 3 shows the EMC removals for both TT1 and TT2 during Phase 2. Consistent and excellent removal of TSS and TP (mean removal > 90%) were noted for both pilot biofilters. The difference in the mean EMC removal percentages for TSS and TP were within 0.3% and 1.2% (both $p > 0.05$) for TT1 between the different periods, whereas the difference in the mean EMC removal percentages for TSS and TP were 0.8% and 0.9% (both $p > 0.05$) for TT2 between the different periods. Similar to TN, EMC removal showed the most distinct difference, with a mean EMC difference of 22.4% for TT1 and 10.2% for TT2 (both $p < 0.05$). The overall EMC removal performance of TT1 and TT2 for the two conditions was reflected in Appendix E. Similar observations were found whereby TN removal was significantly higher in the wet period compared to the dry period when actual stormwater runoff was used for the MBT systems.

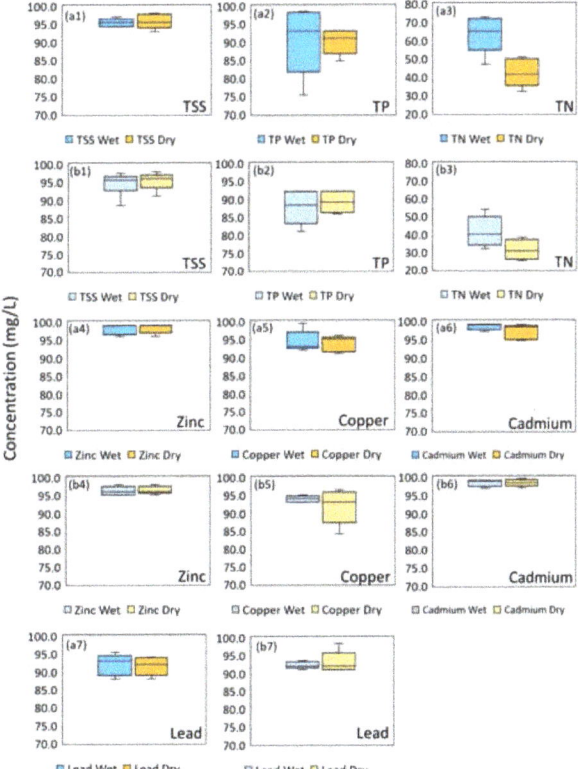

Figure 3. Phase 2 of EMC removal % for (**a1**) TT1 TSS, (**a2**) TT1 TP, (**a3**) TT1 TN, (**a4**) for TT1 Zinc, (**a5**) TT1 Copper, (**a6**) TT1 Cadmium, (**a7**) TT1 Lead, (**b1**) EMC removal % for TT2 TSS, (**b2**) TT2 TP, (**b3**) TT2 TN, (**b4**) TT2 Zinc, (**b5**) TT2 Copper, (**b6**) TT2 Cadmium, and (**b7**) TT2 Lead.

Heavy metal removal in Phase 2 for both biofilters is excellent, with an average EMC removal of more than 90%. Similar to Phase 1, the EMC removal of heavy metals is shown to be independent of the different wet/dry periods whereby there is no significant difference ($p > 0.05$) between the wet and dry period, except cadmium for TT1. Despite this, the effluent concentration of TT1 during wet period dosing is similar to that of dry period dosing (average of 0.05 µg/L and 0.09 µg/L, respectively).

On an overall note, investigating the performance of BRS systems in general, on a laboratory or pilot scale, has the edge of flexible configurations and modes of operations. However, the results may not always correspond to the findings of field application. In a controlled environment and setting (e.g., consistent temperature and routine dosing of water), both Phase 1 and Phase 2 had a similar range of TSS and TP pollutant removal efficiency. The difference in TN removal efficiency could be due to the variability of nitrogen concentration and the dissolved composition of the influent water that was used for the pilot biofilter studies [10,61]. While Phase 1 of the study indicated the most ideal and controlled environment, synthetic stormwater may not be perfect for the actual representation of stormwater runoff pollutants. For instance, TN performance in TT2 using real stormwater runoff is lower compared to synthetic stormwater runoff. This could be due to the lower NO_3^- concentration in the influent of Phase 2. In addition, competition from other pollutants in stormwater runoff and the root growth pattern could limit the process of denitrification in TT2. Other studies also reported differences in BRS performance when synthetic stormwater and actual stormwater runoff were used as influents, with most studies performing laboratory experiments using simulated stormwater followed by a field study [59,67]. Hence, this showed that the usage of simulated stormwater as a screening tool might not necessarily be sufficient.

Figures 4 and 5 show the EMC removal efficiency of the biofilters for Phase 1 and Phase 2, respectively. Comparing the different tree species used in this study, TT1 and TT2 had similar removal efficiencies of TSS (89.8–93.6%), TP (85.7–89.3%), Zn (92.3–98.2%), Cu (84.2–93.8%), Cd (95.0–98.3%), and Pb (84.7–92.2%), whereas TT1 showed a significantly higher removal of TN. This is likely due to the removal mechanism of the MBT. TSS and TP removals were consistent over time for Phase 1 and Phase 2. It was reported that TSS and TP removals were independent of the tree species used and were more influenced by the media used [36,71]. Likewise, heavy metal removals were more dependent on the media used and were less contributed to by different plant species. The bulk of phosphorus pollutants was likely removed via adsorption within the filter media, which in this study, contained WTR as amendments, whereas the differences in the removal of TN could likely be accounted for by the uptake of nitrogenous species by the plants [37] as well as how the tree grows within the soil media. The removal of TN in both phases were variable over time. Landsman and Davis [72] also noted the varied removal efficiency of TN. This could be attributed to diversity of N forms and the multiple internal treatment mechanisms within MBT. It was noted that TN removal for TT1 was better than that of TT2. Deeper plant roots were reported to increase nitrogen retention in the bioretention system [12,73] as well as strengthen microbial activity in the soil [74]. The smaller TN removal variation observed in Figures 4 and 5 compared to previous studies [72,75] could be attributed to the different settings. Phase 1 and Phase 2 of this study were conducted in a controlled environment. Environmental variables and different N species in the field studies could confound the results and could cause larger variation in TN removal.

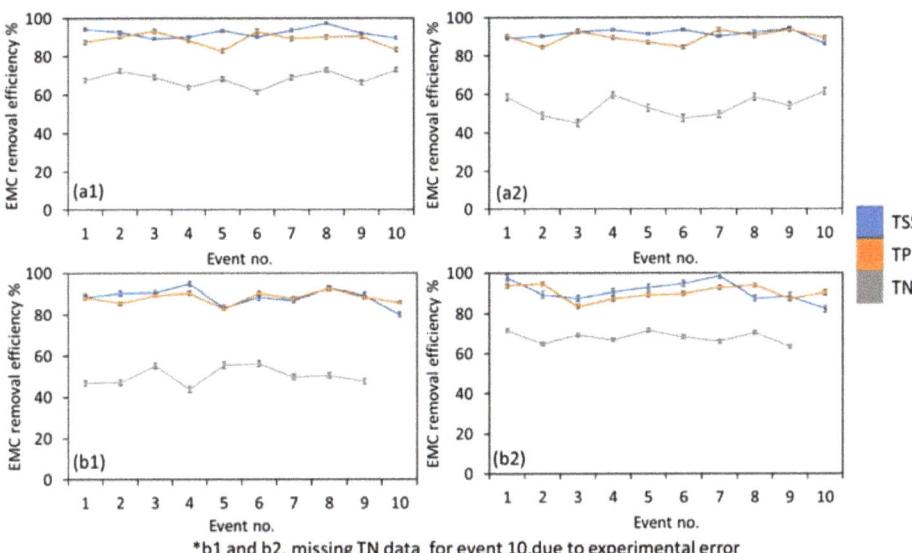

Figure 4. Phase 1 EMC removal efficiency of TSS, TP and TN for (**a1**) TT1 in dry condition, (**a2**) TT2 in dry condition, (**b1**) TT1 in wet condition, (**b2**) TT2 in wet condition.

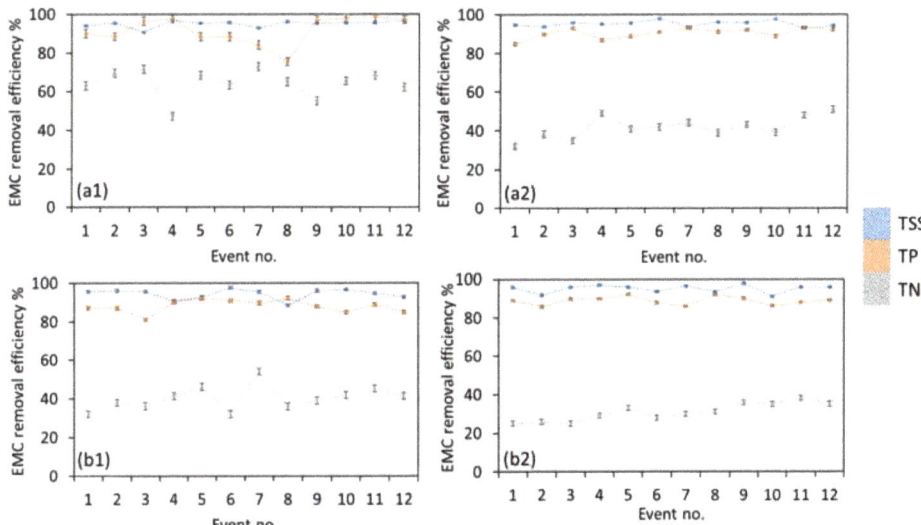

Figure 5. Phase 2 EMC removal efficiency of TSS, TP and TN for (**a1**) TT1 in dry condition, (**a2**) TT2 in dry condition, (**b1**) TT1 in wet condition, (**b2**) TT2 in wet condition.

New root growth in Phase 2 was recorded and summarised in Figure 6. The new root growth for the 7th and 10th month was not recorded due to an obstruction blocking the access to the system caused by other ongoing experiments. This root study provides insights into tree selection based on root growth for BRS applications. The root growth rate of the tree was observed to stabilise from the 3rd month onwards, with no new root growth observed for the first 2 months in TT2. Root growth for TT2 was significantly slower, with the vast majority of root growth occurring in the shallower regions of the system. Throughout the study period, >50% of new root growth was found to be below the 600 mm depth for TT1, with a high percentage (>30%) of the new root growth being observed in the deeper regions (800–1000 mm). The total number of new root tips for TT2 was also lesser than that of TT1, with TT1 having about twice that of TT2. This correlated well with the higher degree of removal of TN for TT1 as compared to TT2. Root depth was found to correlated well to the removal efficiency of nitrogen species, with deeper roots favoring higher removal efficiency [39,40]. Kristian-Thorup-Kristensen [40] reported that subsoil nitrate was well correlated to root intensity and rooting depth. Additionally, McMurtrie et al. [39] reported that root depth and overall root mass had a high correlation to the removal efficiency of nitrogen species. Due to the variation in the environment that the tree species proliferate in, the root structure of the two species shows a significant difference. Coastal plants such as the tree species used in TT1 have deeper roots compared to rainforest plants such as the tree species used in TT2. Hence, in the selection of tree species for a BRS, there needs to be a balance between the depth of the filter media to support plant growth as well as to improve pollutant removal. It may be advantageous to plant a forest species in a BRS with shallower filter depths. On the other hand, when the infiltration rate of the filter media is of concern, coastal tree species may be more valuable, as the continuous high root growth could provide many passageways and pockets for runoff to flow through. In this study, TT1 has better pollutant removal performance with a deeper filter media design, hence the plant species *Talipariti tiliaceum* would be used in the subsequent field study (Phase 3).

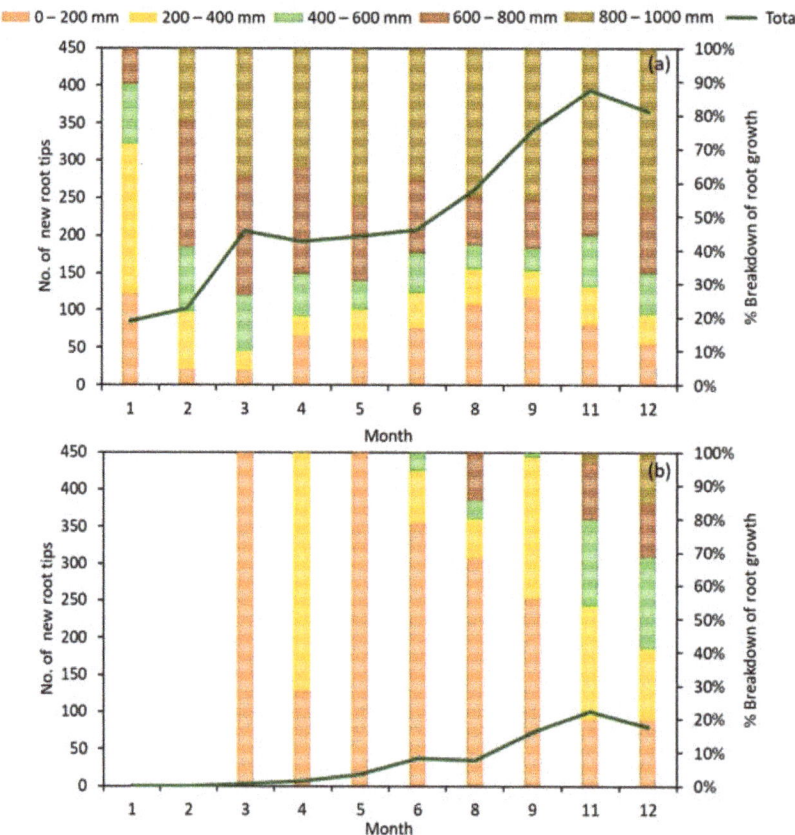

Figure 6. (a) The number of new root tips observed at different depths of the biofilters planted with *Talipalitri tiliaceum*, TT1 and (b) *Sterculia macrophylla*, TT2 in the bioretention system for the pilot study over a period of 12 months.

3.3. Field Study (Phase 3)

The performance of the field-scale MBT and the factors impacting its performance will be discussed in this section.

3.3.1. Tree-Soil Relationship

The *T. tiliaceum* tree in the MBT was monitored for the first 12 months, and the indices for plant growth are reflected in Figure 7a. The height of tree, Fv:Fm ratio, and soil plant analysis development (SPAD) factor remained fairly constant throughout the study. The average height of the tree was noted to be 3.84 m while the average Fv:Fm ratio was 0.823 over 12 months. Tree height remained consistent, with a minute difference in growth height throughout the period of study. The tree in this study depicted a Fv:Fm ratio range from 0.73 to 0.88. This result is similar to other tree phytoremediation studies in the tropics [37,76], which shows that the plants are healthy. Furthermore, SPAD sampled at the top of the tree also had a fairly constant value, with an average of 39.32, indicating that the tree maintained its chlorophyll levels. A demarcated branch near the top of the tree was selected to monitor new leaf growth. Results showed a steady increase in the number of new leaves throughout the study. Based on the definition by Chen et al. [37], the *T. tiliaceum* used in this study was found to have moderate growth with a controlled increase in the number of new leaves. Overall, this suggests that the *T. tiliaceum* tree is suitable for the MBT, due to its controlled growth and good indication of health.

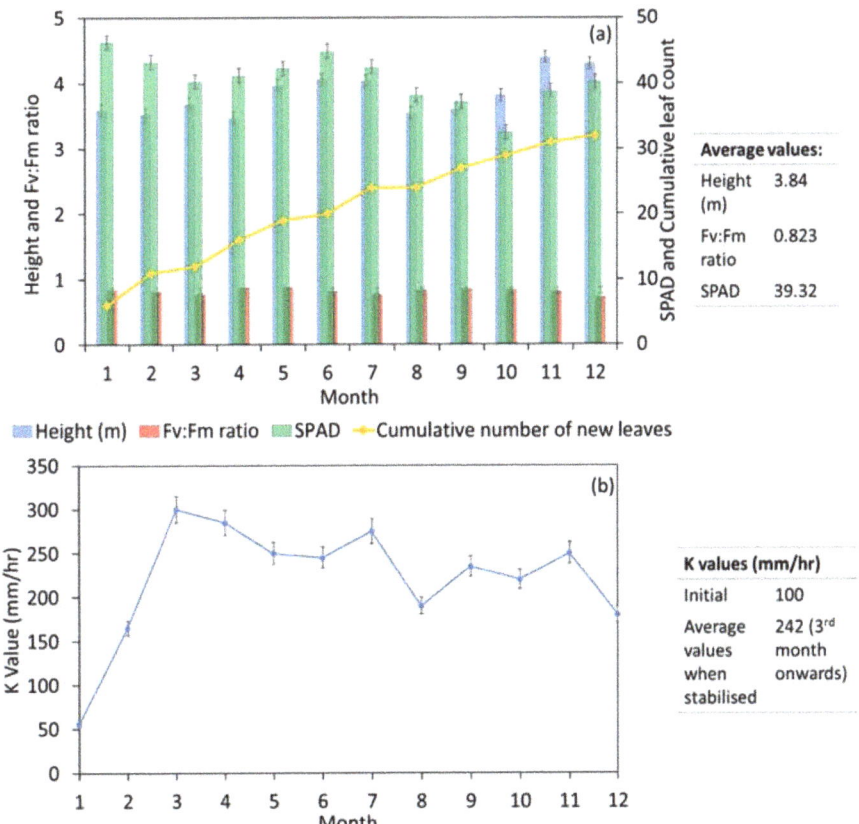

Figure 7. (a) Plant growth indices in the MBT system and (b) Soil conductivity indices.

The hydraulic conductivity of soil is affected by other factors such as the antecedent dry period [77] and plant characteristics, especially root properties [78,79]. Archer et al. [78] found that plants with thick roots can form macropores and can facilitate water to percolate deeper, which leads to higher hydraulic conductivity. The hydraulic conductivity of the soil in the field study was monitored for 12 months to observe key changes to the infiltration rate of the soil due to the growth of the tree in the system (Figure 7b). The first measurement (0 months) of 100 mm/h was taken before the planting of the tree. It can be later seen that the K values at the initial stage decreased slightly and subsequently increased to a stabilised range. The initial decrease in saturated hydraulic conductivity is a typical trait of BRS and was similar to other field studies [46,70], which could be caused by the natural compaction of the filter media layer due to hydraulic loading from stormwater runoff. Furthermore, washed-off sediments that accumulated and were deposited due to the installation of the MBT system could also decrease the K value, which is the primary reason for the surface clogging of such soil-based systems [80,81]. When the hydraulic conductivity is too low, the low infiltration rate might promote high levels of ponding and even overflow, causing runoff to bypass the system and go untreated. Thus, the surface soil layer was removed was conducted after 1 month of operation, resulting in an improvement in the subsequent measurement of hydraulic conductivity. As the tree continued to grow, root growth penetrated the filter media, creating pores and reducing the clogging potential of the BRS [82,83]. The K values stabilised between the range of 180 and 296 mm/h from the 3rd month onwards, which corresponds well with the recommended range (100 mm/h to 300 mm/h) of the technical guidelines in Singapore [84].

3.3.2. Pollutants Removal Performance

A detailed breakdown of the field performance efficiency of the system is shown in Appendix F. The field system generally performed well in TSS removal, with the effluent TSS EMCs (ranged from 0.3 mg/L to 28.4 mg/L) having an average removal efficiency of 43.6%, as shown in Figure 8. This efficiency result was lower compared to other conducted field studies conducted [70,85,86]. Poor removal efficiency was obtained due to low influent EMCs (further discussed in Section 4.2.1). For several events during the monitoring period, the TSS effluent concentration was much higher than the influent TSS concentration. This phenomenon was also observed in another field study [87]. During a prolonged dry period, soil aggregates can break down into fine particles that can migrate to deeper levels or can be washed out in the next storm event. Besides that, antecedent dry weather can also lead to low soil moisture, forming cracks within the filter media due to the shrinkage of soil aggregates. This modifies the preferential flow path of the stormwater runoff, reducing the maximum treatment area of the bioretention system [46]. Effluent TP concentration from the bioretention tree system varied from 0.01 mg/L to 0.13 mg/L. Overall, satisfactory removal of TP was observed with an average of 32.1%. Occasional negative removal efficiency was observed during the monitoring of the bioretention tree system, suggesting that phosphorus in the filter media might be washed out with the effluent. However, the low effluent TP EMC might be due to low phosphorus loading into the bioretention tree system. Roof runoff generally had a lower TP concentration than other catchment types such as lawns and roads. Similar to what was being observed for TSS pollutants, low inflow TP EMC in this study led to a negative removal efficiency for some of the rainfall events [88]. Furthermore, the WTR within the soil mix also aided in the effective removal of TP, as seen by other field studies [89,90]. For nitrogen, effluent TN ranged from 0.08 mg/L to 3.22 mg/L with an average TN removal efficiency of 45.6%. The ability of the bioretention tree system to act as an effective sink for nitrogen could be due to the choice of plant species used. The tree planted in the bioretention tree system is a coastal plant species that required a large amount of nitrogen for growth. As such, a large amount of nitrogen in the stormwater runoff was used by the tree. Additionally, tree species typically have deeper roots and a vast root network that can aid in the uptake of dissolved nitrogen in water [17].

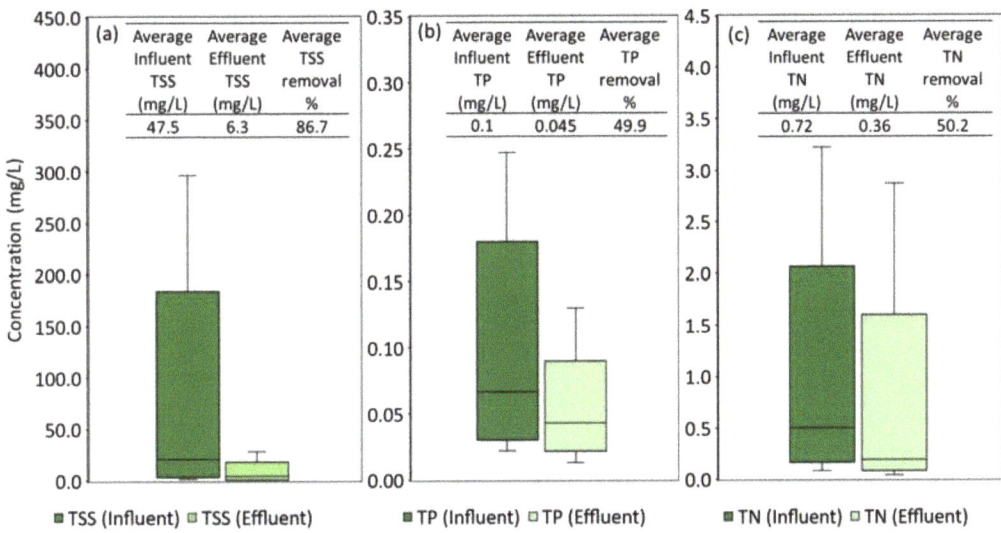

Figure 8. (a) TSS removal performance of field MBT, (b) TP removal performance of field MBT, and (c) TN removal performance of field MBT.

4. Discussion

4.1. Comparison to Other Bioretention Systems in the Tropics

Generally, the effluent of the field MBT system was very clean with average TSS, TP, and TN effluent EMC of 4.8 mg/L, 0.04 mg/L and 0.27 mg/L, respectively. Compared to other conventional BRS systems in the tropics, this MBT showed comparable removal performance, as summarised in Table 1.

The studies by Ong et al. [91] and Wang et al. [92] were conducted in a region with similar rainfall and tropical conditions. It is important to note that both studies monitored a BRS with a submerged zone, which was reported to typically have a much more significant removal of TN [90,92]. This highlights that despite not having a submerged zone, the pollutant removal performance of our MBT was comparable to that of other reported studies, which may be due to the thicker depth of the filter media being used for the MBT. A deeper filter media increases the retention time of the system and contains more sorption sites for nitrogen removal [10]. Compared to other pilot studies of conventional BRS in the tropics, the average effluent of the MBT is also shown to be significantly cleaner. Since there were no reported studies on tree systems in a tropical climate, the results were benchmarked against bioretention tree systems in temperate conditions (Table 2). The performance of the MBT in this study was comparable to other studies.

However, in comparison to Phase 1 and 2, the pollutant removal efficiency of the field study was not remarkable, and this is likely due to various factors that will be discussed in the following sections.

Table 2. Effluent pollutants EMC of the field MBT system and comparison with literature studies.

	Climate Condition	System	Scale of Study	Average Effluent (mg/L)			Average EMC Removal Efficiency (%)		
				TSS	TP	TN	TSS	TP	TN
This Study	Tropical	MBT with engineered media (Phase 1—TT1)	Pilot study	3.8	0.11	1.93	92	89	61
		(Phase 2—TT1)	Pilot study	5.8	0.18	1.62	95	91	52
		(Phase 3—Field study)	Field study	4.8	0.04	0.27	44	32	46
[18]	Temperate	BRS without submerged zone, planted with (A)—Red Maple; (B)—Loblolly Pine; (C)—Pin Oak	Pilot study	5 ± 1 (A) 3 ± 1 (B) 2 ± 1 (C)	5 ± 1 (A) 3 ± 1 (B) 2 ± 1 (C)	0.06 ± 0 (A) 0.06 ± 0 (B) 0.06 ± 0 (C)	-	-	-
[21]	Temperate	Ecosol^{TM} Tree Pit	Field study	-	-	-	95	65 [1]	50 [1]
[22]	Temperate	Stormwater tree pit		-	-	-	85	74	68
[23]	Temperate	StormTree	Field study	-	-	-	>90	>63	>48

[1] Particulate form.

4.2. Environmental Influences on Performance of MBT

4.2.1. Initial Pollutant Concentration

Various environmental factors were explored to have a better understanding of MBT performance in field conditions, and these are summarized in Figure 9. Understanding design and environmental factors is essential to better assess the performance of MBT. This is particularly important for the prediction of the MBT performance and further improvements can be made to optimize this design for other locations with different environmental conditions. Influent pollutant loading is an important factor that affects the removal performance of the BRS [93]. As seen in Figure 9a, as influent pollutant concentration increased, higher removal efficiency was achieved. This is likely due to the strong influence of 'background' pollutants. McNett et al. [94] found that effluent nutrient pollutant concentration was strongly influenced by the 'background' pollutant concentration and less affected by the influent pollutant concentration. The effluent concentration is controlled by the characteristics of the media filter and not by pollutant loading, though this phenomenon is more prominent for phosphorus than nitrogen.

The influent runoff pollutant concentration in this study is significantly lower compared to other studies reported for urban areas [95,96] (Appendix G). Roof runoff generally has a smooth surface, resulting in a cleaner runoff compared to other runoff sources such as pedestrian footpath and road [97]. In the context of Singapore, the high degree of variability in the runoff quality may also be due to the frequency and seasonal severity of storm events [44]. Furthermore, the contributing catchment area in this study is quite small (90 m^2) compared to other field studies e.g., 2200 m^2 [29]; 2400 m^2 [87], leading to a lower pollutant accumulation. This indicates that the field MBT system could potentially manage and treat runoff with higher pollutant concentration, such as industrial and road runoffs. Earlier pilot biofilter studies also noted a much higher pollutant removal efficiency compared to the field study, which is likely due to the higher influent pollutant concentration. Further study to evaluate the performance of MBT systems in a larger catchment area would be recommended. Simulated storm events with higher pollutant inflow levels can be conducted to further validate the performance of the MBT.

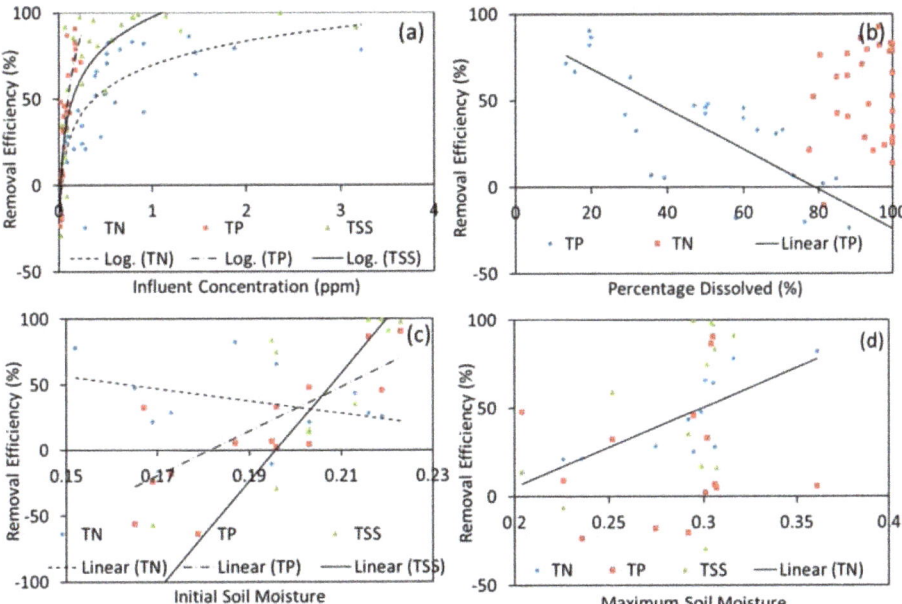

Figure 9. Environmental factors that affect the performance of MBT: (**a**) influent of pollutant, (**b**) % of dissolved pollutant, (**c**) initial soil moisture, and (**d**) maximum soil moisture.

4.2.2. Dissolved Percentage of Influent Pollutants

The dissolved percentage of influent pollutants and their respective removal efficiency are visualized in Figure 7b with more detailed results shown in Appendix H. The removal efficiency of TP was observed to be affected by the dissolved fraction in the influent sample. In this study, the influent dissolved percentage of phosphorus (DP/TP) ranged between 14% to 85%. A higher dissolved fraction of influent phosphorus resulted in the lower phosphorus removal performance of the MBT. As adsorption is the major removal mechanism for phosphorus in BRS [10,36,98], the phosphorus adsorption capacity of the filter media is an important parameter in managing and removing dissolved phosphorus.

On the other hand, little relationship was observed between TN removal efficiency and the dissolved fraction of the pollutants in this study. The difference in observation could be due to the narrow range of the observed results. Most of the observed results had a dissolved nitrogen content ranging from 80% to 100%, with little data below 80%. This led to a skewed result due to the poor spread of the collected data. Similar to most reported studies, the particulate fraction of nitrogen accounted for the lowest fraction of TN in urban stormwater [99,100]. Nitrogen loading is mainly from atmospheric loading, giving rise to a predominantly dissolved percentage of nitrogen in stormwater runoff [60,101]. While dissolved nitrogen forms a large fraction of TN in this study, the removal of TN is shown to have a great degree of variability, due to the relatively complicated TN removal mechanism. The removal of dissolved nitrogen species such as nitrate and nitrite often posed great difficulty, as these nitrogen species are highly soluble and do not readily adsorb into the BRS soil media readily [102].

It is also noted that environmental influences such as weather conditions and the number of antecedent dry days can affect the health of both plants and soil microbial communities, while soil moisture can affect the mobility of the dissolved pollutants. Tropical countries like Singapore have high rainfall, but the time interval between consecutive rainfall events is short [103]. Thus, with varying environmental conditions, the removal of dissolved pollutants may be highly inconsistent and difficult.

4.2.3. Soil Moisture

In this study, soil moisture content is used as an indicator to represent the wet and dry conditions that are experienced by the MBT. The removal efficiency of TP and TSS displayed a positive relationship with the initial volumetric water content in the soil before the rainfall event, while that of TN was more influenced by the maximum soil moisture content during the storm event. This difference in influencing parameters could be due to the different removal mechanisms involved.

TSS and the particulate fraction of phosphorus were removed via a filtration process through the soil column. Higher initial soil moisture prevented the shrinking and cracking of the soil particles to ensure an effective and efficient removal process. TSS and TP have seemingly no relationship with the maximum soil moisture experienced by the MBT during a storm event. Higher maximum soil moisture suggests that the MBT receives a larger volume of stormwater runoff. The field study results coincide with the results of the pilot biofilter studies in which wet and dry conditions did not necessarily affect TSS and TP performance.

On the other hand, nitrogen fate and transport are highly regulated by the moisture content [61,104]. Higher soil moisture would increase the mobility and bioavailability of dissolved nitrogen species. Nitrogen (nitrate species) can be removed through the denitrification process, which is encouraged by an anoxic environment. With higher maximum moisture content in the soil, there is less air in the spaces between soil particles, promoting a temporary anoxic environment for denitrification to occur. Higher initial soil moisture also increases the bioavailability and mobility of dissolved nutrients as well as maintains a healthy microbial consortium to promote nutrient uptake and improves nutrient removal [105].

5. Conclusions and Impact of Studies

Overall, this study has provided insights into the performance of a MBT system in terms of stormwater runoff quality improvements at both the pilot and field scale in the tropical region. MBT was shown to be successful in highly urbanized areas and was able to do so with low construction time while also maintaining high removal performance without the addition of submerged zone layer. Wet or dry dosing conditions had a significant influence on the TN removal performance of both TT1 and TT2 in both phases. While the effective removal of TSS, TP, and heavy metals was found to be independent of wet and dry dosing conditions, variations in TN removal between Phase 1 and 2 was found to be due to the differences in organic speciation found in the stormwater runoff compared to synthetic water. Hence, this study showed that the usage of simulated stormwater as a screening tool might not necessarily be sufficiently representative of field conditions compared to real stormwater runoff. Variation in TN removal was also attributed to the plant species used. Coastal plants such as *Talipalitri tiliaceum* are more suitable for MBT in the tropics due to their moderate growth rate, high root infiltration within the soil media, and high uptake of TN.

The field study showed that the tree retained a healthy growth rate while maintaining the optimal infiltration rate of the system to treat incoming stormwater runoff effectively. Pollutant removal performance of MBT during the field monitoring of 15 months was comparable with other reported studies in the tropics, with the average EMC removal of TSS, TP and TN being 44%, 32% and 46%, respectively. Environmental factors such as low influent pollutant concentrations, soil moisture, and dissolved species were found to affect the performance of MBT in the field. While the present study only explores the simulation of a single rewetting–drying alternative condition based on real rainfall, it might not be fully representative of the removal efficiency for the natural cycle of drying and rewetting caused by climate changes. Future challenge tests with controlled dosing conditions can be further explored in field MBT to monitor the performance of multiple periods of wetting and drying.

Author Contributions: Conceptualization, F.Y.L., H.G., and S.Z.G.; methodology development, F.Y.L. and S.Z.G.; investigation, F.Y.L., S.Z.G.; writing—data analysis and draft preparation, T.H.N. and B.C.Y.L.; supervision and project administration, J.H. and S.L.O., G.S.O., and C.X.L. All authors have read and agreed to the published version of the manuscript.

Funding: This research was funded by PUB, Singapore's National Water Agency under the 'Novel Bioretention Systems Development for Sustainable Stormwater Management' research project, grant number WBS: R-706-000-020-490.

Institutional Review Board Statement: Not applicable.

Informed Consent Statement: Not applicable.

Data Availability Statement: The data presented in this study is available on request from the corresponding author.

Conflicts of Interest: The authors declare no conflict of interest.

Appendix A

Tree Species Screened

Table A1. Summary of tree species screened for suitability for bioretention tree system.

S/N	Trees	Growth Rate	Flow Rate	Nitrate Removal	Phosphate Removal
1	Ardisia elliptica	Slow	No change	Effective	Ineffective
2	Baccaurea minor	Slow	Improved	Effective	Ineffective
3	Barringtonia asiaticum	Moderate	Improved	Ineffective	Ineffective
4	Bhesa paniculata	Slow	No change	Effective	Ineffective
5	Bhesa robusta	Fast	No change	Ineffective	Ineffective
6	Diospyros discolor	Slow	Improved	Ineffective	Ineffective
7	Dipterocarpus kerrii	Slow	No change	Ineffective	Ineffective
8	Elateriospermum tapos	Moderate	Improved	Effective	Ineffective
9	Garcinia cowa	Slow	No change	Ineffective	Ineffective
10	Garcinia subelliptica	Slow	Worsened	Ineffective	Ineffective
11	Gardenia tubifera	Slow	No change	Effective	Ineffective
12	Hopea ferrea	Moderate	Worsened	Effective	Ineffective
13	Kopsia arborea	Slow	Improved	Ineffective	Ineffective
14	Lithocarpus sundaicus	Fast	Improved	Effective	Ineffective
15	Magnolia coco	Slow	Improved	Ineffective	Ineffective
16	Memecylon edule	Slow	No change	Ineffective	Ineffective
17	Magnolia coco	Slow	Improved	Ineffective	Ineffective
18	Memecylon edule	Slow	No change	Ineffective	Ineffective
19	Sterculia macrophylla	Moderate	No change	Ineffective	Ineffective
20	Suregada multiflora	Moderate	Improved	Ineffective	Ineffective
21	Syzygium acuminatissimum	Moderate	Improved	Effective	Ineffective
22	Syzygium gratum	Fast	Improved	Ineffective	Ineffective

Table A1. Cont.

S/N	Trees	Growth Rate	Flow Rate	Nitrate Removal	Phosphate Removal
23	Syzygium leucoxylon	Fast	No change	Effective	Ineffective
24	Syzygium myrtifolium	Moderate	Worsened	Effective	Ineffective
25	Talipalitri tiliaceum (red-leaf variety)	Moderate	Improved	Effective	Ineffective
26	Tristaniopsis whiteana	Moderate	No change	Ineffective	Ineffective

The growth rate and flow rate of the plant systems were determined using various non-destructive techniques such as SPAD-502 readings, new leaf growth and leaf length as well as destructive techniques such as root and shoot analysis. Destructive analysis in this context refers to irreversible damage to the plant after the observation. SPAD-502 chlorophyll meter (Konica Minolta, Japan) can be used to determine leaf colour changes, which have been reported to be related to leaf chlorophyll-a and nitrogen levels.

Appendix B

Synthetic Stormwater Composition and Actual Stormwater Runoff Concentration

Tap water was left overnight in order to dechlorinate the water samples and subsequently dosed with the chemicals shown in the table below.

Table A2. Characteristics of simulated water.

Pollutant	Chemical Used	Target Concentration of the Pollutant (mg/L)
Total Suspended Soils (TSS)	Sediment and sand	100
Phosphate (PO_4^{3-})	H_2KO_4P	1.80
Nitrogen (N)	$MgN_2O_6 \cdot 6H_2O$	0.80
	$C_6H_5NO_2$	1.30
	NH_4Cl	0.40
Copper (Cu)	Cl_2Cu	0.241
Lead (Pb)	Cl_2Pb	0.09025
Zinc (Zn)	Cl_2Zn	1.127
Cadmium (Cd)	$CdN_2O_6 \cdot 4H_2O$	0.00457

The range of the pollutants for the stormwater runoff is shown below.

Table A3. Characteristics of stormwater runoff used for Phase 2.

Pollutant	Min–Max (Mean) Concentration
TSS (mg/L)	78.0–164.0 (121.8)
TP (mg/L)	1.63–3.06 (2.20)
TN (mg/L)	2.24–4.56 (3.45)
NO_{3-} (mg/L)	0.25–0.78 (0.50)
NH_3 (mg/L)	0.12–0.60 (0.33)
Zinc, Zn (µg/L)	666.36–1957.32 (1034.35)
Cadmium, Cd (µg/L)	1.21–6.67 (3.74)
Copper, Cu (µg/L)	87.26–514.98 (232.28)
Lead, Pb (µg/L)	35.21–263.98 (100.26)

Appendix C

Peak Runoff Flow Rate Computation

The peak runoff flow rate resulting from the precipitation event is calculated using the rational formula shown in equation below, where C is the runoff coefficient, I is the average rainfall intensity, and A is the catchment area served.

$$Q = \frac{CIA}{360} \tag{A1}$$

The runoff coefficient is obtained by assuming the catchment area to be comprised of 50% urban areas both fully and closely built-up and 50% residential or industrial areas densely built up. The coefficient was then obtained as 0.85 from the weighted average of the runoff coefficient of these two areas [94].

Appendix D

EMC Removal Efficiency for Pilot Biofilter Study—TT1 and TT2 (Phase 1 and 2)

Table A4. EMC removal efficiency of pilot bioretention tree biofilter study (TT1: *Talipariti tiliaceum*; TT2: *Sterculia macrophylla*) for Phase 1.

Phase 1						
TT1						
	EMC Removal Efficiency (%)					
Pollutant	Mean		Minimum		Maximum	
	Wet	Dry	Wet	Dry	Wet	Dry
TSS	92.5	91.5	89.1	86.4	97.4	94.2
TP	89.0	89.7	83.0	84.3	93.4	93.7
TN	68.4	53.6	61.8	45.1	73.0	61.7
Zn	97.2	96.3	93.8	93.4	98.4	97.4
Cu	89.4	88.6	76.1	76.6	92.1	91.9
Cd [a]	95.3	94.6	65.5	67.2	99.6	99.6
Pb	85.1	84.3	72.1	71.6	90.8	91.0
TT2						
	EMC Removal Efficiency (%)					
Pollutant	Mean		Minimum		Maximum	
	Wet	Dry	Wet	Dry	Wet	Dry
TSS	91.0	88.6	82.4	80.4	98.2	94.8
TP	90.3	88.2	83.3	83.0	94.5	92.6
TN	68.0	50.3	63.5	44.0	71.8	56.2
Zn	92.5	92.1	83.2	85.5	98.8	96.9
Cu	83.9	84.5	74.3	75.3	92.5	91.0
Cd [b]	94.3	96.1	84.1	84.0	99.5	99.6
Pb	87.0	85.9	80.3	78.4	98.5	97.2

[a] only 3 events are detectable, others are below detection limit; [b] only 5 events are detectable, others are below detection limit.

Table A5. EMC removal efficiency of pilot bioretention tree biofilter study (TT1: *Talipariti tiliaceum*; TT2: *Sterculia macrophylla*) for Phase 2.

Phase 2						
TT1						
Pollutant	EMC Removal Efficiency (%)					
	Mean		Minimum		Maximum	
	Wet	Dry	Wet	Dry	Wet	Dry
TSS	95.0	95.3	90.6	92.9	96.9	97.9
TP	91.5	90.3	75.3	84.8	98.4	93.1
TN	94.1	41.7	47.0	32.1	72.8	50.9
Zn	98.2	98.2	96.0	97.0	99.0	99.0
Cu	93.8	93.7	92.0	91.0	99.5	96.0
Cd	98.6	96.9	98.1	94.7	99.1	99.0
Pb	92.0	91.8	88.0	88.0	95.4	94.0
TT2						
Pollutant	EMC Removal Efficiency (%)					
	Mean		Minimum		Maximum	
	Wet	Dry	Wet	Dry	Wet	Dry
TSS	94.3	95.1	88.5	91.2	97.5	97.8
TP	88.0	88.9	81.0	85.9	92.1	92.2
TN	40.1	30.9	31.9	24.9	54.0	38.1
Zn	96.2	96.3	95.0	95.0	98.0	98.0
Cu	93.8	92.3	93.0	84.1	95.0	96.5
Cd	98.3	98.3	96.9	97.0	99.0	99.5
Pb	91.9	92.4	91.0	91.0	93.5	98.3

Appendix E

Statistical Analysis for Phase 1 and Phase 2

Table A6. Paired sample test for Phase 1 of study.

TT1					
	Paired Samples Test				
	Mean	Std Dev	Std Error Mean	df	Sig. (2-tailed)
TT1 (TSS Wet)–TT1 (TSS Dry)	0.953	3.539	1.119	9	0.417
TT1 (TP Wet)–TT1 (TP Dry)	−0.716	4.689	1.483	9	0.641
TT1 (TN Wet)–TT1 (TN Dry)	14.850	6.178	1.954	9	0.000
TT1 (Zinc Wet)–TT1 (Zinc Dry)	−0.018	1.723	0.581	9	0.468
TT1 (Copper Wet)–TT1 (Copper Dry)	0.147	2.264	0.532	9	0.154
TT1 (Cadmium Wet)–TT1 (Cadmium Dry)	-	-	-	-	-
TT1 (Lead Wet)–TT1 (Lead Dry)	0.142	2.701	0.628	9	0.573

Table A6. *Cont.*

TT2					
	Paired Samples Test				
	Mean	Std Dev	Std Error Mean	df	Sig. (2-tailed)
TT2 (TSS Wet)–TT2 (TSS Dry)	2.450	6.077	1.922	9	0.234
TT2 (TP Wet)–TT2 (TP Dry)	2.021	4.633	1.465	9	0.201
TT2 (TN Wet)–TT2 (TN Dry)	17.706	4.044	1.348	9	0.000
TT2 (Zinc Wet)–TT2 (Zinc Dry)	−0.084	1.203	0.416	9	0.457
TT2 (Copper Wet)–TT2 (Copper Dry)	1.552	2.413	0.315	9	0.224
TT2 (Cadmium Wet)–TT2 (Cadmium Dry)	-	-	-	9	-
TT2 (Lead Wet)–TT2 (Lead Dry)	−0.544	1.892	0.621	9	0.654

Table A7. Paired sample test for Phase 2 of study.

TT1					
	Paired Samples Test				
	Mean	Std Dev	Std Error Mean	df	Sig. (2-tailed)
TT1 (TSS Wet)–TT1 (TSS Dry)	−0.333	2.245	0.648	11	0.617
TT1 (TP Wet)–TT1 (TP Dry)	1.150	7.595	2.192	11	0.610
TT1 (TN Wet)–TT1 (TN Dry)	22.408	10.749	3.103	11	0.000
TT1 (Zinc Wet)–TT1 (Zinc Dry)	−0.025	1.781	0.514	11	0.962
TT1 (Copper Wet)–TT1 (Copper Dry)	0.133	2.570	0.742	11	0.861
TT1 (Cadmium Wet)–TT1 (Cadmium Dry)	1.642	1.794	0.518	11	0.009
TT1 (Lead Wet)–TT1 (Lead Dry)	0.158	2.578	0.744	11	0.835

TT2					
	Paired Samples Test				
	Mean	Std Dev	Std Error Mean	df	Sig. (2-tailed)
TT2 (TSS Wet)–TT2 (TSS Dry)	−0.742	3.631	1.048	11	0.494
TT2 (TP Wet)–TT2 (TP Dry)	−0.809	3.351	0.967	11	0.421
TT2 (TN Wet)–TT2 (TN Dry)	9.229	5.749	1.660	11	0.000
TT2 (Zinc Wet)–TT2 (Zinc Dry)	−0.084	1.505	0.435	11	0.851
TT2 (Copper Wet)–TT2 (Copper Dry)	1.552	3.306	0.954	11	0.132
TT2 (Cadmium Wet)–TT2 (Cadmium Dry)	0.024	1.109	0.320	11	0.942
TT2 (Lead Wet)–TT2 (Lead Dry)	−0.544	2.423	0.699	11	0.453

Appendix F

Table A8. Performance of MBT for Phase 3 (Field Study).

Event	TSS			TP			TN		
	Influent (mg/L)	Effluent (mg/L)	Removal Efficiency (%)	Influent (mg/L)	Effluent (mg/L)	Removal Efficiency (%)	Influent (mg/L)	Effluent (mg/L)	Removal Efficiency (%)
1	99.8	2.6	97.4	0.19	0.03	82.4	1.47	0.34	76.9
2	99.8	2.1	97.9	0.19	0.02	90.9	1.47	0.52	64.6
3	296.3	0.4	99.9	0.07	0.04	46.3	0.08	0.06	25.0
4	106.6	0.8	99.2	0.10	0.01	87.0	0.46	0.33	28.3
5	4.8	4.1	14.6	0.03	0.02	46.9	0.10	0.08	20.0
6	6.1	18.7	−206.6	0.04	0.05	−19.0	0.10	0.07	30.0
7	1.4	2.1	−50.0	0.02	0.03	−22.7	0.29	0.23	20.7
8	4.0	10.0	−150.0	0.06	0.05	7.0	0.53	0.09	83.0
9	1.3	0.3	76.9	0.05	0.03	34.0	0.22	0.55	−150.0
10	9.0	7.5	16.7	0.05	0.08	−57.1	0.61	0.32	47.5
11	3.1	4.0	−29.0	0.04	0.04	2.6	0.41	0.14	65.9
12	42.0	6.8	83.8	0.14	0.13	7.1	0.27	0.30	−11.1
13	4.6	3.0	34.8	0.04	0.04	−19.4	0.21	0.12	42.9
14	11.3	12.0	−6.2	0.03	0.03	9.1	3.22	0.70	78.3
15	28.4	0.7	97.5	0.09	0.05	46.0	0.79	0.14	82.3
16	8.4	0.7	91.7	0.03	0.02	50.0	1.88	0.39	79.3
17	84.3	5.7	93.2	0.11	0.04	63.6	2.39	1.42	40.6
18	7.2	6.0	16.7	0.03	0.03	6.3	0.16	0.13	18.8
19	3.1	0.8	74.2	0.19	0.04	78.6	0.25	0.19	24.0
20	11.9	5.3	55.5	0.17	0.05	72.8	0.40	0.14	65.0
21	20.5	3.0	85.4	0.05	0.05	6.0	0.52	0.12	76.9
22	52.3	10.0	80.9	0.11	0.06	46.9	0.52	0.15	71.2
23	10.9	7.3	33.0	0.06	0.04	39.3	0.54	0.04	92.6
24	71.3	11.3	84.2	0.19	0.06	66.8	0.91	0.16	82.4
25	47.5	2.0	95.8	0.05	0.04	30.2	0.50	0.23	54.0
26	144.0	2.7	98.1	0.09	0.05	43.0	1.39	0.19	86.3
27	3.3	0.3	90.9	0.06	0.05	21.7	0.26	0.17	34.6

Appendix G

Stormwater Runoff Water Quality and Comparison with Literature Studies

Table A9. Runoff water quality of the study area and comparison with literature studies.

	Min-Max (Mean)			
	TSS EMC (mg/L)	TP EMC (mg/L)	TN EMC (mg/L)	References
This Study	1.3–296.3 (47.5)	0.02–0.25 (0.09)	0.08–3.22 (0.72)	-
High Urban Areas	155	0.32	2.63	[95]
Ang Mo Ki—commercial, residential, carparks, road	112.07	0.17	1.85	[96]
Pemimpin—residential	31.92	0.07	1.16	
Lower Seletar—parkland	147.34	0.31	3.02	

Appendix H

Dissolved Nutrients Composition

Table A10. Dissolved nutrient composition and respective removal efficiency.

Dissolved Phosphorus (DP)/Total Phosphorus (TP)	TP Removal Efficiency (%)	Dissolved Nitrogen (DN)/Total Nitrogen (TN)	TN Removal Efficiency (%)
13.5	71.4	77.6	21.4
15.8	66.7	78.6	52.2
19.6	90.7	80.4	76.2
19.6	82.1	80.4	76.2
20.0	86.7	81.7	−10.8
29.0	42.1	84.8	63.6
30.3	63.6	85.0	42.6
31.8	32.7	87.8	64.4
35.7	7.1	87.8	76.9
39.2	5.9	87.8	76.9
47.1	47.1	87.9	40.6
50.0	46.2	91.5	86.1
50.0	42.9	91.7	70.7
50.6	48.1	92.6	28.6
57.8	−17.9	93.1	79.3
60.0	46.0	93.1	79.3
60.0	40.2	93.6	47.9
63.7	33.3	94.9	21.3
68.8	31.3	96.4	92.5
70.4	33.3	96.6	82.0
73.3	6.7	98.0	24.3
77.8	22.2	99.3	78.3
81.2	2.5	99.5	83.0
84.8	5.0	99.9	52.7
		100.0	13.9
		100.0	25.5
		100.0	28.2
		100.0	34.8
		100.0	43.8
		100.0	66.0
		100.0	78.7
		100.0	82.5

References

1. Roy-Poirier, A.; Champagne, P.; Filion, Y. Review of Bioretention System Research and Design: Past, Present, and Future. *J. Environ. Eng.* **2010**, *136*, 878–889. [CrossRef]
2. Luell, S.K.; Hunt, W.F.; Winston, R.J. Evaluation of undersized bioretention stormwater control measures for treatment of highway bridge deck runoff. *Water Sci. Technol.* **2011**, *64*, 974–979. [CrossRef] [PubMed]

3. Miller, J.D.; Hutchins, M. The impacts of urbanisation and climate change on urban flooding and urban water quality: A review of the evidence concerning the United Kingdom. *J. Hydrol. Reg. Stud.* **2017**, *12*, 345–362. [CrossRef]
4. Goh, H.W.; Zakaria, N.A.; Lau, T.L.; Foo, K.Y.; Chang, C.K.; Leow, C.S. Mesocosm Study of Enhanced Bioreten-tion Media in Treating Nutrient Rich Stormwater for Mixed Development Area. *Urban Water J.* **2017**, *14*, 134–142. [CrossRef]
5. Chow, W.T.L.; Cheong, B.D.; Ho, B.H. A Multimethod Approach towards Assessing Urban Flood Patterns and Its Associated Vulnerabilities in Singapore. *Adv. Meteorol.* **2016**, *2016*, 1–11. [CrossRef]
6. Gerber, J.D.; Hartmann, T.; Hengstermann, A. *Instruments of Land Policy: Dealing with Scarcity of Land*; Routledge: London, UK, 2018.
7. Lopez-Ponnada, E.V.; Lynn, T.J.; Ergas, S.J.; Mihelcic, J.R. Long-term field performance of a conventional and modified bioretention system for removing dissolved nitrogen species in stormwater runoff. *Water Res.* **2020**, *170*, 115336. [CrossRef]
8. Davis, A.P.; Hunt, W.F.; Traver, R.G.; Clar, M. Bioretention Technology: Overview of Current Practice and Future Needs. *J. Environ. Eng.* **2009**, *135*, 109–117. [CrossRef]
9. Goh, H.W.; Lem, K.S.; Azizan, N.A.; Chang, C.K.; Talei, A.; Leow, C.S.; Zakaria, N.A. A review of bioretention components and nutrient removal under different climates—Future directions for tropics. *Environ. Sci. Pollut. Res.* **2019**, *26*, 14904–14919. [CrossRef]
10. LeFevre, G.; Paus, K.H.; Natarajan, P.; Gulliver, J.; Novak, P.J.; Hozalski, R. Review of Dissolved Pollutants in Urban Storm Water and Their Removal and Fate in Bioretention Cells. *J. Environ. Eng.* **2015**, *141*, 04014050. [CrossRef]
11. Lim, H.S.; Lim, W.; Hu, J.Y.; Ziegler, A.; Ong, S.L. Comparison of Filter Media Materials for Heavy Metal Remov-al from Urban Stormwater Runoff Using Biofiltration Systems. *J. Environ. Manag.* **2015**, *147*, 24–33. [CrossRef]
12. Barrett, M.E.; Limouzin, M.; Lawler, D.F. Effects of Media and Plant Selection on Biofiltration Performance. *J. Environ. Eng.* **2013**, *139*, 462–470. [CrossRef]
13. Angers, D.A.; Caron, J. Plant-induced Changes in Soil Structure: Processes and Feedbacks. *Biogeochemistry* **1998**, *42*, 55–72. [CrossRef]
14. Binkley, D.; Giardina, C. Why Do Tree Species Affect Soils? The Warp and Woof of Tree-Soil Interactions. In *Plant-Induced Soil Changes: Processes and Feedbacks*; Springer: Dordrecht, The Netherlands, 1998; Volume 42, pp. 89–106.
15. Van Breemen, N.; Finzi, A.C. Plant-Soil Interactions: Ecological Aspects and Evolutionary Implications. *Biogeochemistry* **1998**, *42*, 1–19. [CrossRef]
16. Skorobogatov, A.; He, J.; Chu, A.; Valeo, C.; van Duin, B. The impact of media, plants and their interactions on bioretention performance: A review. *Sci. Total Environ.* **2020**, *715*, 136918. [CrossRef]
17. Dagenais, D.; Brisson, J.; Fletcher, T.D. The role of plants in bioretention systems; does the science underpin current guidance? *Ecol. Eng.* **2018**, *120*, 532–545. [CrossRef]
18. Tirpak, R.A.; Hathaway, J.M.; Franklin, J.A. Investigating the Hydrologic and Water Quality Performance of Trees in Bioretention Mesocosms. *J. Hydrol.* **2019**, *576*, 65–71. [CrossRef]
19. Frosi, M.H.; Kargar, M.; Jutras, P.; Prasher, S.O.; Clark, O.G. Street tree pits as bioretention units: Effects of soil organic matter and area permeability on the volume and quality of urban runoff. *Water Air Soil Pollut.* **2019**, *230*, 152. [CrossRef]
20. Elliott, R.M.; Adkins, E.R.; Culligan, P.J.; Palmer, M.I. Stormwater infiltration capacity of street tree pits: Quantifying the influence of different design and management strategies in New York City. *Ecol. Eng.* **2017**, *111*, 157–166. [CrossRef]
21. EcosolTM Tree Pit Technical Specification. Available online: https://urbanassetsolutions.com.au/wp-content/uploads/UAS-Tree-Pit-Technical-Specifications-2018.pdf (accessed on 20 June 2021).
22. Stockholm Tree Pits. Available online: https://stockholmtreepits.co.uk/ (accessed on 20 June 2021).
23. StormTree. Available online: https://www.storm-tree.com/ (accessed on 20 June 2021).
24. Macedo, M.B.D.; Lago, C.A.F.D.; Mendiondo, E.M.; Giacomoni, M.H. Bioretention performance under different rainfall regimes in substropical conditions: A case study in Sao Carlos, Brazil. *J. Environ. Manag.* **2019**, *248*, 109266. [CrossRef]
25. Blecken, G.-T.; Zinger, Y.; Deletetić, A.; Fletcher, T.D.; Viklander, M. Influence of Intermittent Wetting and Drying Conditions on Heavy Metal Removal by Stormwater Biofilters. *Water Res.* **2009**, *43*, 4590–4598. [CrossRef]
26. Rahman, Y.A.; Nachabe, M.H.; Ergas, S.J. Biochar amendment of stormwater bioretention systems for nitrogen and *Escherichia coli* removal: Effect of hydraulic loading rates and antecedent dry periods. *Bioresour. Technol.* **2020**, *310*, 123428. [CrossRef]
27. Manka, B.; Hathaway, J.; Tirpak, R.; He, Q.; Hunt, W. Driving forces of effluent nutrient variability in field scale bioretention. *Ecol. Eng.* **2016**, *94*, 622–628. [CrossRef]
28. Hunt, W.F.; Jarrett, A.R.; Smith, J.T.; Sharkey, L.J. Evaluating Bioretention Hydrology and Nutrient Removal at Three Field Sites in North Carolina. *J. Irrig. Drain. Eng.* **2006**, *132*, 600–608. [CrossRef]
29. Brown, R.A.; Hunt, W.F. Underdrain Configuration to Enhance Bioretention Exfiltration to Reduce Pollutant Loads. *J. Environ. Eng.* **2011**, *137*, 1082–1091. [CrossRef]
30. Kazemi, F.; Golzarian, M.R.; Myers, B. Potential of combined water sensitive urban design systems for salinity treat-ment in urban environments. *J. Environ. Manag.* **2018**, *209*, 169–175. [CrossRef]
31. An Engineered Soil Composition and a Method of Preparing the Same. Singapore Patent Application No. SG2013082722A, 27 August 2013.
32. Barrett, J.E.; Burke, I.C. Potential Nitrogen Immobilization in Grassland Soils across a Soil Organic Matter Gradient. *Soil Biol. Biochem.* **2000**, *32*, 1707–1716. [CrossRef]

33. O'Neill, S.W.; Davis, A.P. Water Treatment Residual as a Bioretention Amendment for Phosphorus. I: Evaluation Studies. *J. Environ. Eng.* **2012**, *138*, 318–327. [CrossRef]
34. McGechan, M.B.; Lewis, D.R. Sorption of Phosphorus by Soil, Part 1: Principles, Equations and Models. *Biosyst. Eng.* **2002**, *82*, 1–24. [CrossRef]
35. Xu, D.; Lee, L.Y.; Lim, F.Y.; Lyu, Z.; Zhu, H.; Ong, S.L.; Hu, J. Water treatment residual: A critical review of its applications on pollutant removal from stormwater runoff and future perspectives. *J. Environ. Manag.* **2020**, *259*, 109649. [CrossRef]
36. Hermawan, A.A.; Talei, A.; Leong, J.Y.C.; Jayatharan, M.; Goh, H.W.; Alaghmand, S. Performance Assessment of a Laboratory Scale Prototype Biofiltration System in Tropical Region. *Sustainability* **2019**, *11*, 1947. [CrossRef]
37. Chen, X.C.; Huang, L.; Chang, T.H.A.; Ong, B.L.; Ong, S.L.; Hu, J. Plant Traits for Phytoremediation in the Tropics. *Engineering* **2019**, *5*, 841–848. [CrossRef]
38. Judd, L.A.; Jackson, B.E.; Fonteno, W.C. Advancements in Root Growth Measurement Technologies and Observation Capabilities for Container-Grown Plants. *Plants* **2015**, *4*, 369–392. [CrossRef] [PubMed]
39. McMurtrie, R.E.; Iversen, C.M.; Dewar, R.C.; Medlyn, B.E.; Näsholm, T.; Pepper, D.A.; Norby, R.J. Plant root distributions and nitrogen uptake predicted by a hypothesis of optimal root foraging. *Ecol. Evol.* **2012**, *2*, 1235–1250. [CrossRef]
40. Thorup-Kristensen, K. Are differences in root growth of nitrogen catch crops important for their ability to reduce soil nitrate-N content, and how can this be measured? *Plant Soil* **2001**, *230*, 185–195. [CrossRef]
41. Barwick, M.; Van der Schans, A.; Claudy, J.B. *Tropical & Subtropical Trees: A Worldwide Encyclopaedic Guide*; Thames & Hudson: London, UK, 2004.
42. Poorter, H.; Bühler, J.; Van Dusschoten, D.; Climent, J.; Postma, J.A. Pot Size Matters: A Meta-Analysis of the Effects of Rooting Volume on Plant Growth. *Funct. Plant Biol.* **2012**, *39*, 839–850. [CrossRef]
43. Qiu, Z.C.; Wang, M.; Lai, W.L.; He, F.H.; Chen, Z.H. Plant Growth and Nutrient Removal in Constructed Monoculture and Mixed Wetlands. *Hydrobiologia* **2011**, *661*, 251–260. [CrossRef]
44. Lim, H.S. Variations in the Water Quality of a Small Urban Tropical Catchment: Implications for Load Estimation and Water Quality Monitoring. In *The Interactions between Sediments and Water*; Springer: Dordrecht, The Netherlands, 2003; pp. 57–63.
45. Chui, P. Characteristics of Stormwater Quality from Two Urban Watersheds in Singapore. *Environ. Monit. Assess.* **1997**, *44*, 173–181. [CrossRef]
46. Zinger, Y.; Prodanovic, V.; Zhang, K.; Fletcher, T.D.; Deletic, A. The effect of intermittent drying and wetting stormwater cycles on the nutrient removal performances of two vegetated biofiltration designs. *Chemosphere* **2021**, *267*, 129294. [CrossRef]
47. American Public Health Association (APHA). *Standard Methods for Examination of Water and Wastewater*, 22nd ed.; APHA: Washington, DC, USA, 2012.
48. Joshi, U.M.; Balasubramanian, R. Characteristics and Environmental Mobility of Trace Elements in Urban Runoff. *Chemosphere* **2010**, *80*, 310–318. [CrossRef]
49. Gorbe, E.; Calatayud, A. Applications of chlorophyll fluorescence imaging technique in horticultural research: A review. *Sci. Hortic.* **2012**, *138*, 24–35. [CrossRef]
50. Makarova, V.; Kazimirko, Y.; Krendeleva, T.; Kukarskikh, G.; Lavrukhina, O.; Pogosyan, S.; Yakovleva, O. Fv/Fm as a Stress Indicator for Woody Plants from Urban-Ecosystem. In *Photosynthesis: Mechanisms and Effects*; Springer: Dordrecht, The Netherlands, 1998; pp. 4065–4068.
51. Netto, A.T.; Campostrini, E.; de Oliveira, J.G.; Bressan-Smith, R.E. Photosynthetic pigments, nitrogen, chlorophyll a fluorescence and SPAD-502 readings in coffee leaves. *Sci. Hortic.* **2005**, *104*, 199–209. [CrossRef]
52. Richardson, A.D.; Duigan, S.P.; Berlyn, G.P. An evaluation of noninvasive methods to estimate foliar chlorophyll content. *New Phytol.* **2002**, *153*, 185–194. [CrossRef]
53. Sepúlveda, P.; Johnstone, D.M. A Novel Way of Assessing Plant Vitality in Urban Trees. *Forests* **2018**, *10*, 2. [CrossRef]
54. FAWB. *Guidelines for Filter Media in Biofiltration Systems (Version 3.01)*; FAWB: Clayton, Australia, 2009; pp. 1–8.
55. Bratieres, K.; Fletcher, T.; Deletic, A.; Zinger, Y. Nutrient and sediment removal by stormwater biofilters: A large-scale design optimisation study. *Water Res.* **2008**, *42*, 3930–3940. [CrossRef]
56. Hatt, B.; Fletcher, T.; Deletic, A. Hydraulic and pollutant removal performance of stormwater filters under variable wetting and drying regimes. *Water Sci. Technol.* **2007**, *56*, 11–19. [CrossRef]
57. Read, J.; Wevill, P.; Fletcher, T.; Deletic, A. Variation among plant species in pollutant removal from stormwater in biofiltration systems. *Water Res.* **2008**, *42*, 893–902. [CrossRef]
58. Lucas, W.C.; Greenway, M. Phosphorus Retention by Bioretention Mesocosms Using Media Formulated for Phosphorus Sorption: Response to Accelerated Loads. *J. Irrig. Drain. Eng.* **2011**, *137*, 144–153. [CrossRef]
59. Ergas, S.J.; Sengupta, S.; Siegel, R.; Pandit, A.; Yao, Y.; Yuan, X.; Davis, A.P.; Shokouhian, M.; Sharma, H.; Minami, C. Water Quality Improvement through Bioretention Media: Nitrogen and Phosphorus Removal. *Water Environ. Res.* **2006**, *78*, 284–293.
60. Taylor, G.D.; Fletcher, T.D.; Wong, T.H.; Breen, P.F.; Duncan, H.P. Nitrogen composition in urban runoff—implications for stormwater management. *Water Res.* **2005**, *39*, 1982–1989. [CrossRef]
61. Zhang, H.; Ahmad, Z.; Shao, Y.; Yang, Z.; Jia, Y.; Zhong, H. Bioretention for removal of nitrogen: Processes, operational conditions, and strategies for improvement. *Environ. Sci. Pollut. Res.* **2021**, *28*, 10519–10535. [CrossRef]
62. Rycewicz-Borecki, M.; McLean, J.E.; Dupont, R.R. Nitrogen and phosphorus mass balance, retention and uptake in six plant species grown in stormwater bioretention microcosms. *Ecol. Eng.* **2017**, *99*, 409–416. [CrossRef]

63. Barron, N.J.; Hatt, B.; Jung, J.; Chen, Y.; Deletic, A. Seasonal operation of dual-mode biofilters: The influence of plant species on stormwater and greywater treatment. *Sci. Total Environ.* **2020**, *715*, 136680. [CrossRef]
64. Eissenstat, D. Root structure and function in an ecological context. *New Phytol.* **2000**, *148*, 353–354. [CrossRef]
65. Read, J.; Fletcher, T.D.; Wevill, P.; Deletic, A. Plant Traits that Enhance Pollutant Removal from Stormwater in Biofiltration Systems. *Int. J. Phytoremed.* **2010**, *12*, 34–53. [CrossRef]
66. Collins, K.A.; Lawrence, T.J.; Stander, E.K.; Jontos, R.J.; Kaushal, S.S.; Newcomer, T.A.; Grimm, N.B.; Ekberg, M.L.C. Opportunities and Challenges for Managing Nitrogen in Urban Stormwater: A Review and Synthesis. *Ecol. Eng.* **2010**, *36*, 1507–1519. [CrossRef]
67. Hsieh, C.-H.; Davis, A.P. Evaluation and Optimization of Bioretention Media for Treatment of Urban Storm Water Runoff. *J. Environ. Eng.* **2005**, *131*, 1521–1531. [CrossRef]
68. Chu, Y.; Yang, L.; Wang, X.; Wang, X.; Zhou, Y. Research on Distribution Characteristics, Influencing Factors, and Maintenance Effects of Heavy Metal Accumulation in Bioretention Systems: Critical Review. *J. Sustain. Water Built Environ.* **2021**, *7*, 03120001. [CrossRef]
69. Muthanna, T.M.; Viklander, M.; Blecken, G.-T.; Thorolfsson, S.T. Snowmelt pollutant removal in bioretention areas. *Water Res.* **2007**, *41*, 4061–4072. [CrossRef] [PubMed]
70. Hatt, B.E.; Fletcher, T.; Deletic, A. Hydrologic and pollutant removal performance of stormwater biofiltration systems at the field scale. *J. Hydrol.* **2009**, *365*, 310–321. [CrossRef]
71. Brix, H. Functions of Macrophytes in Constructed Wetlands. *Water Sci. Technol.* **1994**, *29*, 71–78. [CrossRef]
72. Landsman, M.R.; Davis, A.P. Evaluation of Nutrients and Suspended Solids Removal by Stormwater Control Measures Using High-Flow Media. *J. Environ. Eng.* **2018**, *144*, 04018106. [CrossRef]
73. Gold, A.C.; Thompson, S.P.; Piehler, M.F. Nitrogen Cycling Processes within Stormwater Control Measures: A Review and Call for Research. *Water Res.* **2019**, *149*, 578–587. [CrossRef]
74. Xu, B.; Wang, X.; Liu, J.; Wu, J.; Zhao, Y.; Cao, W. Improving Urban Stormwater Runoff Quality by Nutrient Removal through Floating Treatment Wetlands and Vegetation Harvest. *Sci. Rep.* **2017**, *7*, 1–11. [CrossRef]
75. Li, H.; Davis, A.P. Water Quality Improvement through Reductions of Pollutant Loads Using Bioretention. *J. Environ. Eng.* **2009**, *135*, 567–576. [CrossRef]
76. Chen, X.C.; Huang, L.; Ong, B.L. The Phytoremediation Potential of a Singapore Forest Tree for Bioretention Systems. *J. Mater. Sci. Eng. A* **2014**, *4*, 220–227.
77. Gadi, V.K.; Tang, Y.R.; Das, A.; Monga, C.; Garg, A.; Berretta, C.; Sahoo, L. Spatial and Temporal Variation of Hydraulic Conductivity and Vegetation Growth in Green Infrastructures Using Infiltrometer and Visual Technique. *Catena* **2017**, *155*, 20–29. [CrossRef]
78. Archer, N.A.L.; Quinton, J.N.; Hess, T. Below-ground relationships of soil texture, roots and hydraulic conductivity in two-phase mosaic vegetation in South-East Spain. *J. Arid Environ.* **2002**, *52*, 535–553. [CrossRef]
79. Lu, J.; Zhang, Q.; Werner, A.D.; Li, Y.; Jiang, S.; Tan, Z. Root-induced changes of soil hydraulic properties—A review. *J. Hydrol.* **2020**, *589*, 125203. [CrossRef]
80. Le Coustumer, S.; Fletcher, T.; Deletic, A.; Barraud, S.; Lewis, J.F. Hydraulic performance of biofilter systems for stormwater management: Influences of design and operation. *J. Hydrol.* **2009**, *376*, 16–23. [CrossRef]
81. Le Coustumer, S.; Fletcher, T.; Deletic, A.; Barraud, S.; Poelsma, P. The influence of design parameters on clogging of stormwater biofilters: A large-scale column study. *Water Res.* **2012**, *46*, 6743–6752. [CrossRef]
82. Langergraber, G.; Haberl, R.; Laber, J.; Pressl, A. Evaluation of Substrate Clogging Processes in Vertical Flow Constructed Wetlands. *Water Sci. Technol.* **2003**, *48*, 25–34. [CrossRef]
83. Siriwardene, N.R.; Deletic, A.; Fletcher, T.D. Clogging of Stormwater Gravel Infiltration Systems and Filters: Insights from a Laboratory Study. *Water Res.* **2007**, *41*, 1433–1440. [CrossRef]
84. PUB. *ABC Waters Design Guidelines*; PUB: Singapore, 2018.
85. Hsieh, C.-H.; Davis, A.P.; Needelman, B.A. Bioretention Column Studies of Phosphorus Removal from Urban Stormwater Runoff. *Water Environ. Res.* **2007**, *79*, 177–184. [CrossRef]
86. Yan, Q.; James, B.R.; Davis, A.P. Lab-Scale Column Studies for Enhanced Phosphorus Sorption from Synthetic Ur-ban Stormwater Using Modified Bioretention Media. *J. Environ. Eng.* **2017**, *143*, 04016073. [CrossRef]
87. Davis, A.P. Field Performance of Bioretention: Water Quality. *Environ. Eng. Sci.* **2007**, *24*, 1048–1064. [CrossRef]
88. Zhou, N.; Bi, C.-J.; Chen, Z.-L.; Yu, Z.-J.; Wang, J.; Han, J.-C. Phosphorus loads from different urban storm runoff sources in southern China: A case study in Wenzhou City. *Environ. Sci. Pollut. Res.* **2013**, *20*, 8227–8236. [CrossRef]
89. Liu, J.; Davis, A.P. Phosphorus Speciation and Treatment Using Enhanced Phosphorus Removal Bioretention. *Environ. Sci. Technol.* **2014**, *48*, 607–614. [CrossRef]
90. Qiu, F.; Zhao, S.; Zhao, D.; Wang, J.; Fu, K. Enhanced nutrient removal in bioretention systems modified with water treatment residuals and internal water storage zone. *Environ. Sci. Water Res. Technol.* **2019**, *5*, 993–1003. [CrossRef]
91. Ong, G.S.; Kalyanaraman, G.; Wong, K.L.; Wong, T.H.F. Monitoring Singapore's First Bioretention System: Rain Garden at Balam Estate. In *WSUD 2012: Water Sensitve Urban Design, Proceedings of the Building the Water Sensitve Community, Seventh International Conference on Water Sensitive Urban Design, Melbourne, Australia, 21–23 February 2012*; ACT: Barton, Australia, 2012; pp. 601–608.
92. Wang, M.; Zhang, D.; Li, Y.; Hou, Q.; Yu, Y.; Qi, J.; Fu, W.; Dong, J.; Cheng, Y. Effect of a Submerged Zone and Carbon Source on Nutrient and Metal Removal for Stormwater by Bioretention Cells. *Water* **2018**, *10*, 1629. [CrossRef]

93. Barrett, M.E. Comparison of BMP Performance Using the International BMP Database. *J. Irrig. Drain. Eng.* **2008**, *134*, 556–561. [CrossRef]
94. McNett, J.K.; Hunt, W.F.; Davis, A.P. Influent Pollutant Concentrations as Predictors of Effluent Pollutant Concentrations for Mid-Atlantic Bioretention. *J. Environ. Eng.* **2011**, *137*, 790–799. [CrossRef]
95. Duncan, H.P. *Urban Stormwater Quality: A Statistical Overview*; CRC for Catchment Hydrology: Victoria, Australia, 1999; 134p.
96. Song, H.; Qin, T.; Wang, J.; Wong, T.H.F. Characteristics of Stormwater Quality in Singapore Catchments in 9 Different Types of Land Use. *Water* **2019**, *11*, 1089. [CrossRef]
97. Wei, Z.; Simin, L.; Fengbing, T. Characterization of Urban Runoff Pollution between Dissolved and Particulate Phases. *Sci. World J.* **2013**, *2013*, 964737. [CrossRef] [PubMed]
98. Hermawan, A.A.; Jung, D.Y.; Talei, A. Removal Process of Nutrients and Heavy Metals in Tropical Biofilters. *E3S Web Conf.* **2018**, *65*, 05026. [CrossRef]
99. Jani, J.; Yang, Y.Y.; Lusk, M.G.; Toor, G.S. Composition of Nitrogen in Urban Residential Stormwater Runoff: Concentrations, Loads, and Source Characterization of Nitrate and Organic Nitrogen. *PLoS ONE* **2020**, *15*, e0229715. [CrossRef] [PubMed]
100. Miguntanna, N.P.; Liu, A.; Egodawatta, P.; Goonetilleke, A. Characterising nutrients wash-off for effective urban stormwater treatment design. *J. Environ. Manag.* **2013**, *120*, 61–67. [CrossRef] [PubMed]
101. Müller, A.; Österlund, H.; Marsalek, J.; Viklander, M. The pollution conveyed by urban runoff: A review of sources. *Sci. Total Environ.* **2020**, *709*, 136125. [CrossRef]
102. Clark, S.E.; Pitt, R. Targeting treatment technologies to address specific stormwater pollutants and numeric discharge limits. *Water Res.* **2012**, *46*, 6715–6730. [CrossRef]
103. Chui, T.F.M.; Trinh, D.H. Modelling infiltration enhancement in a tropical urban catchment for improved stormwater management. *Hydrol. Process.* **2016**, *30*, 4405–4419. [CrossRef]
104. Mangangka, I.; Liu, A.; Egodawatta, P.; Goonetilleke, A. Performance characterisation of a stormwater treatment bioretention basin. *J. Environ. Manag.* **2015**, *150*, 173–178. [CrossRef]
105. Hsieh, C.-H.; Davis, A.P.; Needelman, B.A. Nitrogen Removal from Urban Stormwater Runoff Through Layered Bioretention Columns. *Water Environ. Res.* **2007**, *79*, 2404–2411. [CrossRef]

Article

Evaluation of Active, Beautiful, Clean Waters Design Features in Tropical Urban Cities: A Case Study in Singapore

Teck Heng Neo [1], Dong Xu [1], Harsha Fowdar [2], David T. McCarthy [2], Enid Yingru Chen [3], Theresa Marie Lee [3], Geok Suat Ong [3], Fang Yee Lim [1], Say Leong Ong [1] and Jiangyong Hu [1,*]

1 Department of Civil & Environmental Engineering, Faculty of Engineering, National University of Singapore, Block E1A, #07-01, 1 Engineering Drive 2, Singapore 117576, Singapore; teckheng.neo@u.nus.edu (T.H.N.); erixd@nus.edu.sg (D.X.); rlimtony@gmail.com (F.Y.L.); ceeongsl@nus.edu.sg (S.L.O.)
2 Cooperative Research Centre for Water Sensitive Cities, Melbourne, VIC 3800, Australia; harsha.fowdar@monash.edu (H.F.); David.McCarthy@monash.edu (D.T.M.)
3 PUB, Singapore's National Water Agency, 40 Scotts Road #09-01 Environment Building, Singapore 228231, Singapore; Enid_CHEN@pub.gov.sg (E.Y.C.); Theresa_Marie_LEE@pub.gov.sg (T.M.L.); ONG_Geok_Suat@pub.gov.sg (G.S.O.)
* Correspondence: ceehujy@nus.edu.sg

Citation: Neo, T.H.; Xu, D.; Fowdar, H.; McCarthy, D.T.; Chen, E.Y.; Lee, T.M.; Ong, G.S.; Lim, F.Y.; Ong, S.L.; Hu, J. Evaluation of Active, Beautiful, Clean Waters Design Features in Tropical Urban Cities: A Case Study in Singapore. *Water* **2022**, *14*, 468. https://doi.org/10.3390/w14030468

Academic Editor: Maria Mimikou

Received: 29 December 2021
Accepted: 26 January 2022
Published: 4 February 2022

Publisher's Note: MDPI stays neutral with regard to jurisdictional claims in published maps and institutional affiliations.

Copyright: © 2022 by the authors. Licensee MDPI, Basel, Switzerland. This article is an open access article distributed under the terms and conditions of the Creative Commons Attribution (CC BY) license (https:// creativecommons.org/licenses/by/ 4.0/).

Abstract: In Singapore, active, beautiful, clean waters design features (ABCWDFs), such as rain gardens and vegetated swales, are used as a sustainable approach for stormwater management. Field monitoring studies characterising the performance of these design features in the tropical region are currently limited, hampering the widespread implementation of these systems. This study characterised the performance of individual ABCWDFs in the tropical climate context by monitoring a rain garden (FB7) and a vegetated swale (VS1) that were implemented in a 4-ha urban residential precinct for a period of 15 months. Results showed that total suspended solids (TSS), total phosphorus (TP) and total nitrogen (TN) concentrations were low in the new residential precinct runoff, leading to poor removal efficiency despite the effluent concentrations of individual ABCWDFs that were within the local stormwater treatment objectives. Average TSS, TP and TN EMCs of four sub-catchment outlets were lower (23.2 mg/L, 0.11 mg/L and 1.00 mg/L, respectively) when compared to the runoff quality of the major catchments in Singapore, potentially demonstrating that the ABCWDFs are effective in improving the catchment runoff quality. Findings from this study can help to better understand the performance of ABCWDFs receiving low influent concentrations and implications for further investigations to improve stormwater runoff management in the tropics.

Keywords: urban stormwater runoff management; field monitoring; ABC Waters design features; water quality; bioretention; swales

1. Introduction

In an urban landscape, the high percentage of impervious surfaces often results in a greater volume of surface runoff [1,2]. Flood risks are higher due to an intense storm that generates a higher volume of surface runoff that exceeds the drainage design capacity. Deterioration of stormwater runoff quality inevitably occurs as catchments become more developed and stormwater runoff washes accumulated pollutants deposited on the impervious surfaces [3]. With increasing future urban developments, the level of imperviousness is likely to increase and deposition of pollutants would also increase with more development activities. Thus, it is essential to have an effective stormwater management solution that is built into the developments to capture, detain and treat the runoff before channeling it to downstream water bodies through drains or canals in order to maintain good water quality in the reservoirs.

Key blue–green infrastructures such as bioretention systems and vegetated swales are widely used in stormwater management solutions due to their effectiveness in the management of stormwater peak flow, runoff volume and stormwater pollution [4,5]. Bioretention systems are effective solutions for the removal of pollutants in stormwater runoff, such as suspended solids [6,7], nutrients [8–10] and heavy metals [11,12]. Vegetated swales were also reported to improve stormwater runoff quality [13–15]. However, the performance of vegetated swales in nutrient removal varies in different studies. Yu et al. [15] reported a decrease in stormwater phosphorus and nitrogen mass loadings after going through the grassed swale, while others reported an increase in nutrient concentration [14,16,17]. This variation can somewhat be attributed to the different characteristics of the runoff in terms of speciation and concentration in these studies.

In Singapore, the Active, Beautiful, Clean Waters (ABC Waters) Programme is a sustainable stormwater management approach that seamlessly integrates the environment, water bodies and the community. This creates new community spaces and lifestyles around developments where runoff is generated during wet weather whilst improving the water quality of urban runoff [18]. Where appropriately designed, ABC Waters design features (ABCWDFs) utilise blue–green systems that could manage stormwater quality and peak runoff before it is discharged into the downstream drainage system [18,19]. ABCWDFs use natural systems such as plants and soil to detain and treat stormwater runoff before discharging the cleansed runoff into the drainage system. Biodiversity and living environment are also supplemented with the implementation of ABCWDFs, which include bioretention basin (rain garden), bioretention swales, vegetated swales, constructed wetlands and sedimentation basins. The ABC Waters design guidelines were originally referenced from temperate countries such as Australia [20] and were adapted to Singapore's context. However, Singapore being a tropical country has different meteorological conditions such as rainfall, temperature, humidity and evapotranspiration rate [21]. Guidelines applicable to temperature contexts might not be suitable in Singapore. The difference in rainfall characteristics implies systems need to be sized larger in terms of surface area and storage capacity in tropical climates. Pollutant generation differs between the two rainfall regimes with a high annual mean rainfall typically leading to lower stormwater runoff pollutants concentration [22]. The runoff pollutant concentration can affect the assessment of the performance of stormwater management features [23,24]. Lintern et al. [23] reported that the influent pollutant (nutrients) concentration was found to be strongly correlated with the effluent coming out from the stormwater management systems. Thus, the results from these temperate field studies might not be adopted as the actual performance of stormwater management features in such tropical countries [21]. Present field-scale studies were also evaluated based on actual rain events. Few studies provided a simulation of water features under extreme weather conditions. With the advent of global climate changes, extreme weather conditions such as prolonged dry and wet periods would be increasingly common. Harsh conditions such as prolonged antecedent dry period (ADP) were reported to affect runoff retention and pollutants removal performance [25–27]. Batalini de Macedo et al. [25] found that soil moisture largely affected runoff retention efficiency during a dry period while rainfall depth and intensity is the primary factor during the wet period. On a smaller laboratory scale, Zinger et al. [10] reported that ADP affects changes in hydraulic conductivity of bioretention systems that affect solids and nutrient removal.

Herein, this paper aims to evaluate the performance of the ABCWDFs in a tropical urban context. A residential precinct-scale study spanning for a period of 15 months was conducted in Singapore and used as a case study, facilitating further improvements to existing stormwater management systems. To keep up with global climate changes, the performance of the ABC design was also assessed under challenging operational conditions (e.g., simulated events with prolonged dry periods and higher pollutant concentrations). Hence, the effectiveness and feasibility of these design features can be better understood in terms of managing stormwater runoff from more polluted sources. This paper serves to supplement the knowledge with regards to the field performance of ABCWDFs (in terms of water quantity and quality improvements) in the tropical setting. This field monitoring study also aims to determine whether water quality targets can be achieved at the catchment scale with the implementation of ABCWDFs.

2. Materials and Methods

2.1. Study Area

The study area is a 4-ha pilot urban residential project (Figure 1) named Waterway Ridges, in collaboration with Singapore's Water Agency (PUB) and the Housing Development Board of Singapore (HDB). The study area is an urbanised residential precinct of around 4 ha. Various ABCWDFs were built in the entire precinct (21 bioretention basins or rain gardens, 4 vegetated swales and 2 gravel swales) and the precinct is divided into 4 sub-catchment areas (Figure 1) with 4 distinct sampling outlets (SP1–SP4). Detailed design for each design feature can be found in the hydraulic modelling study by Yau et al. [28]. For the purpose of this study, a representative rain garden and vegetated swale was monitored. The rain garden (FB7) was selected due to its close proximity to the central location of the precinct whereas the vegetated swale (VS1) was selected due to its close proximity to the main road. The 4 catchment outlets of the residential precinct were also monitored for water quantity and quality. Detailed design characteristics of the monitored ABCWDFs and precinct outlets are summarised in Table 1.

Table 1. Characteristics of the monitored ABCWDFs and the 4 catchment outlets of the monitored precinct.

	Monitored ABCWDFs			
	FB7		VS1	
Construction completion	April 2017		April 2017	
Drainage area	572 m^2 (236 m^2—roof, 12 m^2—other impervious, 244 m^2 pervious)		806 m^2 (442 m^2—roof, 81 m^2—other impervious, 163 m^2 pervious)	
Surface area	80 m^2		120 m^2	
Media depth	0.40 m		$-$ [a]	
Number of inlet points	1		2	
	Monitored Sub-Catchments			
	SP1	SP2	SP3	SP4
Catchment area	13,833 m^2	15,713 m^2	2176 m^2	8105 m^2
Impervious fraction	0.69	0.71	0.53	0.69
Area treated (percentage of total area)	32%	57%	62%	88%
ABCWDFs present [b]	Bioretention system (5) Swale system (2)	Bioretention system (11) Swale system (4)	Bioretention system (1)	Bioretention system (4)

[a] there is no filter media depth for vegetated swale. [b] bioretention system includes both bioretention basin and bioretention lawn; swale system includes both vegetated swale and gravel swale.

FB7 is a bioretention basin/rain garden in catchment 2 with a surface area of 80 m^2 and a soil-based filter media depth of 400 mm. The bottom of the rain garden is lined with an impermeable liner and the overflow pit allows a maximum detention depth of 200 mm. This feature received runoff predominantly from the roof of surrounding residential buildings. Both the inlet (*FB7-Inlet*) and outlet of FB7 (*FB7-Outlet*) were monitored for water quantity and water quality.

VS1 is a vegetated swale that is gravel-lined and vegetated with short grass located in catchment 1. It does not have an impermeable liner at the bottom of the gravel-lined layer. It has two inlet points for stormwater runoff. The first inlet point (*VS1-1-Inlet*) is located at the top of the slope while the second inlet point (*VS1-2-Inlet*) is in the middle of the swale. However, only *VS1-1-Inlet* was monitored for water quantity and quality due to instrumental error at *VS1-2-Inlet*. As the catchment characteristics for both inlet points are similar (roof catchment with similar area), the water quality characteristics of both inlet points are assumed to be the same. For a more holistic and accurate assessment of the entire VS1, the water quantity information for *VS1-2-Inlet* was obtained from calibrated model simulation (Model for Urban Stormwater Improvement Conceptualisation, MUSIC V6, for simulation of hydrological model and Water Sensitive Urban Design (WSUD) systems) [29].

The outlet of VS1 (*VS1-Outlet*) was also monitored for both water quantity and quality, where it captures effluent of VS1 that travels through the surface of the swale.

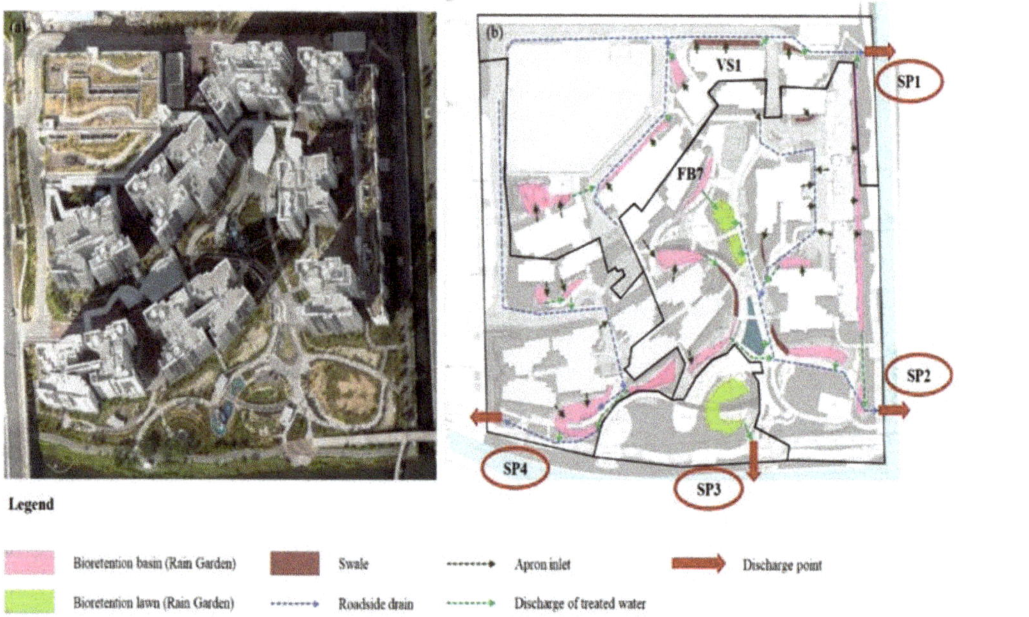

Figure 1. Satellite image of monitored precinct (**a**); Location of monitored ABC water design features (FB7—bioretention system, VS1—swale system) and catchment outlet stormwater drains (SP1–SP4) (**b**).

2.2. Natural Storm Events Monitoring Protocol and Sampling Methodology

The monitoring study spanned a period of 15 months that corresponded to the wet and dry periods in Singapore caused by local monsoon. A pressure transducer (4–20 mA, Heron, ON, Canada) and 90° V-notch weir plate were used to continuously monitor flow depth (interval of 15 s) at the inlet and outlet of the ABCWDFs. Discharge (L/s) and total volume (L) were calculated by incorporating the use of the stage–discharge rating curve Measurements of water level and flow rate at each monitoring point were taken to generate the reliable stage–discharge rating curve. The curve is further refined and calibrated twice throughout the monitoring period). For the discharge at the catchment outlets, area–velocity (AV) sensors (2150, ISCO, Lincoln, NE, USA) were used. Automatic water samplers (3700, ISCO, Lincoln, NE, USA) were set to collect a maximum of 24 water samples using a time-based discrete sampling method. A total of 1 L of stormwater runoff samples were collected at a user-defined interval. Rainfall depth was monitored using a tipping bucket rain gauge (TB3, Hydrological Services, Lakeworth, FL, USA), installed at the high open space within the precinct. Rainfall information from the rain gauge was logged at a 1 min interval.

Water quality samples were collected within 24 h of each storm event and transported in an icebox to SINGLAS-accredited laboratory for water quality testing. For each event, 10 samples were chosen out of the 24 samples for testing based on the hyetograph of the event. The selection of the 10 samples reflected the full spread of rainfall and aimed to capture all the rising and falling limbs of the rainfall. More priority was given to the first few samples to focus on "first flush" concentrations.

Over the monitoring period, water quality data were collected for 12 events. The characteristics of the monitored storm events are given in the supporting document (Table S1).

2.3. Challenge Test Framework and Sampling Methodology

The challenge test framework developed for this study was adapted from prior challenge test study by Zhang et al. [27]. The two ABCWDFs were artificially spiked with synthetic stormwater, with a target pollutants concentration representing the 95th percentile of Singapore's stormwater EMC. Detailed chemicals and target pollutant concentration can be found in Table S2 of supplementary information. Water from the waterway beside the study area was pumped into a mixing tank (1 m^3) and prepared chemicals were mixed using an internal recirculation pump. The mixed-dosed water was then released to the ABCWDFs. The pumping and mixing steps were repeated to attain the event volume for the tests. For sampling, three inflow samples were taken from the outlet hose of the mixing tank and were then composited into a 1 L sample. For the outflow samples, volume-based sampling was conducted (Table S2 in supplementary information). Once the water samples were collected, they were stored in an icebox and delivered to the same laboratory at the end of the challenge test within 8 h for water quality testing.

A total of 4 challenge tests for each ABCWDF (summarised in Table 2) was conducted to study the effects of various antecedent dry period (ADP). Two scenarios were simulated: extreme wet and extreme dry conditions. For extreme wet simulation (simulating back-to-back storm events), 12 h were selected. For extreme dry simulation, the longest antecedent dry period was represented by the 95th percentile of the dry period experienced by the study area, which was found to be 6 days (144 h). This value was estimated using 8 years of rainfall data from the nearest meteorological station from the study area.

Table 2. Summary of challenge test conditions.

ABCWDFs	Challenge Test ID	Date	Antecedent Dry Period (h)	Event Volume (m^3)
FB7	Challenge Test 1 (FB-CT1)	28 February 2019	240	14
	Natural Storm Events	a	-	-
	Challenge Test 2 (FB-CT2)	12 March 2019	16	14
	Challenge Test 3 (FB-CT3)	13 March 2019	12	14
	Challenge Test 4 (FB-CT4)	20 March 2019	144	15
VS1	Challenge Test 1 (VS-CT1)	25 February 2019	168	9
	Challenge Test 2 (VS-CT2)	26 February 2019	12	12
	Challenge Test 3 (VS-CT3)	5 March 2019	144	15
	Natural Storm Events	b	-	-
	Challenge Test 4 (VS-CT4)	28 March 2019	144	12

[a] 1 rainfall event observed on 11 March 2019 (26.2 mm). [b] 2 rainfall events observed on 11 March 2019 (26.2 mm) and 21 March 2019 (19.6 mm).

For the event volume for extreme wet simulation, 1.1 times the pore volumes of the ABCWDFs were selected as the event volume (Table S2 in supplementary information). For extreme dry simulation, a larger event volume for the challenge tests was needed as more water was needed to fill up the drier pores of the filter media. As such, a MUSIC model was set up for the catchment and the 95th percentile of runoff volume received by the ABCWDFs (simulated by MUSIC) was selected as the event volume. Details of the calculation can be found in Table S2 of supplementary information.

2.4. Water Quality Analysis

For both water samples from natural storm events and challenge tests, testing was conducted in accordance with the Standard Method for the Examination of Water and Wastewater [30]. The tested parameters are listed in Table S4 (supplementary information).

2.5. Performance Assessment and Analysis

In this study, the water quantity and quality data were used to compute the Event Mean Concentration (EMC) of TSS, TP and TN. When the pollutant concentration of the water samples was below the detection limit, half of the lowest detection limit was selected to estimate the EMC [31]. To calculate the efficiency of pollutant removal in ABCWDFs, the efficiency ratio (ER) method was used to assess the performance of these design features [32]. Student *t*-test was conducted to demonstrate the significance of results. A *p*-value of <0.05 is considered significant. The performance of the ABCWDFs was compared with the stormwater treatment objectives in Singapore [18], listed in Table 3 below.

Table 3. Stormwater treatment objectives for Singapore [18].

Pollutants	Stormwater Treatment Objectives
Total Suspended Solids (TSS)	80% Removal or less than 10 mg/L
Total Nitrogen (TN)	45% Removal or less than 1.2 mg/L
Total Phosphorus (TP)	45% Removal or less than 0.08 mg/L

3. Results

3.1. Characteristics of Influent Stormwater Runoff

For FB7-Inlet, water quality data from nine events were measured. TSS influent EMCs ranged from 7.4 to 23.8 mg/L (average of 11.3 mg/L), TP influent EMCs ranged from 0.03 to 0.42 mg/L (average of 0.12 mg/L) and TN influent EMCs ranged from 0.38 to 1.50 mg/L (average of 0.92 mg/L). For VS1-1-Inlet, water quality data from eight events were gathered. TSS influent EMCs for this point ranged from 21.1 to 190.6 mg/L (average of 88.4 mg/L), TP influent EMCs ranged from 0.04 to 0.43 mg/L (average of 0.27 mg/L) and TN influent EMCs ranged from 1.19 to 2.84 mg/L (average of 2.02 mg/L).

Overall, the precinct stormwater runoff (represented by the average of FB7-Inlet and VS1-1-Inlet) in this study had a much lower EMC for TSS, TP and TN when compared to the world data for high urban areas reported by Duncan [22] (Table 4). It is important to note that stormwater runoff concentration depends heavily on land uses [22]. For the precinct in this study, the monitored catchments are predominantly hard roof catchments, which typically have lower levels of pollutant generation compared to other surfaces such as road, pedestrian footpaths and parkland. The TP and TN runoff concentrations observed in this study were also comparable to the observations from another study in Queenstown [33], another urban residential area in Singapore.

Table 4. Runoff water quality of the precinct and comparison with literature studies.

	Runoff Pollutants EMC		
	Average TSS EMC (mg/L)	Average TP EMC (mg/L)	Average TN EMC (mg/L)
This study			
FB7-Inlet (9 events)	11.3	0.12	0.92
VS1-1-Inlet (8 events)	88.4	0.27	2.02
Average	**49.9**	**0.2**	**1.47**
Literature			
(Duncan, 1999)			
High Urban Areas	155	0.32	2.63
Roofs	35	0.13	
(Lim, 2003)			
Queenstown high urban residential	100	0.13	1.25

Typically, as the antecedent dry period increases, more pollutants will accumulate and this leads to higher loading or concentration in the stormwater runoff. The first flush analysis was conducted to further verify the relationship between runoff pollutants loading/concentration and ADP. Bertrand-Krajewski [34] proposed that a significant first flush phenomenon arises when 80% of pollutant loads were transported within the first 30% of runoff volume. This signifies that the pollutants were disproportionately high discharge in the beginning stage of a storm event [35]. Using the established criteria from Bertrand-Krajewski [34], there was no presence of the first flush phenomenon for TSS, TP and TN. The dimensionless cumulative pollutant load against the dimensionless cumulative runoff volume (MV curve) for the storm event is shown in Figure S1. From

Figure 2, it is interesting to note that there was no trend being observed between ADP and the EMC/loadings of the stormwater runoff (low R^2 values). This could be attributed to the high frequency and intensity of storm events of the tropical meteorological condition in Singapore. Stormwater pollutants' wash-off happened more often and led to a decrease in pollutant accumulation [36]. Furthermore, the monitored precinct is rather new (construction was completed 1 year prior to the field monitoring), which may lead to a lower accumulation of stormwater pollutants. Overall, this observation of low stormwater runoff concentration was obtained based on 17 samples (11 storm events) with small catchment areas. The results should be further validated by conducting a larger-scale monitoring study with more storm events. Despite Singapore being an urbanised developed country with a high impervious percentage, the runoff quality observation in this study might not be representative of other countries in the tropical region. Furthermore, effective pollutant source management solutions such as strict erosion control practices and routine drainage system maintenance from PUB minimised the presence of urban pollutants within the catchment in Singapore [37].

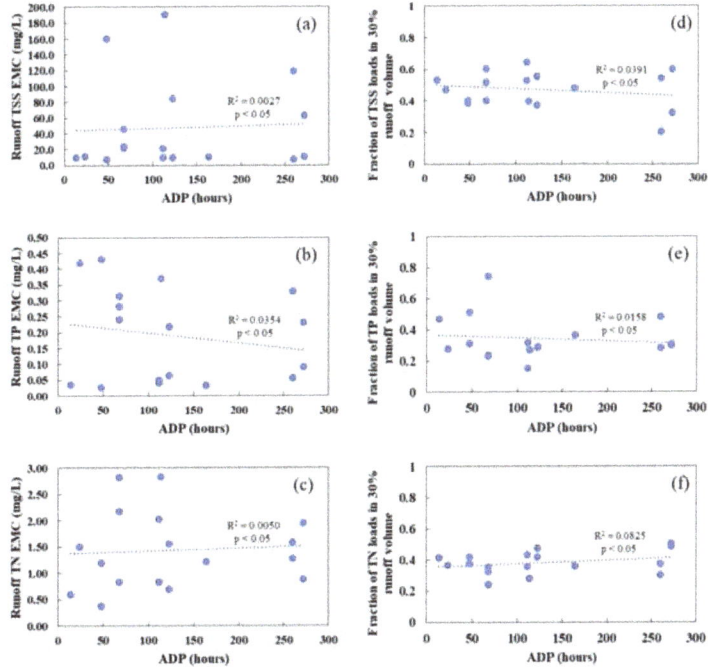

Figure 2. Inflow pollutants EMC against ADP (**a**–**c**); Fraction of pollutants loads in 30% of runoff volume against ADP (**d**–**f**).

3.2. Treatment Performance of Rain Garden

3.2.1. Natural Storm Events

A total of six natural storm events were monitored for the FB7 system (Table 5). Effluent TSS EMCs ranged from 6.3 to 20.7 mg/L (average of 13.4 mg/L), effluent TP EMCs ranged from 0.07 to 0.58 mg/L (average of 0.19 mg/L) while effluent TN EMCs ranged from 0.39 to 1.15 mg/L (average of 0.73 mg/L). Altogether, variation in the inflow concentrations over an event was observed due to the initial spike in pollutant concentration and this was observed for most monitored storm events. On the other hand, pollutant concentrations of the effluent were relatively steadier with minimal fluctuations. An example can be seen from the inflow and outflow TSS, TP and TN pollutographs of FB7 for Event 4 (Figure 3).

Table 5. Pollutant EMC and loads removal for FB7 and VS1 during monitored storm events and challenge tests.

ID	Pollutant EMC								
	TSS			TP			TN		
	Influent (mg/L)	Effluent (mg/L)	Removal (%)	Influent (mg/L)	Effluent (mg/L)	Removal (%)	Influent (mg/L)	Effluent (mg/L)	Removal (%)
Rain garden (FB7) during monitored storm events									
Event 3 [a]	10.9	20.3	−86.4	0.03	0.07	−116.0	1.22	1.15	5.4
Event 4	23.8	6.3	73.4	0.28	0.14	51.6	0.83	0.39	52.8
Event 8	9.8	12.4	−25.7	0.06	0.15	−131.5	0.71	0.73	−3.4
Event 9	7.7	20.7	−168.0	0.06	0.12	−112.7	1.29	0.55	56.9
Event 11	11.7	10.7	8.3	0.42	0.58	−39.1	1.50	0.97	35.6
Event 12	9.8	10.1	−2.5	0.04	0.09	−164.3	0.60	0.59	0.8
Average	12.3	13.4	−33.5	0.15	0.19	−85.3	1.02	0.73	24.7
Rain garden (FB7) during challenge tests									
FB-CT1	32.1	2.3	92.9	0.18	0.04	76.7	3.19	1.93	39.4
FB-CT2	25.1	3.0	87.9	0.23	0.06	73.5	3.20	1.62	49.5
FB-CT3	40.4	6.0	85.2	0.23	0.06	74.5	3.19	1.86	41.6
FB-CT4	44.7	4.2	90.7	0.29	0.07	75.0	3.11	1.88	39.4
Average	35.6	3.9	89.2	0.23	0.06	74.9	3.17	1.82	42.5
Vegetated swale (VS1) during monitored storm events [b]									
Event 7	61.8	145.4	−135.5	0.25	0.12	53.0	2.55	0.89	65.0
Event 8	86.4	70.5	18.4	0.22	0.11	51.9	1.70	0.81	52.4
Event 9	116.8	46.4	60.3	0.44	0.17	60.9	2.00	0.71	64.4
Event 10	180.0	44.8	75.1	0.36	0.07	80.1	2.76	1.25	54.8
Average	111.2	76.8	4.6	0.32	0.12	61.5	2.25	0.92	59.1
Vegetated swale (VS1) during challenge tests									
VS-CT1	39.7	17.8	55.2	0.24	0.23	3.9	3.06	2.78	9.3
VS-CT2	34.5	9.6	72.2	0.15	0.12	16.7	2.93	2.55	12.9
VS-CT3	32.9	13.2	59.9	0.14	0.13	5.3	2.69	2.58	4.2
VS-CT4	44.5	5.8	86.9	0.24	0.21	11.9	3.40	3.17	6.6
Average	37.9	11.6	68.6	0.19	0.17	9.5	3.02	2.77	8.3

[a] accumulation of sediments at the weir plate, causing some inaccurate water level readings. [b] assumed that water quality information for VS1-1-Inlet and VS1-2-Inlet were the same.

FB7 generally performed well in TSS removal with the average effluent TSS EMCs similar to the treatment objectives of 10.0 mg/L as stipulated in the ABC Waters design guidelines. However, if the system is assessed based on the percentage removal efficiency, poor efficiency (−34%) was obtained due to the low influent TSS EMCs [38]. Various field studies in the temperate region reported a similar range of bioretention effluent TSS concentration [39,40], but positive removal efficiency due to higher influent TSS concentration. Similar to the TSS removal performance of another rain garden in the tropics, previous field monitoring studies of Balam estate rain garden [19] showed lower TSS EMC removal efficiency of 57%, caused by the low influent TSS concentration. Both effluent concentration and removal efficiency ratios should be considered for practical and effective stormwater runoff management practices. This also highlights the importance of considering inflow pollutants concentration when assessing the performance of ABCWDFs, especially in a catchment with low pollutant generation sources.

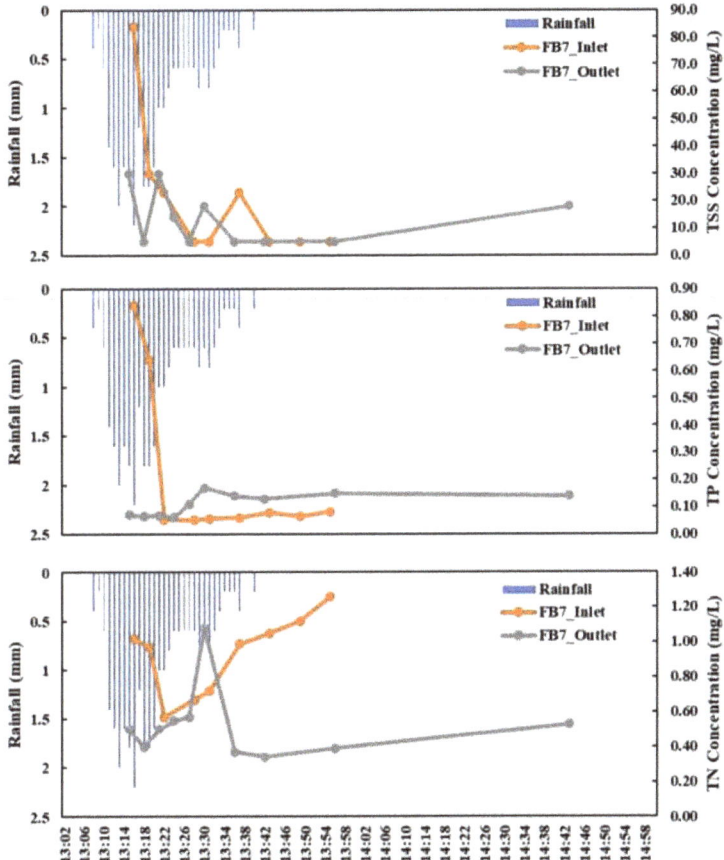

Figure 3. FB7 TSS, TP and TN pollutographs for Event 4.

Similarly, for TP, negative percentage removal efficiency (−85%) was obtained due to low influent TP EMCs. A total of four out of six monitored events had TP influent EMC that was lower than the stormwater quality objectives of 0.08 mg/L. The low influent TP EMC could be due to little phosphorus loading from the rooftop catchment areas (which are not accessible to the general public and hence have fewer anthropogenic activities), in addition to the fact that the development is relatively new. A similar observation was found in the monitoring studies of the Balam estate rain garden [19], where only 27% of TP removal was noted for the rain garden due to the low TP influent EMC.

Likewise, due to the low influent TN EMCs, the average removal efficiency of 25% was observed although the effluent TN EMCs were below the stormwater quality objectives of 1.2 mg/L for all monitored events. TN removal efficiency fluctuated between −3% and 57%. Given that FB7 does not have designed components for denitrification such as a submerged layer or carbon source additives, the TN removal was expected and similar to some of the bioretention studies (without submerged zone and additives) in the literature [40,41].

3.2.2. Challenge Tests

When subjected to higher pollutant loading (in comparison with the natural storm events), clear pollutant reduction patterns for the bioretention system were observed for all four challenge tests. The pollutant removal efficiency of TSS, TP and TN were comparable between the four challenge tests conducted (Table 5). For TSS and TP, all four challenge tests for FB7 showed good EMC removal efficiency for TSS and TP (average of 89% and 75%, respectively). The high TSS performance of the system ($p < 0.05$) demonstrates efficient physical filtration. As phosphorus is closely associated with sediment removal of stormwater runoff (Wu et al., 1998), the high and significant TP removal ($p < 0.05$) is not surprising. High TP removal was also observed in other biofiltration studies [39,42]. Effluent TSS and TP EMCs for all four tests were below the stormwater quality objectives, highlighting the effectiveness of the rain garden in removing stormwater pollutants.

The removal mechanisms of nitrogen are much more complicated compared to phosphorus or suspended solids removal. Particulate nitrogen can be trapped by the soil media via sedimentation and filtration, while dissolved nitrogen could be removed via adsorption or biological uptake by plants [43,44]. From the four TN pollutographs of the challenge tests (Figure 4), it was observed that there was a sudden decrease, followed by a gradual increase in the TN pollutant concentration. This is a characteristic of bioretention systems whereby the pore water within the FB7 soil media, which has a better water quality, was purged out by the influent challenge test water. The lower initial TN concentration could also be due to the sorption of nitrate to the active sites of the soil particles when the challenge test water first percolated through the system. When the active sites were all occupied, nitrate was not removed through sorption. Thus, subsequent samples contained higher nitrate concentration (not shown) and TN concentration. All four tests showed a decent average removal efficiency (42.5%), with 39.4% for FB-CT1, 49.5% for FB-CT2, 41.6% for FB-CT3 and 39.4% for FB-CT4 ($p < 0.05$). The nitrogen removal of the challenge tests is similar except for FB-CT2, which had a lower effluent TN EMC (Table 5). This might be caused by the antecedent natural storm event (rainfall depth of 21.0 mm) that occurred before FB-CT2. As mentioned in the previous section, stormwater runoff in the study area has low pollutants concentration and this may lead to cleaner FB7 pore water that reduces the overall effluent TN concentration of subsequent test (FB-CT2).

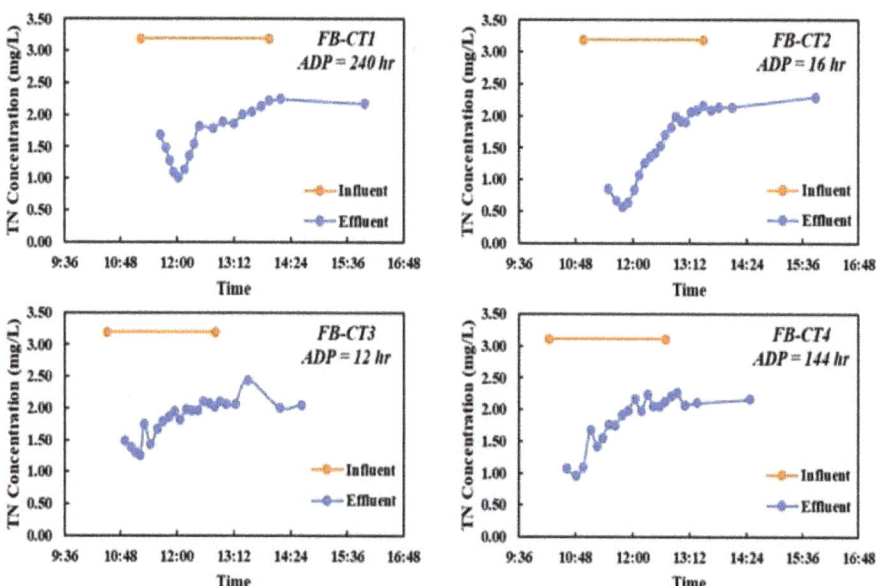

Figure 4. TN pollutographs for FB7 challenge tests.

Hatt et al. [26] found that bioretention effluent nitrate concentration was increased as ADP increased due to the synergistic effect of temperature and moisture (influenced by seasonality and ADP), which may affect the biological processing of nitrogen. Besides that, alternating wetting and drying conditions were shown to affect the soil moisture of the bioretention systems, causing implications to the nitrogen removal performance [44]. However, a tropical country such as Singapore has a minimal seasonality effect. For example, Chui and Tringh [45] found that the average time intervals between consecutive storm events in Singapore were short and this is distinctive as compared to temperate countries. More monitoring studies can be conducted to investigate the relationship between ADP or soil moisture and nitrogen removal in a tropical context.

3.3. Treatment Performance of Vegetated Swale

3.3.1. Natural Storm Events

A total of four water quality datasets for VS1 were obtained. In order to provide a better assessment of VS1, the calculation of inflow pollutant EMC was carried out by dividing the summation of total pollutants load from both VS1-1-Inlet (measured) and VS1-2-Inlet (modelled) with the total event volume.

Influent TSS EMCs ranged from 61.8 to 180.0 mg/L with an average of 111.2 mg/L. Most effluent samples had a lower TSS concentration than the influent, displaying the effectiveness of the swale in reducing solid particles from stormwater runoff (Figure 5). The primary mechanism of TSS removal in the vegetated swale is the sedimentation and filtration process. The removal is influenced by the flow path length, roughness and influent particle size distribution [14]. However, effluent TSS EMCs ranged from 44.8 to 145.4 mg/L (average of 76.8 mg/L), which is higher than the stormwater quality objective of 10.0 mg/L. Throughout the monitoring study, small gravels were found accumulating before the weir plate of VS1-2-Inlet during and after a storm event. Thus, these small gravels were washed off from the landscaping areas (within the contributing catchment) to the swale surface and ended up in the effluent collected at VS1-Outlet.

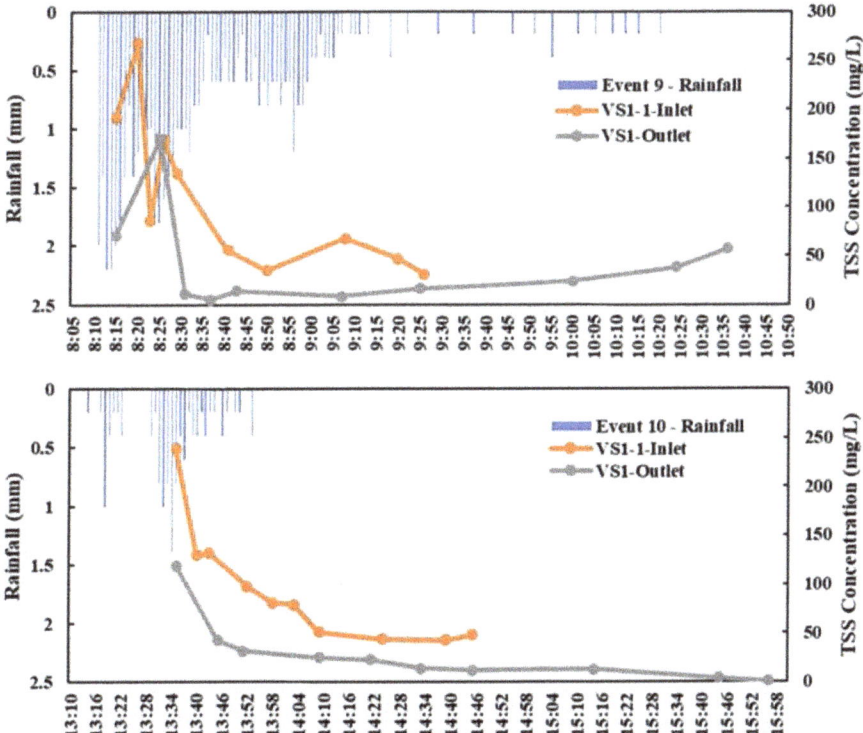

Figure 5. VS1 TSS pollutographs for Event 9 and Event 10.

For phosphorus and nitrogen, influent TP EMCs ranged from 0.22 to 0.36 mg/L (average of 0.32 mg/L) while TN EMCs ranged from 1.70 to 2.76 mg/L (average of 2.25 mg/L). On the other hand, the swale effluent TP EMCs ranged from 0.07 to 0.17 mg/L (average of 0.12 mg/L) while effluent TN ranged from 0.71 to 1.25 mg/L (average of 0.92 mg/L), which were comparable to the stormwater quality objectives of 0.08 mg/L and 1.2 mg/L, respectively (both $p < 0.05$). The low effluent nutrients concentration could again be attributed to two points: (1) the low influent pollutant concentration of the catchment in general; (2) VS1 is unlined and the exfiltration leads to the loss of some TP and TN pollutants, causing a "cleaner" effluent. Fardel et al. [46] analysed swale performance in the literature and found that inflow pollutants concentration was strongly correlated with the pollutants efficiency ratios of the swale system.

3.3.2. Challenge Tests

The four challenge tests for VS1 showed decent removal of TSS with an average removal efficiency of 69.0%, with a range of 55.2% to 86.9% ($p < 0.05$). The effluent of the challenge tests (average of 11.6 mg/L) was just slightly higher than the stormwater quality objectives for TSS. The long-vegetated swale typically conveys the stormwater at a slower speed compared to concrete stormwater drains and this allows better settlement of sediments. Other runoff simulation experiments also reported similar TSS EMC reduction (61%–86% for Deletic and Fletcher [47]; 79%–98% for Bäckström [48]).

For TP and TN, all four tests showed poor removal efficiency with an average of 9.4% and 8.2%, respectively. In terms of effluent concentration, the four tests displayed an average concentration of 0.17 mg/L and 2.77 mg/L for TP and TN, respectively (both $p < 0.05$). The absolute difference in concentration between challenge tests influent and effluent is, however, very small. The results are not surprising as TN and TP removal require adequate retention time (affected by flow velocity) for better pollutant removal [15,49]. It is likely that the nutrient reduction observed in the VS1 challenge tests is attributed to the removal of sediment particles, to which the nutrients are attached [50].

Another thing to take note of is that there were no obvious differences in VS1 pollutant treatment performance between the various operating conditions. The dry duration between consecutive challenge test events did not affect the pollutants removal efficiency since the major factor that affects the water quality improvements of vegetated swale is most likely the swale length and slope ratio [14].

3.4. Water Quality of Entire Precinct

The water quality performance of the entire precinct was monitored and summarised in Table 6.

Table 6. EMCs for the catchment outlets.

Catchment Outlets	EMC		
	Min–Max (Mean)		
	TSS (mg/L)	TP (mg/L)	TN (mg/L)
SP1	16.2–50.3 (32.5)	0.07–0.18 (0.10)	0.63–2.09 (1.26)
SP2	17.4–55.5 (30.7)	0.06–0.22 (0.12)	0.77–1.60 (1.08)
SP3	3.8–48.1 (12.8)	0.04–0.16 (0.10)	0.30–1.02 (0.67)
SP4	8.8–37.3 (16.6)	0.04–0.17 (0.10)	0.69–1.38 (0.99)
Average	**23.2**	**0.11**	**1.00**

SP1 and SP2 are larger catchments in the precinct, and where most residential buildings and common green areas are located. This might explain the higher pollutants concentration for these two catchments. Some of the runoffs from these large catchments (68% of the total catchment area for SP1 and 43% for SP2) were untreated and entered the drainage system directly. On the other hand, SP3 and SP4 are smaller and have a higher percentage of the treated area (62% and 88% respectively), the pollutant concentrations were generally lower. With a higher runoff treatment percentage by ABCWDFs, better water quality can be obtained at the catchment outlets, which is important for downstream urban waterbodies protection.

In terms of meeting stormwater quality objectives, TSS and TP water quality at the catchment outlets were slightly above the guideline value for most monitored events while TN was generally below the guideline value. However, when compared with the runoff pollutant concentration of residential catchment without installation of ABCWDFs in Singapore (Table 4), the three pollutants average EMC at the catchment outlets were either similar or lower, indicating that the ABCWDFs are effective in improving the overall catchment runoff quality. Larger improvements can be seen if compared with worldwide high urban land uses [22].

4. Discussion

This study presents an important field study of ABCWDFs that can guide urban catchment managers for effective stormwater runoff management in tropical countries. Overall, the effluent of both FB7 and VS1 are comparable to the literature studies (Table 7) and these systems can be employed in tropical climates for effective stormwater runoff management. When comparing with other field studies in Singapore, similar information was obtained. Lim et al. [38] observed poor pollutant removal efficiency for a modular bioretention tree system due to low influent runoff pollutant concentration, which may be caused by the high frequency and seasonal severity of storm events. Compared with another ABCWDF by Ong et al. [19], the effluent concentrations were slightly lower in that study. Lower nitrogen concentration in the effluent could be due to the incorporation of the submerged zone in their ABCWDF design. Moreover, the differences in the effluent quality could also be attributed to the difference in volume and runoff characteristics. However, it should be noted that the effluent pollutant EMCs were in accordance with those of previous studies.

Table 7. Comparison of effluent pollutant EMC of this study and literature studies.

Study	Climate	Scope	TSS Effluent (mg/L)	TP Effluent (mg/L)	TN Effluent (mg/L)
Rain Gardens					
This study, FB7	Tropical	Field monitoring of rain garden in residential precinct, Singapore	13.9	0.19	0.73
Lim et al. [38]	Tropical	Field monitoring of modular bioretention tree system with engineered media, Singapore	4.8	0.04	0.27
Wang et al. [36]	Tropical	Field monitoring of rain garden with submerged zone, Singapore	2.0	0.04	0.61
Ong et al. [19]	Tropical	Field monitoring of rain garden with submerged zone, Singapore	11.3	0.11	0.66
Lopez-Ponnada et al. [9]	Temperate	Field monitoring of modified rain garden, USA	-	-	0.74
Brown and Hunt [41]	Temperate	Field monitoring of rain garden with submerged zone, USA	16.9	0.09	0.43
Davis [39]	Temperate	Field monitoring of rain garden, USA	18, 13	0.15, 0.17	-
Swales					
This study, VS1	Tropical	Field monitoring of vegetated swale in residential precinct, Singapore	76.8	0.12	0.92
Leroy et al. [17]	Temperate	Field monitoring of vegetated swale in treating road runoff, France	178	-	-
Stagge et al. [14]	Temperate	Field monitoring of vegetated swale, USA	16, 18	0.29, 0.20	2.12, 2.63

With higher inflow pollutants concentration (challenge tests), FB7 could still produce a treated effluent with TSS and TP concentration that achieves the stormwater quality objectives. Long-term TSS accumulation at the rain garden could decrease hydraulic conductivity and affect pollutant uptake capability over time [51,52], but the impact might be lessened due to periodic tropical rain that reduces pollutant build-up. To further enhance phosphorus removal, background TP concentration can be reduced by using media with absorbent to phosphorus such as engineered soil or stringent use of fertilisers for the maintenance of the plants in the ABCWDFs.

No significant difference was noted for the TN removal in the challenge test study (different operational phases). A similar observation was made by Wang et al. [53] where the number of antecedent dry days did not affect TN performance in bioretention column studies. Despite not having a submerged zone [10,54], the TN removal efficiency of this study was still comparable to regional studies of rain gardens. In this study, the rain gardens were able to meet the stormwater quality objectives imposed. Other tropical regions with different runoff characteristics might require more thorough removal of TN, which could be enhanced with submerged zone incorporated rain gardens. Other possible considerations for effective nitrogen removal include additives that could promote denitrification within or below the filter media [55–57]. Moreover, studies have shown that plant species can influence pollutants removal [58,59]. Plant species with dense and fibrous root architecture can perform better in terms of nitrogen removal [60]. Loh [61] also showed that tropical plants, especially native species, could improve the nitrate removal performance of the rain gardens.

The performance of VS1 is similar to the literature studies of temperate climate, where the vegetated swale is effective in slowing down the flow of stormwater runoff as well as removing coarse solids or sediments. In an urban setting, a vegetated swale is a versatile feature that can be installed easily in small catchment areas to convey stormwater runoff as part of a designed drainage system or as a pretreatment feature for other ABCWDFs, such as a rain garden or wetland, which are more expensive and harder to construct in urban environments [49]. The usage of vegetated swale would be justified for residential land use with low stormwater runoff pollutants concentration. From our study, VS1 is able to handle stormwater runoff with low influent pollutant concentrations and produce effluent that meets stormwater treatment objectives. Combining the different functions of these ABCWDFs allows better and innovative ways of managing hydrological and treatment performance during normal and extreme conditions [62]. Besides that, the design of vegetated swales can also be modified further to include check dams, gravel layers (infiltration swales), engineered media (bioretention swales) to enhance the overall pollutants removal performance [63–65]. These modifications have the potential to be further explored.

Future monitoring of ABCWDFs in other land use of Singapore (e.g., industry area, park or business districts) will also provide insights on the performance of the ABCWDFs in removing certain stormwater pollutants such as heavy metals, faecal coliform, TOC, etc., which are more prevalently found in non-residential catchments. Bioretention can be optimised to treat stormwater runoff from commercial, residential and industrial areas [66]. For instance, a previous study was carried out to model the removal of TSS, TN, TP and heavy metal from an industry area [67]. Biochar was recommended in biofilters for treating wastewaters containing hazardous contaminants [68]. Nevertheless, the field application of bioretention or ABCWDFs in industry and business areas is still lacking. Currently, more focus is on the optimisation of engineered media on a laboratory scale to study the factors affecting the removal of industry-associated pollutants, such as organic pollutants and heavy metals [69–71]. Other key design considerations such as catchment characteristics, rainfall pattern and intensities, size of the feature, type of vegetation, soil additives, etc. should be further explored and elucidated to better understand the factors affecting the removal rates of these pollutants.

5. Conclusions

In this study, selected ABCWDFs in an urban residential precinct in Singapore were monitored and assessed in terms of achieving water quality and peak runoff management objectives. The features were monitored for 15 months. During wet events, ABCWDFs such as rain gardens and vegetated swale were able to produce effluent that is within local stormwater treatment objectives. Results showed that total suspended solids (TSS), total phosphorus (TP) and total nitrogen (TN) concentrations were low in the runoff from this new residential precinct. Effluent pollutant (TSS, TP and TN) concentrations at the outlet of the individual treatment systems were within the local stormwater treatment objectives over the monitored storm events. The effluent TSS, TP and TN concentrations are consistent with other field studies in the literature. Through challenge tests with synthetic runoff, a higher removal efficiency of pollutants was observed. Challenge test results displayed that the rain garden performed well even during extreme conditions with clear pollutant reduction. Removal efficiency based on Event Mean Concentration (EMC) in the order of 89%, 75% and 43% for TSS, TP and TN, respectively, were achieved in several simulated storm events, using higher pollutant influent levels. This study provides evidence that ABCWDFs can provide sustainable management of urban stormwater runoff quality in tropical climates.

Supplementary Materials: The following supporting information can be downloaded at: https://www.mdpi.com/article/10.3390/w14030468/s1, Table S1: Summary of monitored natural storm events; Table S2: Target pollutants concentration and dosing volume of challenge tests; Table S3: Number of samples for challenge test events; Table S4: Test methods for water quality parameters; Figure S1: First flush (MV curve) for TSS, TP and TN of FB7 (a–c); VS1 (d–f)

Author Contributions: Conceptualisation, T.H.N. and J.H.; methodology, T.H.N., D.X., H.F. and D.T.M.; investigation, T.H.N., D.X. and H.F.; Writing—original draft preparation, T.H.N.; Writing—review and editing, F.Y.L., J.H. and T.M.L.; visualisation, T.H.N. and F.Y.L.; supervision, J.H., S.L.O. and D.T.M.; project administration, J.H. and S.L.O.; funding acquisition, G.S.O., E.Y.C. and T.M.L. All authors have read and agreed to the published version of the manuscript.

Funding: This research was funded by the Public Utilities Board (PUB), grant number WBS: R-347-000-267-490.

Institutional Review Board Statement: Not applicable.

Informed Consent Statement: Not applicable.

Data Availability Statement: The data presented in this study is available on request from the corresponding author.

Acknowledgments: The authors of this paper would like to thank the PUB, Singapore's National Water Agency, for funding this project "ABC Waters Evaluation and Modelling Project: Punggol New Town—C39 Precinct". The authors would also like to thank Shie-Yui Liong, Mengjie Liew, Kim Dong Eon and Jiandong Liu from the NUS Tropical Marine Science Institute (TMSI) for their assistance in the project.

Conflicts of Interest: The authors declare no conflict of interest.

References

1. Miller, J.D.; Kim, H.; Kjeldsen, T.R.; Packman, J.; Grebby, S.; Dearden, R. Assessing the impact of urbanization on storm runoff in a peri-urban catchment using historical change in impervious cover. *J. Hydrol.* **2014**, *515*, 59–70. [CrossRef]
2. Sun, S.; Barraud, S.; Castebrunet, H.; Aubin, J.-B.; Marmonier, P. Long-term stormwater quantity and quality analysis using continuous measurements in a French urban catchment. *Water Res.* **2015**, *85*, 432–442. [CrossRef] [PubMed]
3. Zhang, S.; Guo, Y. Stormwater Capture Efficiency of Bioretention Systems. *Water Resour. Manag.* **2014**, *28*, 149–168. [CrossRef]
4. Mahmoud, A.; Alam, T.; Yeasir, A.; Rahman, M.; Sanchez, A.; Guerrero, J.; Jones, K. Evaluation of field-scale stormwater bioretention structure flow and pollutant load reductions in a semi-arid coastal climate. *Ecol. Eng.* **2019**, *142*, 100007. [CrossRef]
5. Vijayaraghavan, K.; Biswal, B.; Adam, M.; Soh, S.; Tsen-Tieng, D.; Davis, A.; Chew, S.; Tan, P.; Babovic, V.; Balasubramanian, R. Bioretention systems for stormwater management: Recent advances and future prospects. *J. Environ. Manag.* **2021**, *292*, 112766. [CrossRef] [PubMed]
6. Alam, T.; Bezares-Cruz, J.; Mahmoud, A.; Jones, K. Nutrients and solids removal in bioretention columns using recycled materials under intermittent and frequent flow operations. *J. Environ. Manag.* **2021**, *297*, 113321. [CrossRef] [PubMed]
7. Søberg, L.; Al-Rubaei, A.; Viklander, M.; Blecken, G. Phosphorus and TSS Removal by Stormwater Bioretention: Effects of Temperature, Salt, and a Submerged Zone and Their Interactions. *Water Air Soil Pollut.* **2020**, *231*, 270. [CrossRef]
8. Hatt, B.E.; Fletcher, T.D.; Deletic, A. Hydrologic and pollutant removal performance of stormwater biofiltration systems at the field scale. *J. Hydrol.* **2009**, *365*, 310–321. [CrossRef]

9. Lopez-Ponnada, E.V.; Lynn, T.J.; Ergas, S.J.; Mihelcic, J.R. Long-term field performance of a conventional and modified bioretention system for removing dissolved nitrogen species in stormwater runoff. *Water Res.* **2020**, *170*, 115336. [CrossRef]
10. Zinger, Y.; Prodanovic, V.; Zhang, K.; Fletcher, T.D.; Deletic, A. The effect of intermittent drying and wetting stormwater cycles on the nutrient removal performances of two vegetated biofiltration designs. *Chemosphere* **2021**, *267*, 129294. [CrossRef]
11. Caldelas, C.; Gurí, R.; Araus, J.; Sorolla, A. Effect of ZnO nanoparticles on Zn, Cu, and Pb dissolution in a green bioretention system for urban stormwater remediation. *Chemosphere* **2021**, *282*, 131045. [CrossRef] [PubMed]
12. Lange, K.; Österlund, H.; Viklander, M.; Blecken, G. Metal speciation in stormwater bioretention: Removal of particulate, colloidal and truly dissolved metals. *Sci. Total Environ.* **2020**, *724*, 138121. [CrossRef] [PubMed]
13. Boger, A.; Ahiablame, L.; Mosase, E.; Beck, D. Effectiveness of roadside vegetated filter strips and swales at treating roadway runoff: A tutorial review. *Environ. Sci. Water Res. Technol.* **2018**, *4*, 478–486. [CrossRef]
14. Stagge, J.H.; Davis, A.P.; Jamil, E.; Kim, H. Performance of grass swales for improving water quality from highway runoff. *Water Res.* **2012**, *46*, 6731–6742. [PubMed]
15. Yu, S.L.; Kuo, J.T.; Fassman, E.A.; Pan, H. Field test of grassed-swale performance in removing runoff pollution. *J. Water Resour. Plan. Manag.* **2001**, *127*, 168–171. [CrossRef]
16. Barrett, M.E. Performance comparison of structural stormwater best management practices. *Water Environ. Res.* **2005**, *77*, 78–86. [CrossRef] [PubMed]
17. Leroy, M.-C.; Portet-Koltalo, F.; Legras, M.; Lederf, F.; Moncond'huy, V.; Polaert, I.; Marcotte, S. Performance of vegetated swales for improving road runoff quality in a moderate traffic urban area. *Sci. Total Environ.* **2016**, *566–567*, 113–121. [CrossRef]
18. Public Utilities Board (PUB). *ABC Waters Design Guidelines*, 4th ed.; Public Utilities Board (PUB): Singapore, 2018.
19. Ong, G.S.; Kalyanaraman, G.; Wong, K.L.; Wong, T.H. Monitoring Singapore's first bioretention system: Rain garden at balam estate. In Proceedings of the WSUD 2012—7th International Conference on Water Sensitive Urban Design: Building the Water Sensitive Community, Melbourne, Australia, 21–23 February 2012.
20. Facility for Advancing Water Biofiltration (FAWB). *Adoption Guidelines for Stormwater Biofiltration Systems*; Monash University: Melbourne, Australia, 2008.
21. Goh, H.W.; Lem, K.S.; Azizan, N.A.; Chang, C.K.; Talei, A.; Leow, C.S.; Zakaria, N.A. A review of bioretention components and nutrient removal under different climates—future directions for tropics. *Environ. Sci. Pollut. Res.* **2019**, *26*, 14904–14919. [CrossRef]
22. Duncan, H. *Urban Stormwater Quality: A Statistical Overview*; Report 99/3; Cooperative Research Centre for Catchment Hydrology: Victoria, Australia, 1999.
23. Lintern, A.; McPhillips, L.; Winfrey, B.; Duncan, J.; Grady, C. Best Management Practices for Diffuse Nutrient Pollution: Wicked Problems Across Urban and Agricultural Watersheds. *Environ. Sci. Technol.* **2020**, *54*, 9159–9174. [CrossRef]
24. McNett, J.K.; Hunt, W.F.; Davis, A.P. Influent Pollutant Concentrations as Predictors of Effluent Pollutant Concentrations for Mid-Atlantic Bioretention. *J. Environ. Eng.* **2011**, *137*, 790–799. [CrossRef]
25. Batalini de Macedo, M.; Ambrogi Ferreira do Lago, C.; Mendiondo, E.; Giacomoni, M. Bioretention performance under different rainfall regimes in subtropical conditions: A case study in São Carlos, Brazil. *J. Environ. Manag.* **2019**, *248*, 109266. [CrossRef] [PubMed]
26. Hatt, B.E.; Fletcher, T.D.; Deletic, A. Hydraulic and Pollutant Removal Performance of Fine Media Stormwater Filtration Systems. *Environ. Sci. Technol.* **2008**, *42*, 2535–2541. [CrossRef] [PubMed]
27. Zhang, K.; Randelovic, A.; Page, D.; McCarthy, D.T.; Deletic, A. The validation of stormwater biofilters for micropollutant removal using in situ challenge tests. *Ecol. Eng.* **2014**, *67*, 1–10. [CrossRef]
28. Yau, W.K.; Radhakrishnan, M.; Liong, S.-Y.; Zevenbergen, C.; Pathirana, A. Effectiveness of ABC Waters Design Features for Runoff Quantity Control in Urban Singapore. *Water* **2017**, *9*, 577. [CrossRef]
29. eWater CRC. *Model for Urban Stormwater Improvement Conceptualisation (MUSIC Version 6.0)*; eWater Cooperative Research Centre: Canberra, Australia, 2014.
30. Rice, E.W.; Baird, R.B.; Eaton, A.D.; Clesceri, L.S. *Standard Methods for the Examination of Water and Wastewater*; American Public Health Association (APHA): Washington, DC, USA, 2012.
31. Dombeck, G.D.; Perry, M.W.; Phinney, J.T. Mass balance on water column trace metals in a free-surface-flow-constructed wetlands in Sacramento, California. *Ecol. Eng.* **1998**, *10*, 313–339. [CrossRef]
32. Maniquiz, M.C.; Choi, J.-Y.; Lee, S.-Y.; Cho, H.-J.; Kim, L.-H. Appropriate Methods in Determining the Event Mean Concentration and Pollutant Removal Efficiency of a Best Management Practice. *Environ. Eng. Res.* **2010**, *15*, 215–223. [CrossRef]
33. Lim, H.S. Variations in the water quality of a small urban tropical catchment: Implications for load estimation and water quality monitoring. In *The Interactions between Sediments and Water*; Springer: Dordrecht, The Netherlands, 2003.
34. Bertrand-Krajewski, J.-L.; Chebbo, G.; Saget, A. Distribution of pollutant mass vs volume in stormwater discharges and the first flush phenomenon. *Water Res.* **1998**, *32*, 2341–2356. [CrossRef]
35. Geiger, W.F. Flushing effects in combined sewer systems. In Proceedings of the 4th International Conference Urban Drainage, Lausanne, Switzerland, 31 August–4 September 1987.
36. Wang, J.; Chua, L.H.C.; Shanahan, P. Evaluation of pollutant removal efficiency of a bioretention basin and implications for stormwater management in tropical cities. *Environ. Sci. Water Res. Technol.* **2017**, *3*, 78–91. [CrossRef]

37. Lim, M.H.; Leong, Y.H.; Tiew, K.N.; Seah, H. Urban stormwater harvesting: A valuable water resource of Singapore. *Water Pr. Technol.* **2011**, *6*, 1–2. [CrossRef]
38. Lim, F.; Neo, T.; Guo, H.; Goh, S.; Ong, S.; Hu, J.; Lee, B.; Ong, G.; Liou, C. Pilot and Field Studies of Modular Bioretention Tree System with *Talipariti tiliaceum* and Engineered Soil Filter Media in the Tropics. *Water* **2021**, *13*, 1817. [CrossRef]
39. Davis, A.P. Field performance of bioretention: Water quality. *Environ. Eng. Sci.* **2007**, *24*, 1048–1064. [CrossRef]
40. Hunt, W.F.; Smith, J.T.; Jadlocki, S.J.; Hathaway, J.M.; Eubanks, P.R. Pollutant Removal and Peak Flow Mitigation by a Bioretention Cell in Urban Charlotte, N.C. *J. Environ. Eng.* **2008**, *134*, 403–408. [CrossRef]
41. Brown, R.A.; Hunt, W.F., III. Impacts of Media Depth on Effluent Water Quality and Hydrologic Performance of Undersized Bioretention Cells. *J. Irrig. Drain. Eng.* **2011**, *137*, 132–143. [CrossRef]
42. Davis, A.P.; Shokouhian, M.; Sharma, H.; Minami, C. Water Quality Improvement through Bioretention Media: Nitrogen and Phosphorus Removal. *Water Environ. Res.* **2006**, *78*, 284–293. [CrossRef]
43. Biswal, B.; Vijayaraghavan, K.; Adam, M.; Lee Tsen-Tieng, D.; Davis, A.; Balasubramanian, R. Biological nitrogen removal from stormwater in bioretention cells: A critical review. *Crit. Rev. Biotechnol.* **2021**, 1–23. [CrossRef]
44. Zhang, H.; Ahmad, Z.; Shao, Y.; Yang, Z.; Jia, Y.; Zhong, H. Bioretention for removal of nitrogen: Processes, operational conditions, and strategies for improvement. *Environ. Sci. Pollut. Res.* **2021**, *28*, 10519–10535. [CrossRef]
45. Chui, T.F.M.; Trinh, D.H. Modelling infiltration enhancement in a tropical urban catchment for improved stormwater management. *Hydrol. Process.* **2016**, *30*, 4405–4419. [CrossRef]
46. Fardel, A.; Peyneau, P.E.; Béchet, B.; Lakel, A.; Rodriguez, F. Analysis of swale factors implicated in pollutant removal efficiency using a swale database. *Environ. Sci. Pollut. Res.* **2019**, *26*, 1287–1302. [CrossRef]
47. Deletic, A.; Fletcher, T.D. Performance of grass filters used for stormwater treatment—A field and modelling study. *J. Hydrol.* **2006**, *317*, 261–275. [CrossRef]
48. Bäckström, M. Sediment transport in grassed swales during simulated runoff events. *Water Sci. Technol.* **2002**, *45*, 41–49. [CrossRef]
49. Ekka, S.; Rujner, H.; Leonhardt, G.; Blecken, G.; Viklander, M.; Hunt, W. Next generation swale design for stormwater runoff treatment: A comprehensive approach. *J. Environ. Manag.* **2021**, *279*, 111756. [CrossRef]
50. Lucke, T.; Nichols, P.W.B. The pollution removal and stormwater reduction performance of street-side bioretention basins after ten years in operation. *Sci. Total Environ.* **2015**, *536*, 784–792. [CrossRef]
51. Le Coustumer, S.; Fletcher, T.D.; Deletic, A.; Barraud, S.; Lewis, J.F. Hydraulic performance of biofilter systems for stormwater management: Influences of design and operation. *J. Hydrol.* **2009**, *376*, 16–23. [CrossRef]
52. Li, H.; Davis, A.P. Heavy metal capture and accumulation in bioretention media. *Environ. Sci. Technol.* **2008**, *42*, 5247–5253. [CrossRef]
53. Wang, F.; Wang, H.; Sun, C.; Yan, Z. Conventional bioretention column with Fe-hydrochar for stormwater treatment: Nitrogen removal, nitrogen behaviour and microbial community analysis. *Bioresour. Technol.* **2021**, *334*, 125252. [CrossRef]
54. Wang, M.; Zhang, D.; Li, Y.; Hou, Q.; Yu, Y.; Qi, J.; Fu, W.; Dong, J.; Cheng, Y. Effect of a Submerged Zone and Carbon Source on Nutrient and Metal Removal for Stormwater by Bioretention Cells. *Water* **2018**, *10*, 1629. [CrossRef]
55. Guo, H.; Lim, F.Y.; Zhang, Y.; Lee, L.Y.; Hu, J.Y.; Ong, S.L.; Yau, W.K.; Ong, G.S. Soil column studies on the performance evaluation of engineered soil mixes for bioretention systems. *Desalination Water Treat.* **2015**, *54*, 3661–3667. [CrossRef]
56. Goh, H.W.; Zakaria, N.A.; Lau, T.L.; Foo, K.Y.; Chang, C.K.; Leow, C.S. Mesocosm study of enhanced bioretention media in treating nutrient rich stormwater for mixed development area. *Urban Water* **2017**, *14*, 134–142. [CrossRef]
57. Tirpak, R.A.; Afrooz, A.N.; Winston, R.J.; Valenca, R.; Schiff, K.; Mohanty, S.K. Conventional and amended bioretention soil media for targeted pollutant treatment: A critical review to guide the state of the practice. *Water Res.* **2021**, *189*, 116648. [CrossRef]
58. Bratieres, K.; Fletcher, T.D.; Deletic, A.; Zinger, Y. Nutrient and sediment removal by stormwater biofilters: A large-scale design optimisation study. *Water Res.* **2008**, *42*, 3930–3940. [CrossRef] [PubMed]
59. Skorobogatov, A.; He, J.; Chu, A.; Valeo, C.; van Duin, B. The impact of media, plants and their interactions on bioretention performance: A review. *Sci. Total Environ.* **2020**, *715*, 136918. [CrossRef] [PubMed]
60. Monash University Water for Liveability Centre. Vegetation Guidelines for Stormwter Biofilters in the South-West of Western Australia. Available online: https://watersensitivecities.org.au/wp-content/uploads/2016/07/381_Biofilter_vegetation_guidelines_for_southwestWA.pdf (accessed on 13 December 2021).
61. Loh, B. *A Selection of Plants for Bioretention Systems in the Tropics*; Research Technical Note of Centre for Urban Greenery and Ecology: Singapore, 2012.
62. Bernhardt, E.S.; Palmer, M.A. Restoring streams in an urbanizing world. *Freshw. Biol.* **2007**, *52*, 738–751. [CrossRef]
63. Hunt, W.; Davis, A.; Traver, R. Meeting Hydrologic and Water Quality Goals through Targeted Bioretention Design. *J. Environ. Eng.* **2012**, *138*, 698–707. [CrossRef]
64. Line, D.; Jennings, G.; Shaffer, M.; Calabria, J.; Hunt, W. Evaluating the Effectiveness of Two Stormwater Wetlands in North Carolina. *Trans. ASABE* **2008**, *51*, 521–528. [CrossRef]
65. Winston, R.; Powell, J.; Hunt, W. Retrofitting a grass swale with rock check dams: Hydrologic impacts. *Urban Water J.* **2018**, *16*, 404–411. [CrossRef]
66. USEPA. Storm Water Technology Fact Sheet Bioretention. *Off. Water* **1999**, *12*, 1–8.
67. Gao, J.; Wang, R.; Huang, J.; Liu, M. Application of BMP to urban runoff control using SUSTAIN model: Case study in an industrial area. *Ecol. Model.* **2015**, *318*, 177–183. [CrossRef]

68. Deng, S.; Chen, J.; Chang, J. Application of biochar as an innovative substrate in constructed wetlands/biofilters for wastewater treatment: Performance and ecological benefits. *J. Clean. Prod.* **2021**, *293*, 126156. [CrossRef]
69. Xu, T.; Yu, J.; Cai, D.; You, Z.; Shah, K.J. Removal of Cd (II) Ions from Bioretention System by Clay and Soil Wettability. *Water* **2021**, *13*, 3164. [CrossRef]
70. Mehmood, T.; Lu, J.; Liu, C.; Caurav, G.K. Organics removal and microbial interaction attributes of zeolite and ceramsite assisted bioretention system in copper-contaminated stormwater treatment. *J. Env. Manage.* **2021**, *292*, 112654. [CrossRef]
71. Xiong, J.; Li, G.; Zhu, J.; Li, J.; Yang, Y.; An, S.; Liu, C. Removal characteristics of heavy metal ions in rainwater runoff by bioretention cell modified with biochar. *Environ. Technol.* **2021**, 1–13. [CrossRef] [PubMed]

Article

Flood Control and Aquifer Recharge Effects of Sponge City: A Case Study in North China

Bo Meng [1,2], Mingjie Li [1,2], Xinqiang Du [1,2] and Xueyan Ye [1,2,*]

[1] Laboratory of Groundwater Resources and Environment, Ministry of Education, Jilin University, Changchun 130021, China; mengbo19@mails.jlu.edu.cn (B.M.); lh141120@163.com (M.L.); duxq@jlu.edu.cn (X.D.)
[2] College of New Energy and Environment, Jilin University, Changchun 130021, China
* Correspondence: Yexy@jlu.edu.cn

Abstract: Sponge City is an integrated urban stormwater management approach and practice to tackle waterlogging, flooding, water scarcity, and their related problems. Despite many positive effects of Sponge City on flood control that have been investigated and revealed, the effect on aquifer recharge is still less known. Considering maximizing the function of natural elements such as surface water bodies and subsurface storage space, to minimize the use of a gray drainage system, a Sponge City design was proposed to substitute the planning development scheme in the study area. The stormwater management model of SWMM (storm water management model) and the groundwater flow model of MODFlow (Modular Three-dimensional Finite-difference Groundwater Flow Model) were adopted to evaluate the flood-control effect and aquifer-recharge effect, respectively. Compared with the traditional planning scenario, the peak runoff is approximately 92% less than that under the traditional planning scenario under the condition of a 5-year return period. Due to the increase in impervious areas of urban construction, the total aquifer recharge from precipitation and surface water bodies was decreased both in the present planning scenario and the Sponge City design scenario. However, the Sponge City design has a positive impact on maintaining groundwater level stabilization and even raises the groundwater level in some specific areas where stormwater seepage infrastructure is located.

Keywords: Sponge City; aquifer recharge; urban stormwater; green infrastructure

Citation: Meng, B.; Li, M.; Du, X.; Ye, X. Flood Control and Aquifer Recharge Effects of Sponge City: A Case Study in North China. *Water* **2022**, *14*, 92. https://doi.org/10.3390/w14010092

Academic Editors: Haifeng Jia, Jiangyong Hu, Tianyin Huang, Albert S. Chen and Yukun Ma

Received: 15 November 2021
Accepted: 29 December 2021
Published: 4 January 2022

Publisher's Note: MDPI stays neutral with regard to jurisdictional claims in published maps and institutional affiliations.

Copyright: © 2022 by the authors. Licensee MDPI, Basel, Switzerland. This article is an open access article distributed under the terms and conditions of the Creative Commons Attribution (CC BY) license (https://creativecommons.org/licenses/by/4.0/).

1. Introduction

With rapid urbanization and global climate change, urban flooding has become a major issue in China [1–3]. The government of China promoted the application of the "Sponge City" approach to reduce urban flood risk and improve the environment in these cities [4]. The concept of "Sponge City" is closely related to stormwater management strategies and practices, such as the "Sustainable Draining System" of the United Kingdom (UK), "Best Management Practice" and "Low Impact Development" from the United States and New Zealand, and "Water Sensitive City" from Australia [2,5,6]. In addition, the idea of "Sponge City" is consistent with the strategies of flood mitigation through natural and ecological approaches, e.g., to achieve flood control and aquifer recharge by using riparian woodland planting along the riverbed in the floodplain of the UK [7]. The Sponge City concept aims to: (i) control urban peak runoff and to temporarily store, recycle and purify stormwater; (ii) to upgrade the drainage systems using more green infrastructures and (iii) to integrate natural water bodies and encourage multi-functional objectives within drainage design [2]. Thus, Sponge City is an integrated urban stormwater-management approach with multiple purposes of ecological, architectural, hydrologic, hydrogeological, and economic aspects [5,8,9]. In October 2014, following the publication of "Sponge City Construction Technology Guidelines", the Ministries of Finance, Housing and Urban–Rural Development, and Water Resources, collectively initiated a Sponge City pilot program [10]. With the strong national policy push and huge financial investment, Sponge City construction has been widely carried out in China. There are 30 "Sponge City" pilot sites that have been developed since 2015 [11], and it was planned to refit 80% of urban areas in China by 2030 to absorb and reuse at least 70% rainwater. However, the challenges and gaps in technology, finance, management, and even the idea still exist in current urban Sponge City practices [12–17], and a more comprehensive understanding and a collaborative work platform from various professionals are needed eagerly.

Like other integrated stormwater-management approaches, such as LID (Low Impact Development) and BMPs (Best Management Practices) [18,19], it is also expected that the adoption and implementation of Sponge City will provide a large opportunity to recharge groundwater in urban aquifers [17]. Urban aquifer recharge can alleviate the pressure on sewers, urban streams, and wastewater treatment plants [20]. Hydrogeology conditions and groundwater resources in urban areas were thought of as the key influencing factors for Sponge City implementation [5,21–23]. Xu et al. (2018) [12] recommended that the dispersed ponds and the ditches made of durable and permeable materials in the drainage system should be adopted in Sponge City construction to increase the amount of infiltration, storage, and utilization of rainwater. Kang et al. (2019) [24] summarized the quantification methods of groundwater recharge in Sponge City construction and introduced the basic principles and application conditions of each method. Jin et al. (2021) [20] stated that the urban aquifers should be involved in the Sponge City approach and proposed hydrogeological criteria to improve the strategy. Sun et al. (2020) [25] discussed the influences of Sponge City construction on spring discharge in Jinan City of China, and the result indicated that the extent of spring discharge recovery was not evident in a short time frame based on a numerical model of groundwater flow.

However, compared with many pieces of research of Sponge Cities' effect on control runoff and mitigation of flooding [26–30], less is known regarding their effects on aquifer recharge. Based on a case of Sponge City in North China, the flood control and aquifer recharge effects were analyzed, and positive effects were expected under the scientific designed Sponge City schemes.

2. Materials and Methods

2.1. Study Area

The study area is a planning urban construction area in Zhengzhou City, the capital of Henan Province in North China, with a total area of 91 km^2. The study area has a warm and semi-arid climate, with obvious continental monsoon climate characteristics. The average annual precipitation is 631.76 mm, and the annual precipitation is mostly concentrated between July and September.

The study area belongs to the piedmont alluvial–proluvial plain (Figure 1). Since the Cenozoic era, Neogene and Quaternary strata have been successively deposited in the study area, with a thickness of more than 800 m. The surface of the study area is widely covered with Holocene light-yellow silt with a sedimentary thickness of 2.1~26.0 m (Figure 2). The aquifer is mainly composed of Quaternary Holocene, Upper Pleistocene, and Middle Pleistocene alluvium. The lithology is mainly fine sand and fine silty sand, and the groundwater type is phreatic water and weakly confined water. The groundwater depth in the area is generally more than 5 m, and groundwater evaporation is weak. The rivers in the region are seasonal tributaries with low flow and low permeability of river bed. Thus, the natural recharge of rivers to groundwater is little, which has very weak influence on the groundwater flow field in most of a year (Figure 2).

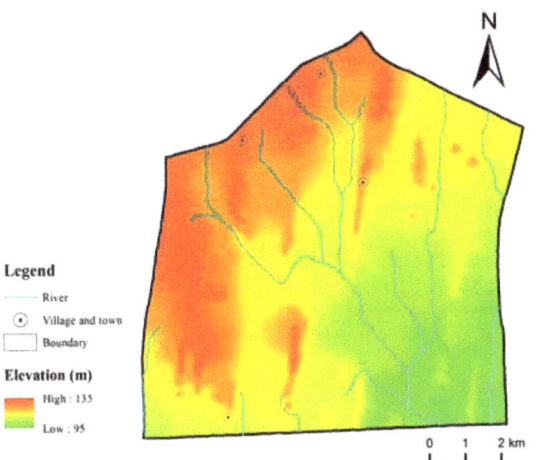

Figure 1. Sketch map of the study area.

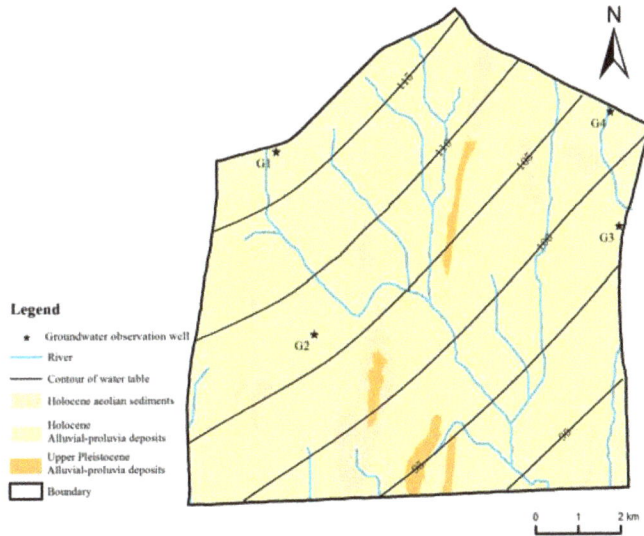

Figure 2. Geology and groundwater flow field.

An advanced manufacturing district is planned in the study area, and the land-use types under the original planning scenario are shown in Figure 3. The design stormwater intensity was synthesized using the local rainstorm intensity formula (Equation (1)) [31]:

$$q = \frac{2387(1 + 0.257 \lg P)}{(t + 10.605)^{0.792}}. \tag{1}$$

where q represents rainfall intensity (L/(s·ha)), P represents the return period (year), and t represents rainfall duration (minute). Since the Chicago storm profile can be easily adapted to China [32], it was used to develop design rainfall hyetographs for this study. Four rainfall return periods of 0.5, 2, 5, and 10 years were selected, and the rainfall lasted for 3 h. The single rainfall is 37.286, 43.531, 47.659, and 50.784 mm under 0.5-, 2-, 5-, and 10-year return periods, respectively (Figure 4).

Figure 3. The originally planned land-use types.

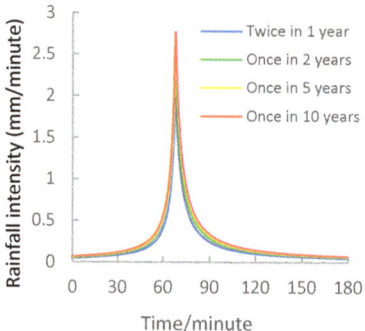

Figure 4. Typical rainfall intensity curve.

2.2. Methods

To compare the flood-control and aquifer-recharge effects of Sponge City and the traditional city development, the Sponge City design scheme was firstly proposed by using the following principles: based on urban construction planning, retain the natural river network, plan LID facilities based on geological and hydrogeological conditions, and minimize the use of gray urban drainage system. This is a coupled Sponge City idea with natural and artificial elements to solve urban stormwater management.

2.2.1. SWMM

The Storm Water Management Model (SWMM) is a widely adopted dynamic hydrologic and hydraulic model often used to estimate runoff quantity and quality in urban drainage systems [33,34]. SWMM was adopted to simulate urban stormwater processes in study area under the present, the original planning, and the Sponge City design scenarios.

The natural catchment area divided by ArcGIS software (Version10.2) is used as the model sub-catchment area, the natural river channel is used as the drainage pipeline, and the drainage nodes are arranged on the river channels. The impermeability rate of each sub-catchment area was calculated (Equation (2)). Under the present scenario, the impermeability rate of sub-catchments is between 5% and 20%, and the average impermeability rate of the study area is 10.3%. Under the planning scenario, the maximum impermeability rate of the sub-catchment is 90%, the minimum is 10%, and the average is 56.54%. Under the design scenario, the land use of each sub-catchment area is based on the planning conditions; the maximum impermeability rate of sub-catchment is 81%, the minimum is 10%, and the average is 56.38%. The infiltration process is simulated by Horton model, the surface runoff is calculated by Manning's formula, and the hydraulic model of discharge routing of drainage system adopts kinematic wave model. The related main hydrological parameters are set with the reference value ranges and the actual conditions (Table 1) [35]:

$$R = (\sum A_j C_j)/A \qquad (2)$$

where R represents the impermeability rate of the sub-catchment area; A_j represents the area of land-use type j in the sub-catchment area; C_j is the impermeability rate corresponding to the land-use type; and A is the total area of the sub-catchment area.

Table 1. The hydrologic parameters.

Manning Roughness Coefficient			Depression Storage/mm		Coefficients for Horton Formula		
Impervious Area	Permeable Area	River Channels	Impervious Area	Permeable Area	Infiltration Rate/mm·h^{-1}		Attenuation Coefficient/h^{-1}
					Maximum	Minimum	
0.015	0.20	0.02	2.80	5.10	14.67–193.22	1.22–98.44	3–4

Parameter calibration method for urban rainfall-runoff model based on runoff coefficient is used to complete the calibration and verification of the model [36]. Taking the runoff coefficient as the objective function of model parameter calibration, compare the values of urban empirical comprehensive runoff coefficient and the runoff coefficient calculated by simulation (Table 2), the parameters are calibrated by using the design rainfall process with a return period of two years, and verified by using the design rainstorm process with a return period of ten years.

Table 2. Empirical value of runoff coefficient [37].

Development Intensity	Proportion of Impervious Area	Empirical Runoff Coefficient
Upper-middle	>70%	0.6~0.8
Middle	50~70%	0.5~0.7
Low-middle	30~50%	0.4~0.6
Low	<30%	0.3~0.5

2.2.2. MODFlow

The groundwater flow model of the study area using Visual MODFLOW is established to evaluate the recharge effect of the Sponge City scheme on the groundwater resources.

The groundwater numerical model domain is consistent with the study area, with a total area of 91 km^2. According to the hydrogeological conditions, the aquifer is divided into two layers; the 1st layer is a shallow unconfined aquifer with a thickness of 80~100 m, and the 2nd layer is a deep, confined aquifer with a thickness of about 220 m. The lithology of aquifers is mainly interbedded by fine sand, silty sand, and silty clay. The aquifers are generalized as heterogeneous and isotropy, and the hydraulic conductivity and specific yield of the 1st layer are 0.6 m/day and 0.05, respectively; the hydraulic conductivity and specific storage of the 2nd layer are 0.3 m/d and 1×10^{-5}/m. There is a certain amount of water exchange inside and outside the boundary of the study area, which is generalized as the general head boundary. The groundwater is mainly recharged by precipitation. The bottom of confined water aquifer is the lower boundary of the model and generalized as no-flow boundary. The characteristics of groundwater movement are generalized as three-dimensional transient flow in accordance with Darcy's law. According to the existing groundwater level data, the model calibration period is from 5 July 2014, to 5 January 2015, and the verification period is from 5 February 2015, to 5 June 2015.

3. Results

3.1. Sponge City Design Scheme

The Sponge City comprised three parts of the natural hydrographic network, LID facilities, and a drainage system. This scheme aims to maximize the function of the natural hydrographic network and minimize the use of the gray urban drainage system.

3.1.1. Natural Hydrographic Network Design

- River

In actual urban construction, because of the overall design of the city, the original river course is often changed or straightened. Its drainage effect and safety are far less than the natural river channel formed by the natural evolution of rainwater runoff. Based on the DEM (Digital levation Model) data generated through ArcGIS, the depression filling, flow direction analysis, and flow calculation were carried out. Then, according to the accumulated water volume of each grid, the river network is divided into four grades: IV—main stream with large flow, III—main stream, II—primary tributary, and I—secondary tributary. River Sections I-5, I-8, and I-13 need to accept the water quantity outside the study area. I-12 and I-13 are connected outside the area. I-1, II-5, I-7, and I-8 are not designed as river channels but are still reserved as drainage ditches because of the small river flow of the original water system. The main stream III-1 is divided into three catchment areas (upper, middle, and lower) due to its long length. The final designed river channels and their catchment areas are shown in Figure 5.

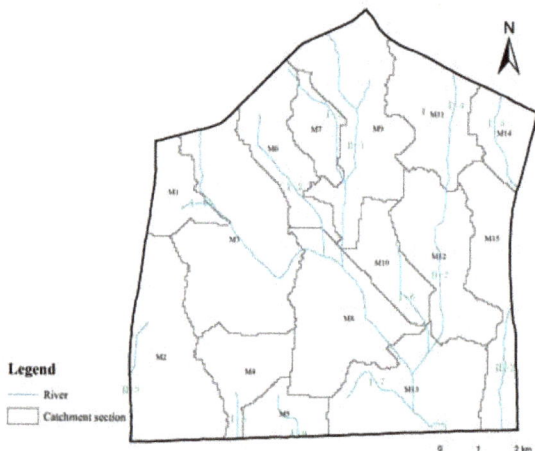

Figure 5. The river and its corresponding catchment area.

The blue line of the river is the boundary of the protection scope of river engineering areas, which include rivers, sandbars, beaches, and areas reserved for river widening, regulation, ecological landscape, greening, and other purposes. The design width of the blue line is set to 60 m for the main stream, 50 m for the primary tributary, and 30 m for the secondary tributary (Figure 6, Table 3).

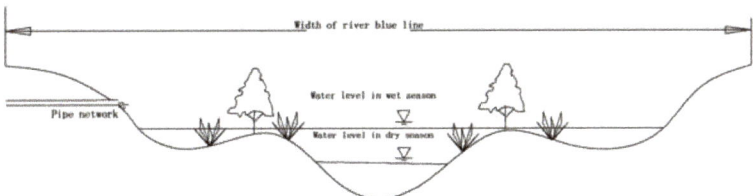

Figure 6. Schematic diagram of the river profile.

Table 3. Parameters of river sections.

River Segment	Length /km	Catchment Area /km²	Rate of Flow /m³s⁻¹	Width of River Blue Line /m
III-1, IV	13.33	30.12	150.23	60
II-1	7.37	9.68	57.42	50
II-2, I-4	7.57	12.38	69.61	50
I-1	0.49	2.81	9.45	30
I-2	4.32	4.91	29.96	30
I-3	3.6	2.53	16.91	30
I-6	3.09	3.99	23.67	30
III-2, I-5	6.31	9.15	51.45	50
II-3	3.22	8.96	41.92	30
I-8	1.22	4.11	17.78	30
I-9	1.42	2.51	13.75	30

- Lake and water corridor

The construction of a small water reservoir has been proved to be an effective and significant measure in water management, especially for runoff control [38,39]. Considering the river network, the location where the tributary flows into the higher-grade river has larger runoff, which is the key point of runoff control. Meanwhile, III-1 is the main drainage channel of the entire area, with a large drainage pressure. Therefore, two small artificial lakes were set before the tributaries flowed into III-1, and the urban water corridor was set at the starting point where I-2 and II-2 flow into III-1. The locations of the lakes and water corridors are shown in Figure 7.

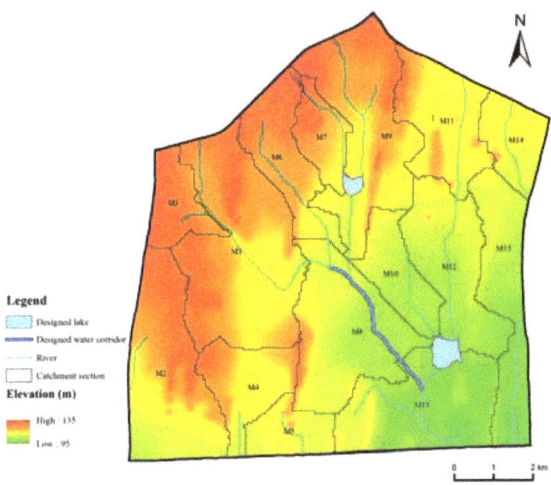

Figure 7. Locations of designed lakes and water corridor.

The areas of the two lakes are bounded by the topographic contour lines as 0.20 km² and 0.46 km². According to previous research, when the width of the water corridor is 80–100 m, the loss of sediment and the flood can be better controlled [40]. Combined with the existing design experience of a plain area [41], the average width of the designed water corridor is 90 m, the length is 4.0 km, and the area is 0.36 km².

3.1.2. LID Facility Design

The selection of LID facilities is related with land use type, local terrain and topography, average annual rainfall, total annual rainfall volume control rate and urban development intensity, etc. [42]. Based on the geological and hydrogeological conditions of the study area, combined with land-use types, the water storage capacity of facilities required is calculated with the goal of controlling 70% of the 3-h rainfall once in five years.

The main LID facilities used are rainwater gardens, permeable pavement, and green roofs. The green roof is suitable for buildings with roof load, waterproof, the height less than 30m and roof slope less than 10°, and the buildings in the study area will be processed to meet the requirements of green roof construction. The scale of the facilities in each sub-region is shown in Table 4.

Table 4. LID facilities' parameters of water catchments.

	Area /km²	Proportion of Land Use Types/%				Green Space /km²	Permeable Pavement /km²	Green Roof /km²	Rainwater Garden /km²	Water Storage Depth /m
		Industrial	Residential	Commercial	Greenspace					
M1	2.81	1.1	0.7	0.0	98.1	2.35	0.23	0.02	0.47	0.2
M2	8.96	61.3	0.0	0.0	38.7	3.22	0.28	2.20	0.64	0.3
M3	12.87	22.8	35.7	5.4	36.1	4.79	1.56	2.98	0.96	0.2
M4	4.11	46.0	24.3	0.0	29.7	1.28	0.34	1.08	0.26	0.3
M5	2.51	17.0	49.4	0.0	33.6	0.92	0.37	0.57	0.18	0.2
M6	4.91	6.3	47.2	1.2	45.3	2.25	0.74	0.89	0.45	0.2
M7	2.53	0.0	64.8	0.0	35.2	1.00	0.46	0.52	0.20	0.2
M8	8.76	27.4	16.0	14.2	42.4	4.00	0.78	2.01	0.80	0.2
M9	9.68	3.3	39.9	7.9	48.9	4.62	1.40	1.73	0.92	0.2
M10	3.51	69.7	15.1	1.5	13.7	0.61	0.17	1.17	0.12	0.3
M11	5.88	0.0	56.6	0.0	43.4	2.67	1.00	1.06	0.53	0.2
M12	6.98	15.6	40.2	20.6	23.6	1.88	0.98	2.02	0.38	0.3
M13	8.5	35.6	26.6	0.0	37.8	2.73	0.80	1.93	0.55	0.3
M14	2.98	0.0	41.9	0.0	58.1	1.66	0.44	0.40	0.33	0.2
M15	6.16	54.2	0.0	4.5	41.3	2.33	0.24	1.47	0.47	0.3

3.1.3. Drainage System Design

Part of the rainfall can be stored or infiltrated to recharge groundwater in time through surface water and LID facilities. However, there are still areas where the building density is too high, so it is difficult to build large-scale green rainwater storage and utilization facilities. It is still necessary to design an urban rainwater pipe network to ensure the discharge of rainwater runoff in these areas. The calculated rainfall flood volume of each area is less than 30% after treatment by LID facilities. However, to ensure the safety of urban drainage and improve the drainage capacity, the rainfall-control capacity of the rainwater network is 40% of the total rainfall.

The catchment area of the river channel is the most suitable drainage area. The entire area is divided into 15 drainage areas, which are the same as the catchment areas. The layout of the rainwater pipe network is based on the urban road, combined with the location of the river channel, and arranged beside the main urban road. The rainwater in each area was discharged into the rivers through the drainage pipe network (Figure 8).

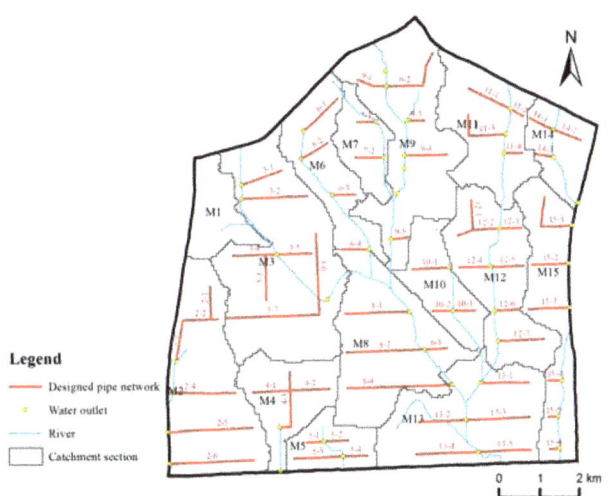

Figure 8. Layout of rainwater pipe network.

3.1.4. Sponge City Scheme

The design of rivers and lakes, LID facilities, and pipe networks are superimposed to form the overall scheme of stormwater control and utilization. In the rainfall process, LID facilities are used to store the rainwater as the bridge between rain and underground space. In the storage process, some rainwater can infiltrate locally to supplement groundwater. The rainwater that cannot be stored by LID facilities will be discharged into the urban rainwater pipe network. This ensures the safety of the city under severe stormwater conditions. Finally, the rainwater discharged by pipe network drains into rivers and lakes, which are the major drainage channels accepting rainfall and water from the urban drainage system. The comprehensive Sponge City design is shown in Figure 9.

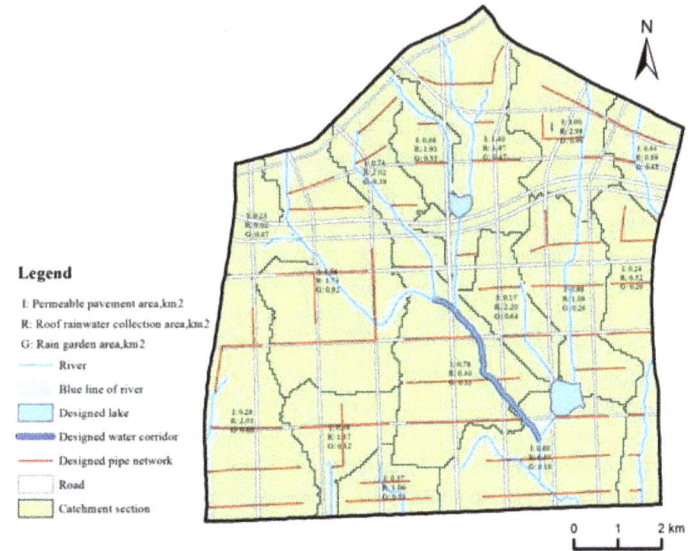

Figure 9. Sponge City scheme.

3.2. Stormwater Control Effect

SWMM is used to evaluate the stormwater-runoff-control ability. In the scheme of stormwater control and utilization, river channels are the major drainage channels of rainwater, and the outlets of the pipe network are free of river channels. According to the design scheme, the study area is generalized into 121 sub-catchment areas, 115 pipe sections, 115 nodes, and 63 outlets, as shown in Figure 10.

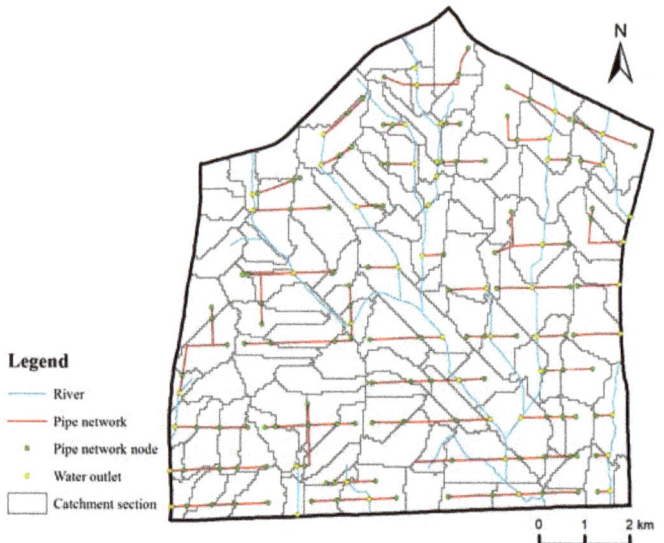

Figure 10. Generalized model of surface runoff under Sponge City design scenario.

The mean comprehensive runoff coefficient of different development intensity areas (Table 5) simulated by the model is within the empirical value range of urban comprehensive runoff coefficient (Table 2). Therefore, the model could reasonably simulate the surface runoff in the study area.

Table 5. Simulated mean comprehensive runoff coefficient.

Development Intensity	Upper-Middle	Middle	Lower-Middle	Low
Simulated runoff coefficient	0.742	0.638	0.441	0.325

Under the Sponge City design scenario, the land-use situation of each sub-catchment area is based on the original planning scenario, the maximum impervious rate is 81%, the minimum is 10%, and the average is 56.38%.

Under the five-year rainstorm condition, the maximum ponding time under the planning and design scenarios at each node is shown in Figure 11. It can be seen that under the planning scenario, the area most prone to ponding in the study area is mainly located in the plain area in the southeast. Due to the increase in the impervious area caused by urban construction in these areas, it is difficult for rainwater to infiltrate over time. Under the design conditions, there are almost no ponding points in the whole study area. Even at the maximum ponding time, there are only seven ponding nodes in the whole area, the maximum ponding is 0.87 m^3/s, and the ponding risk is far lower than that in the design scenario.

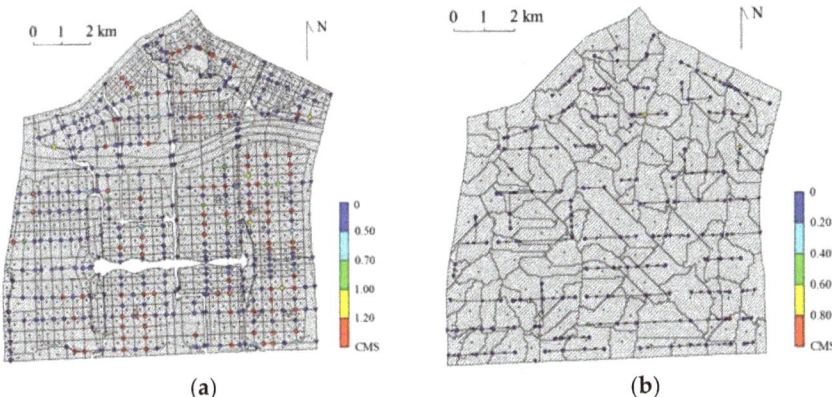

Figure 11. Node ponding circumstances of a rainstorm once in five years under planning and Sponge city scenarios. (**a**) Planning scenario. (**b**) Sponge city design scenario.

A comparison of the runoff simulation results for the planning and Sponge City design scenarios is shown in Table 6, and the runoff process under the planning and the Sponge City design scenarios for each rainfall return period is shown in Figure 12. Compared with the planning scenario, the runoff under the proposed Sponge City design scheme is significantly reduced, but the reduction shows a decreasing trend with an increase in the rainfall return period. The runoff coefficient is significantly reduced under all rainfall return periods, including the 10-year return period rainfall condition; the runoff coefficient is reduced below 0.3, and the shorter the return period, the smaller the runoff coefficient. The results show that the proposed Sponge City stormwater control and utilization scheme has an obvious effect on runoff control, which can meet the requirement of 70% rainfall utilization.

Table 6. Runoff simulation results under different rainfall return periods.

Return Period /year	Rainfall /mm	Scenario	Runoff Amount /mm	Runoff Coefficient	Runoff Reduction Rate
0.5	37.286	planning	16.673	0.447	65.5%
		Sponge City	5.762	0.154	
2	43.53	planning	20.111	0.462	59.5%
		Sponge City	8.137	0.187	
5	47.659	planning	22.456	0.471	56.3%
		Sponge City	9.848	0.206	
10	50.784	planning	24.268	0.478	54.6%
		Sponge City	11.209	0.221	

According to the runoff process curves (Figure 12), compared with the planning scenario, the peak discharge under the Sponge City design conditions was significantly reduced, and the reduction proportion decreased with an increase in the return period. Under the Sponge City design condition of 5-year return periods, the peak runoff is 75.28 m^3/s, which is approximately 92% less than that under the planning scenario, indicating that the Sponge City scheme has an obvious effect on reducing the peak runoff. Compared with the planning scenario, the peak time under the Sponge City design scenario is also significantly delayed, and the delay time decreases with the increase in the rainfall period. The delay time is 48 min under the condition of a 5-year return period rainfall, which indicates that the sensitivity of the runoff response to rainfall is significantly reduced under the Sponge City design scheme.

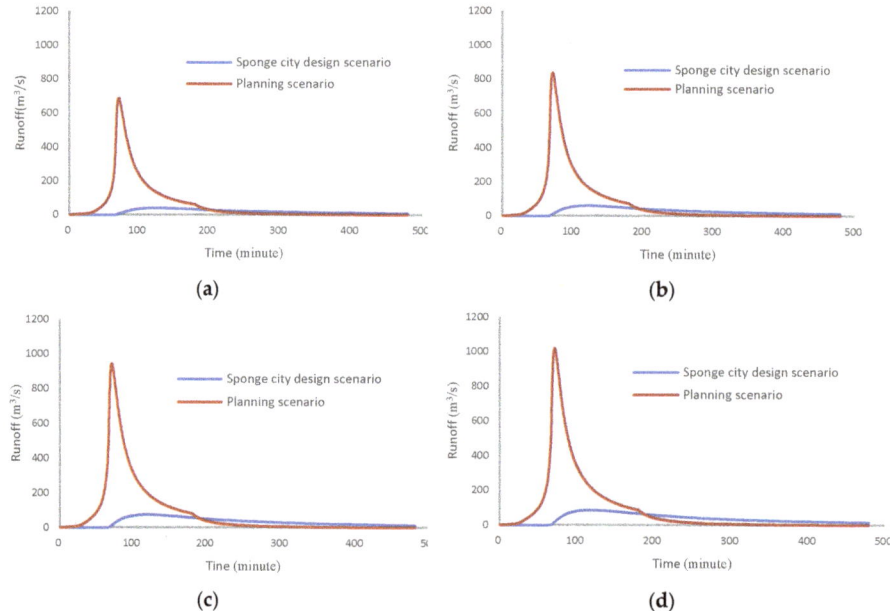

Figure 12. Runoff process curves under different rainfall return periods. (**a**) Twice in 1 year. (**b**) Once in 2 years. (**c**) Once in 5 years. (**d**) Once in 10 years.

3.3. Groundwater-Recharge Effect

The purpose of the stormwater runoff control and utilization system is not only to control surface runoff but also to restore the supplement of precipitation to groundwater under natural conditions. Visual MODFLOW is used to simulate the impact of stormwater-runoff-control and utilization schemes on the groundwater regime.

The groundwater mathematical model established in the study area has good water level fitting in the calibration and verification periods (as shown in Figure 13). The residual mean error is −0.294 m, and the absolute residual mean is 0.541 m in the calibration period. The residual mean is −0.191 m, and the absolute residual mean is 0.498 m in the verification period.

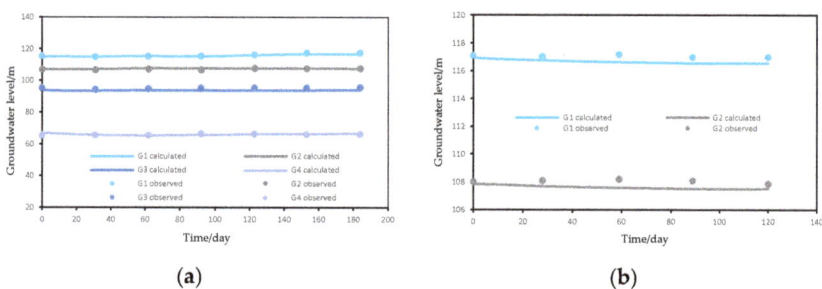

Figure 13. Groundwater level fitting diagram during calibration and verification period. (**a**) Calibration period. (**b**) Verification period.

The change in groundwater level in the study area in the present scenario, planning scenario, and Sponge City design scenario in the next 10 years is predicted (Table 7).

Table 7. Settings of model scenarios.

Condition Setting	Present Scenario	Planning Scenario	Sponge City Design Scenario
Rainfall, evaporation		Annual average	
Groundwater exploitation intensity		Present exploitation intensity	
Proprotion of impermeable area (%)	10.3	56.54	41.61

Modeling results of the three schemes (Figure 14) show that in the present scenario, the groundwater level of each observation well is increased, mainly because the average annual rainfall is greater than that of the current year.

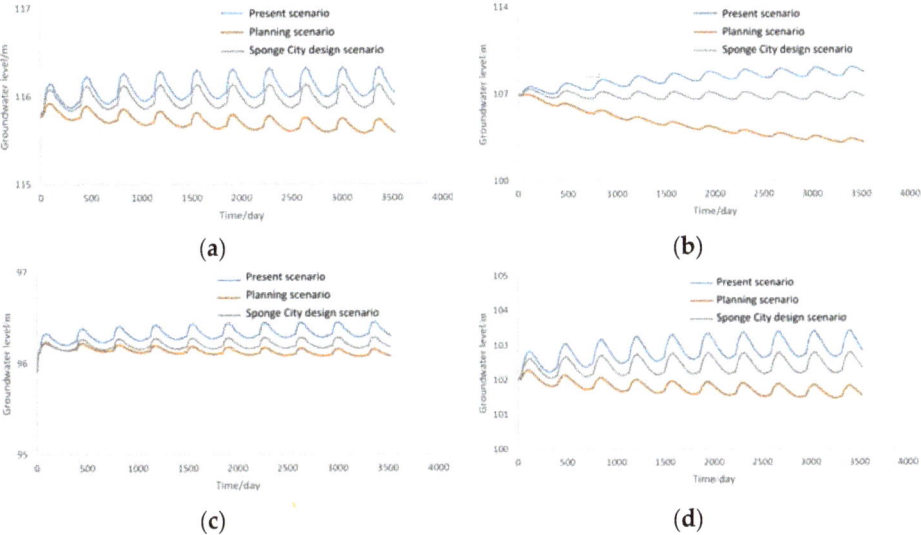

Figure 14. Groundwater level prediction in the next 10 years. (**a**) Observation well G1. (**b**) Observation well G2. (**c**) Observation well G3. (**d**) Observation well G4.

Under the original planning scenario, due to the increase in impervious areas of urban construction, only a small amount of rainfall could replenish groundwater, and canalization of rivers led to the extinguishment of surface water recharge (Table 7). Thus, the groundwater level continues to decline, and the maximum drawdown of the observation well can reach 3.8 m. Although the groundwater level under the Sponge City design scenario is lower than that in the present scenario, it can still maintain the stable state of groundwater level because the LID facilities induced more water returning into the aquifer (Table 8). Figure 13 also revealed different changing ranges in different observation wells; the main reason was the distance to infiltration infrastructure with different types and scales. Therefore, compared with the planning conditions, the Sponge City design scheme can make more precipitation and surface water recharge the subsurface aquifer, thus having a significant positive impact on groundwater resources.

Table 8. The groundwater budget under different scenarios. Unit: 10^8 m^3.

Groundwater Balance Items		Present Scenario	Planning Scenario	Design Scenario
Discharge	Exploitation	0.683	0.683	0.683
	Evaporation	0.233	0.0017	0.2705
	Lateral runoff	0.4832	0.1497	0.3749
Recharge	Precipitation	1.8477	0.5862	0.7007
	LID	0	0	0.9764
	River	0.2104	0	0.1554
	Lateral runoff	0.0495	0.1471	0.0154
Total balance		0.7084	−0.1011	0.5195

4. Conclusions

Based on the dual requirements of stormwater control and utilization in the study area, a Sponge City design scheme that takes full advantage of natural elements is proposed. Combined with the current river distribution, the main drainage channels in the study area are constructed, and the catchment areas corresponding to each drainage channel are divided; according to the geological and hydrogeological conditions of the study area, the analytic hierarchy process was used to determine the suitable types of LID facilities in different blocks of the study area. According to the rainfall and flood volume in each catchment area, the area and volume of LID facilities were calculated according to the green space area of each catchment area, with the goal of 70% of LID facilities' consumption. To ensure the safety of urban drainage under the condition of rainfall, the municipal pipe network drainage system is constructed to discharge 40% of the rainfall. Therefore, the Sponge City scenario showed a better effect in reducing stormwater peak flow, delaying stormwater peak time, and recharging groundwater than that under the traditional city development mode.

According to this research, the coupling of SWMM and MODFlow could be an effective method to analyze and examine the function of flood control and aquifer recharge of a Sponge City. The limitations of this case study are as follows. (1) There was no validation of the hydrological model because there were no available data for the seasonal rivers; (2) the design scenario is only a brief scheme of Sponge City; there has been no assessment of the runoff quantity control and recharge quantity of groundwater in LID facilities on a community scale.

Author Contributions: Conceptualization, X.Y. and X.D.; methodology and formal analysis, M.L.; writing—original draft preparation, B.M.; writing—review and editing, X.Y. and X.D. All authors have read and agreed to the published version of the manuscript.

Funding: This research was funded by the National Natural Science Foundation of China (41672231).

Data Availability Statement: Data is contained within the article.

Conflicts of Interest: The authors declare no conflict of interest.

References

1. Research Group of Control and Countermeasure of Flood. Control and countermeasure of flood in China. *China Flood Drought* **2014**, *3*, 46–48.
2. Chan, F.K.S.; Griffiths, J.; Higgitt, D.; Xu, S.Y.; Zhu, F.F.; Tang, Y.T.; Xu, Y.Y.; Thorne, C. "Sponge City" in China—A breakthrough of planning and flood risk management in the urban contex. *Land Use Policy* **2018**, *76*, 772–778. [CrossRef]
3. Lyu, H.M.; Sun, W.J.; Shen, S.L.; Arulrajah, A. Flood risk assessment in metro systems of mega-cities using a GIS-based modeling approach. *Sci. Total Environ.* **2018**, *626*, 1012–1015. [CrossRef]
4. Zhang, S.Y.; Zevenbergen, C.; Rabe, P.; Jiang, Y. Influences of Sponge City on Property Values in Wuhan, China. *Water* **2018**, *10*, 766. [CrossRef]
5. Lancia, M.; Zheng, C.; He, X.; Lerner, D.N.; Tian, Y. Hydrogeological constraints and opportunities for "Sponge City" development: Shenzhen, southern China. *J. Hydrol. Reg. Stud.* **2020**, *28*, 772–778. [CrossRef]
6. Geiger, W.F. Sponge city and LID technology vision and tradition. *Landsc. Archit. Front.* **2015**, *3*, 10–20.

7. Tazioutzios, C.; Kastridis, A. Multi-Criteria Evaluation (MCE) method for the management of woodland plantations in floodplain areas. *Int. J. Geo-Inf.* **2020**, *9*, 725. [CrossRef]
8. Bertrand-Krajewski, J.L. Intergrated urban stormwater management: Evolution and multidisciplinary perspective. *J. Hydro-Environ. Res.* **2021**, *38*, 72–83. [CrossRef]
9. Zhu, Y.F.; Xu, C.Q.; Yin, D.K.; Xu, J.X.; Wu, Y.Q.; Jia, H.F. Environmental and economic cost-benefit comparison of Sponge City construction in different urban functional regions. *J. Environ. Manag.* **2022**, *304*, 114230. [CrossRef]
10. Tu, X.M.; Tian, T.N. Questions towards a sponge city report on power of public policy: Sponge city and the trend of landscape architecture. *Landsc. Archit. Front.* **2015**, *3*, 22–30.
11. Yin, D.K.; Chen, Y.; Jia, H.F.; Wang, Q.; Chen, Z.X.; Xu, C.X.; Li, Q.; Wang, W.L.; Yang, Y.; Fu, G.T.; et al. Sponge City Practice in China: A Review of Construction, Assessment, Operational and Maintenance. *J. Clean. Prod.* **2021**, *280*, 124963. [CrossRef]
12. Xu, Y.S.; Shen, S.L.; Lai, Y.; Zhou, A.N. Design of sponge city: Lessons learnt from an ancient drainage system in Ganzhou, China. *J. Hydrol.* **2018**, *563*, 900–908. [CrossRef]
13. Chen, S.Y.; Van de Ven, F.H.M.; Zevenbergen, C.; Verbeeck, S.; Ye, Q.H.; Zhang, W.J.; Wei, L. Revisiting China's Sponge City planning approach: Lessons from a case study on Qinhuai District, Nanjing. *Front. Environ. Sci.* **2021**, *9*, 748231. [CrossRef]
14. Wang, Y.T.; Sun, M.X.; Song, B.M. Public perceptions of and willingness to pay for sponge city initiatives in China. *Resour. Conserv. Recycl.* **2017**, *122*, 11–20. [CrossRef]
15. Xie, X.H.; Qin, S.Y.; Gou, Z.H.; Yi, M. Engaging professionals in urban stormwater management: The case of China's Sponge City. *Build. Res. Inf.* **2020**, *48*, 719–730. [CrossRef]
16. Zhang, C.H.; He, M.Y.; Zhang, Y.S. Urban sustainable development based on the framework of sponge city: 71 case study in China. *Sustainability* **2019**, *11*, 1544. [CrossRef]
17. Yang, M.; Sang, Y.F.; Sivakumar, B.; Chan, F.K.S.; Pan, X. Challenges in urban stormwater management in Chinese cities: A hydrologic perspective. *J. Hydrol.* **2020**, *591*, 125314. [CrossRef]
18. Mooers, E.W.; Jamieson, R.C.; Hayward, J.L.; Drage, J.; Lake, C.B. Low-impact development effects on aquifer recharge using coupled surface and groundwater models. *J. Hydrol. Eng.* **2018**, *23*, 04018040. [CrossRef]
19. Newcomer, M.E.; Gurdak, J.J.; Sklar, L.S. Urban recharge beneath low impact development and effects of climate varibility and change. *Water Resour. Res.* **2014**, *50*, 1716–1734. [CrossRef]
20. Jin, M.X.; Lancia, M.; Tian, Y.; Viaroli, S.; Andrews, C.; Liu, J.; Zheng, C. Hydrogeological criteria to improve the sponge city strategy of China. *Front. Environ. Sci.* **2021**, *9*, 700463. [CrossRef]
21. Lancia, M.; Zheng, C.M.; Yi, S.P.; Lerner, D.N.; Andrews, C. Analysis of groundwater resources in densely populated urban watersheds with a complex tectonic setting: Shenzhen, southern China. *Hydrogeol. J.* **2019**, *27*, 183–194. [CrossRef]
22. Lancia, M.; Su, H.; Tian, Y.; Xu, J.T.; Andrews, C.; Lerner, D.N.; Zheng, C.M. Hydrogeology of the Pearl River Delta, southern China. *J. Maps* **2020**, *16*, 388–395. [CrossRef]
23. Su, Y.; Li, T.X.; Cheng, S.K.; Wang, X. Spatial distribution exploration and driving factor identification for soil salinisation based on geodetector models in coastal area. *Ecol. Eng.* **2020**, *156*, 105961. [CrossRef]
24. Kang, H.Z.; Chen, L.; Guo, Q.Z.; Lian, J.J.; Hou, J. An overview of quantification of groundwater recharge in sponge city construction. *Earth Sci. Front.* **2019**, *26*, 58–65.
25. Sun, K.N.; Hu, L.T.; Liu, X.M. The influences of sponge city construction on spring discharge in Jinan city of China. *Hydrol. Res.* **2020**, *51*, 959–975. [CrossRef]
26. Ji, M.C.; Bai, X. Construction of the sponge city regulatory detailed planning index system based on the SWMM model. *Environ. Technol. Innov.* **2021**, *23*, 101645. [CrossRef]
27. Zhou, J.J.; Liu, J.H.; Shao, W.W.; Yu, Y.D.; Zhang, K.; Wang, Y.; Mei, C. Effective evaluation of infiltration and storage measures in sponge city construction: A case study of Fenghuang City. *Water* **2018**, *10*, 937. [CrossRef]
28. Zhou, Y.X.; Sharma, A.; Masud, M.; Gaba, G.S.; Dhiman, G.; Ghafoor, K.Z.; AlZain, M.A. Urban rain flood ecosystem design planning and feasibility study for the enrichment of smart cities. *Sustainability* **2021**, *13*, 5205. [CrossRef]
29. Yang, Y.Y.; Li, J.; Huang, Q.; Xia, J.; Li, J.K.; Liu, D.F. Performance assessment of sponge city infrastructure on stormwater outflows using isochrone and SWMM models. *J. Hydrol.* **2021**, *597*, 126151. [CrossRef]
30. Yin, D.K.; Evans, B.; Wang, Q.; Chen, Z.X.; Jia, H.F.; Chen, A.S.; Fu, G.T.; Ahmad, S.; Leng, L.Y. Integrated 1D and 2D model for better assessing runoff quantity control of low impact development facilities on community scale. *Sci. Total Environ.* **2020**, *720*, 137630. [CrossRef]
31. Zhengzhou Urban and Rural Planning Bureau. *Technical Provisions on Urban Management of Zhengzhou City (2018 Revised Version)*; Office of Zhengzhou Urban and Rural Planning Bureau: Zhengzhou, China, 2018. (In Chinese)
32. Wang, Y.T.; Sun, M.X.; Song, B.M. A framework to support decision making in the selection of sustainable drainage system design alternatives. *J. Environ. Manag.* **2017**, *201*, 145–152. [CrossRef]
33. Tang, S.J.; Jiang, J.; Zheng, Y.; Hong, Y.; Chung, E.S.; Shamseldin, A.Y.; Wei, Y.; Wang, X.H. Robustness analysis of storm water quality modeling with LID infrastructures from natural event-based field monitoring. *Sci. Total Environ.* **2020**, *753*, 142007. [CrossRef] [PubMed]
34. Luan, Q.H.; Fu, X.R.; Song, C.P.; Wang, H.C.; Liu, J.H.; Wang, Y. Runoff Effect Evaluation of LID through SWMM in Typical Mountainous, Low-Lying Urban Areas: A Case Study in China. *Water* **2017**, *9*, 439. [CrossRef]

35. Rossman, L.A. *Storm Water Management Model User's Manual Version 5.1*; United States Environmental Protection Agency: Cincinnati, OH, USA, 2015.
36. Liu, X.P. Parameter calibration method for urban rainfall-runoff model based on runoff coefficient. *Water Wastewater Eng.* **2009**, *35*, 213–217. (In Chinese)
37. Beijing Municipal Design and Research Institute. *Concise Drainage Design Manual*; China Construction Industry Press: Beijing, China, 1990; p. 241.
38. Ashraf, M.; Kahlown, M.A.; Ashfaq, A. Impact of small dams on agriculture and groundwater development: A case study from Pakistan. *Agric. Water Manag.* **2007**, *92*, 90–98. [CrossRef]
39. Kastridis, A.; Stathis, D. The effect of small earth dams and reservoirs on water management in North Greece (Kerkini municipality). *Silva Balc.* **2015**, *16*, 71–84.
40. Zhu, Q.; Yu, K.J.; Li, D.H. The width of ecological corridor in landscape planning. *ACTA Ecol. Sin.* **2005**, *25*, 2406–2412. (In Chinese)
41. Hou, F. Research on the Urban Ecological Corridor Planning Based on Sponge City—Green Corridor of Fengxi New City as a case study. Master's Thesis, Chang'an University, Xi'an, China, June 2017. (In Chinese).
42. Xu, C.Q.; Shi, X.M.; Jia, M.Y.; Han, Y.; Zhang, R.R.; Ahmad, S.; Jia, H.F. China Sponge City database development and urban runoff source control facility configuration comparison between China and the US. *J. Environ. Manag.* **2022**, *304*, 114241. [CrossRef] [PubMed]

MDPI
St. Alban-Anlage 66
4052 Basel
Switzerland
Tel. +41 61 683 77 34
Fax +41 61 302 89 18
www.mdpi.com

Water Editorial Office
E-mail: water@mdpi.com
www.mdpi.com/journal/water

www.ingramcontent.com/pod-product-compliance
Lightning Source LLC
LaVergne TN
LVHW070445100526
838202LV00014B/1667